D0856860

BIOMECHANICS
— OF —
DISTANCE RUNNING

Peter R. Cavanagh, PhD
Penn State University

Human Kinetics Books
Champaign, Ilinois

Library of Congress Cataloging-in-Publication Data

Biomechanics of distance running / editor, Peter R. Cavanagh.
    p.   cm.
   Includes bibliographical references.
   ISBN 0-87322-268-7
   1. Running--Physiological aspects. 2. Biomechanics.
 I. Cavanagh, Peter R.
 QP310.R85B56 1990
612'.044--dc20   89-29302
                                             CIP

ISBN: 0-87322-268-7

Developmental Editors: Holly Gilly, Marie Roy
Copyeditor: Barbara Walsh
Assistant Editors: Valerie Hall, Timothy Ryan
Proofreader: Wendy Nelson
Production Director: Ernie Noa
Typesetter: Angela K. Snyder
Text Design: Keith Blomberg
Text Layout: Tara Welsch
Cover Design: Jack Davis
Printer: Braun-Brumfield, Inc.

Printed in the United States of America

10 9 8 7 6 5 4 3 2 1

Human Kinetics Books
A Division of Human Kinetics Publishers, Inc.
Box 5076, Champaign, IL 61825-5076
1-800-747-4HKP

Contents

Preface

For nearly 4 million years bipedal running has been critical to human survival. Yet in the 150 or so years that we have had the tools to study human locomotion, no book has been devoted exclusively to the scientific study of running mechanics. Lest this observation seem pretentious, I hasten to add that this volume is very much a small stroke on the surface of a large canvas. The pace of study has quickened in the last 10 years, paralleling the popularity of running as a sport; articles on the biomechanics of running are now to be found in many journals and in the proceedings of many conferences. Realizing how much we have yet to learn about the biomechanics of running, it seemed appropriate to make a statement of our initial knowledge of the subject.

This volume collects the contributions of individuals who have been experimenting in the field of running mechanics for many years. It is particularly gratifying to me that over two-thirds have been associated in some way with the program at Penn State. All are colleagues whose work is well known and well respected. *The Biomechanics of Distance Running* is intended for serious students of biomechanics who are seeking a broad perspective on the topic. I expect that not 5 years will pass before a complete rewrite is needed to accommodate new knowledge on the subject, and it is my earnest hope that this book will stimulate ideas from which that new knowledge will grow.

Peter R. Cavanagh

List of Contributors

R. McNeill Alexander, FRS
Department of Pure and Applied
  Biology
Leeds University
Leeds, England LS2 9JT

Anne E. (Betty) Atwater, PhD
The McKale Center
University of Arizona
Tucson, Arizona 85721

Peter R. Cavanagh, PhD
Center for Locomotion Studies
Penn State University
University Park, Pennsylvania
  16801

Christopher J. Edington, MS
Converse, Inc.
North Reading, MA 01864

E.C. Frederick, PhD, President
Exeter Research, Inc.
Brentwood, New Hampshire 03833

Richard N. Hinrichs, PhD
The Department of Health and
  Physical Education
Exercise and Sport Research
  Institute
Arizona State University
Tempe, Arizona 85287

Stanley L. James, MD
Orthopedic and Fracture Clinic
  of Eugene, P.C.
Eugene, Oregon 97401

Donald C. Jones, MD
Orthopedic and Fracture Clinic
  of Eugene, P.C.
Eugene, Oregon 97401

Rodger Kram, MS
Concord Field Station
Harvard University
Bedford, MA 01730

Mark J. Lake, MS
The Center for Locomotion Studies
Penn State University
University Park, Pennsylvania 16802

Irene S. McClay, MS, LPT
School of Life and Health Sciences
Physical Therapy Program
  of Delaware
University of Delaware
Newark, Delaware 19716

Doris I. Miller, PhD
The Center for Locomotion Studies
Penn State University
University Park, Pennsylvania 16802

Monica J. Milliron, MS
National Institute of Occupational
  Safety and Health
Cincinnati, Ohio 45226

Gordon A. Valiant, PhD
Nike Sport Research Laboratory
Beaverton, Oregon 97005

Keith R. Williams, PhD
Physical Education Department
University of California–Davis
Davis, California 95615

Chapter 1

The Mechanics of Distance Running: A Historical Perspective

Peter R. Cavanagh

Twenty years ago the notion that the mechanics of distance running was a valid topic of scientific study would probably have been disputed by many people. Distance running was something that a few hardy ecto-morphs did to demonstrate the limits of human endurance, and their ex-ploits were to be wondered at rather than studied. As running for exercise rather than for competition has become popular, more and more scrutiny has been brought to bear on this classical form of human locomotion.

Locomotion as a Field of Study

Human gait, in its various forms, has always been in the forefront of bio-mechanical study. Ironically, it has been the ravages of war rather than the peace of exercise that has motivated much of the work in the past. Some of Marey's work (Carlet, 1872; Marey, 1895/1972) was sponsored by the French War Office, and Braune and Fischer (1889, 1895–1904) con-ducted studies on gait and posture for the Prussian Army at the height of the militaristic era of the Prussian Empire. Amar (1920) used an early force platform to study the limbless veterans of World War I, and the monumental effort to understand the mechanics of human gait that the University of California studies represent (Klopsteg & Wilson, 1954) was supported by the Committee on Artificial Limbs of the National Research Council of the National Academy of Sciences. Despite the fact that the tools developed during these various programs were precisely those re-quired to study running, the early literature on running mechanics is remarkably sparse.

When 20th-century scientific studies on running did appear they gener-ally had two features in common. First, they were the ''hobby'' activities of scientists who had distinguished themselves in other fields but for whom running held a fascination. Thus Nobel laureate A.V. Hill published on running in the hallowed *Proceedings of the Royal Society* (Furusawa, Hill, & Parkinson, 1927a, 1927b; Hill, 1928) and muscle physiologist Wallace Fenn authored two early and important papers in the *American Journal of Physiology* (Fenn, 1929, 1930). Herbert Elftman, an anatomist, made fundamental contributions with a series of papers between 1939 and 1941 (Elftman, 1939a, 1939b, 1940, 1941). The second characteristic that most of these and other early papers shared was their focus on sprinting. Animals and people that run fast are held in special esteem by most societies. There is something primevally satisfying to many people in watching a boxer demonstrate his superiority over all comers in a rapid, brutal, and punishing way, and the enjoyment of seeing sprinters battle in a few blurred seconds of competition is not altogether dissimilar. Distance running is hardly as intense or as dramatic, and this surely contributed to its past popular and scientific neglect.

Increased Interest in Distance Running

The explosive growth of participation in distance running that occurred in the United States in the late 1960s and early 1970s (Cavanagh, 1981) changed this situation in three important ways. First, questions ultimately related to running mechanics, such as footwear choice and injury prevention, suddenly became relevant to more people. With an estimated 30 million Americans running for exercise there was now a large lobby of healthy, relatively affluent consumers; running mechanics was no longer a minority interest. The second and perhaps most important issue that facilitated research was an economic one. The athletic shoe companies began to reap huge profits from running shoe sales, and many of them proved willing to spend part of their profits on research to understand the nature of this goose that had laid their golden egg. Thus, funding to study distance running, which had never been available from conventional federal sources, suddenly became available to many investigators.

The third factor that facilitated the growth in the scientific study of distance running was the increased incidence of injury that the running boom left in its wake. The effects of this trend have been far-reaching on sports medicine in general and need some examination here. Before distance running became popular, the overwhelming number of injuries to participants in sports in the United States were those from team sports, typically acute traumatic injuries such as the "terrible triad" in the football player's knee, or the leg fractured sliding home in baseball. Thus orthopedics became synonymous with sports medicine, and surgery became synonymous with treatment. Running injuries are usually of a completely different character (Clement, Taunton, Smart, & McNicol, 1981; James

& Jones, this volume). They are subtle problems—often in soft tissue—caused by repetitive application of relatively small forces and not by single acute applications of large forces. Treatment is often problematic, and many injuries resist healing with the same tenacity that runners typically resist surgery. Thus medical practitioners considered experts in "sports medicine" as it had been, found themselves ill equipped and ill trained to diagnose and treat running injuries. The result was a gradual and remarkable acceptance of conservative approaches to running and other sports injuries. A further interesting by-product was the legitimization and acceptance by athletes of podiatry, which, by virtue of the restrictions placed on the profession by conventional medical specialties (such as the lack of surgical privileges at many hospitals), was by nature principally conservative in its approach. This conservative approach required an understanding of the underlying mechanics responsible for the injury, and a great deal of attention has been focused on this area. It is ironic that the center of current thought in podiatry is focused on surgical approaches to the solution of foot problems.

This brings us to the present situation where almost every issue of a sport science journal contains articles on running mechanics. Spanning slightly more than 15 years, the change from oblivion to acceptance has been swift, and distance running mechanics is now studied both for its own sake and because it provides an excellent model of a steady-state endurance activity in which muscle action is repeatable, ground reaction forces are measurable, and changes in mechanics can cause changes in physiological factors such as energy cost.

The remainder of this chapter will briefly review some of the early literature on the mechanics of distance running to set the stage for the contributions that follow.

The Greek Influence—Aristotle

The early approaches to the study of running were by definition kinematic: where instantaneous postures from running were reflected in works of art. There are many illustrations of both sprinting and distance running among the rich treasury of vase paintings that remain from the age of classical Greece (the fifth century B.C.). It is clear that the artist whose work is depicted in Figure 1.1 recognized some of the differences between sprinting and distance running. Despite the strict adherence to frontalism with its unopposed arm and leg movement, the high arm carriage and the restricted hip and knee joint movements are in clear contrast to the extravagant and extensive movements shown for sprinters in Figure 1.2. Of further interest in Figure 1.1, variations in running styles are exhibited,

Figure 1.1.   Greek vase painting (circa 525 B.C.) of distance runners. *Note.* From *Athletics in the Ancient World* (p. 138) by E.N. Gardner, 1967, Oxford, Great Britain: Clarendon. Copyright 1967. Reprinted by permission.

Figure 1.2.   Greek vase painting (circa 470 B.C.) of sprinters. *Note.* From *Athletics in the Ancient World* (p. 134) by E.N. Gardner, 1967, Oxford, Great Britain: Clarendon. Copyright 1967. Reprinted by permission. Photo courtesy of The Metropolitan Museum of Art, Rogers Fund, 1912.

with one runner having extended elbows and another showing more forward lean than the rest of the group. The extent to which this was artistic license or astute observation will never be known.

Approximately 100 years after the vases shown in Figures 1.1 and 1.2 were executed, Aristotle (384–322 B.C.) studied locomotion for more than artistic reasons. His astonishing mind, which had been applied to logic, ethics, politics, mathematics, psychology, biology, and law (among other topics), also turned to human and animal movement in *De Motu Animalium* (*On the Movement of Animals*—Nussbaum, 1978) and to locomotion in particular ("On the Gait of Animals"—*Encyclopedia Britannica*, 1952). Describing the different types of gait observed in the animal kingdom was an extension of his earlier attempts at biological classification. Among Aristotle's many astute observations was an intuitive understanding of the role that ground reaction forces played in progression, anticipating Newton's third law by some 1,900 years:

> Further, the forces of that which causes movement and of that which remains still must be made equal. . . . For just as the pusher pushes, so the pusher is pushed—i.e. with similar force. (*Encyclopedia Britannica*, 1952, p. 234)

While reflecting on the relationship of structure to function, Aristotle observed that flexion of the knee joint is necessary to allow progression and to minimize vertical motion.

> If a man were to walk parallel to a wall in sunshine, the line described (by the shadow of his head) would not be straight but zig-zag, becoming lower as he bends and higher when he stands and lifts himself up. (*Encyclopedia Britannica*, 1952, p. 247)

The European Intellectual Revival— Leonardo da Vinci

As was the case with other branches of knowledge and civilized pursuits, the eclipse of the Grecian society and the decline of its successor, the Roman Empire, led to the virtual abandonment of scientific inquiry in locomotion. With one or two notable exceptions, this lasted until the European intellectual revival of the 14th and 15th centuries. Because a main component of the Renaissance was a revival of Greek and Roman ideas and ideals, the Aristotelian concepts of motion were as well regarded as they had been almost 2,000 years earlier. Science and art were also intimately associated, and no one individual expressed the spirit of those times more completely than Leonardo da Vinci (1452–1519). He displayed interests and abilities that were more expansive than a present-day scholar could possibly embrace. His fascination with human movement was rooted in the desire to accurately represent movement in his paintings, and because no stop-action photography was available, the only substitute to ensure accuracy was painstaking observation and drawing of human movement. Da Vinci reflected his awareness of the complexity of human movement and of the inadequacy of the eye when he wrote in his treatise on painting (McMahon, 1956, p. 138), "It is impossible for any memory to preserve all the aspects and changes of the parts of the body." In this volume, he described and illustrated the principles of motion that were intended to help students of painting with accurate representation of a variety of human locomotor activities (Figure 1.3).

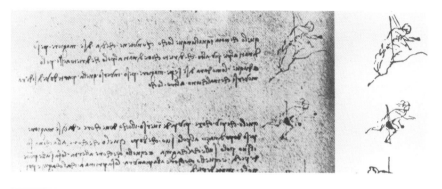

Figure 1.3.   A fragment from Leonardo da Vinci's treatise on painting illustrating a variety of postulates on running. *Note.* From *Leonardo da Vinci's Elements of the Science of Man* (p. 175) by K.D. Keele, 1983, New York: Academic Press. Copyright 1983. Reprinted by permission.

The treatise on painting (McMahon, 1956) contains many postulates referring to running under different conditions, among which the following are particularly interesting:

> 346. *Of motion created by the destruction of balance.* Motion is created by the destruction of balance, that is, of equality of weight, for nothing can move by itself which does not leave its state of balance, and that thing moves most rapidly which is furthest from its balance. (pp. 133-134)

> 347. *Of the motion in man and other animals when running.* When a man or other creature moves either rapidly or slowly, that side which is above the leg that sustains the body will always be lower than the opposite side (p. 134)

> 352. *Of the motion and course of animals.* That figure will appear swiftest in its course, which is about to fall forwards. (p. 135)

The above statements demonstrate remarkable insights into the mechanics of running on topics such as the initiation of movement, the frontal plane movements of the pelvis (what is now known as the positive Trendelenberg sign; Inman et al., 1981), the forward lean against the force of air resistance, and the variations in trunk lean as a function of speed. In Figure 1.3, Leonardo's speculation concerning the distribution of weight under the foot during grade running is shown in his characteristic mirror writing alongside the illustrations. This reads as follows:

He who runs down a slope has his axis on his heels; and he who runs uphill has it on the toes of his feet; and a man running on the level ground has it first on his heels and then on the toes of his feet. (Keele, 1983, p. 175)

This observation strikes a personal chord, because as shown in Figure 1.4, our own experiments (Dick & Cavanagh, 1987), performed some 500 years after Leonardo made these observations, provide quantitative proof of his theory.

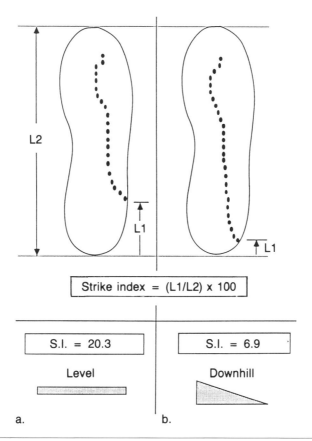

Figure 1.4. Schematic diagram of center of pressure measurements from our own laboratory during (a) running level and (b) downhill running, confirming Leonardo's speculations.

The 17th-Century Scientific Revolution—
Newton and Borelli

Although it is unlikely that any thoughts concerning the mechanics of running ever crossed the mind of Isaac Newton (1642–1727), it would be a mistake to omit mention of him in this brief account. Newton is often described as the culminating figure of the 17th-century scientific revolution, and, in a pleasing piece of scientific continuity, he was born in the year that Galileo died. Despite the advances in the understanding of the world made by individuals such as Galileo, Kepler, and Copernicus, academia was still the domain of Platonists and Aristotelians. Newton's skepticism of the classical explanations that were still revered by his professors at Trinity College are clearly expressed by the following short piece of graffiti written in the margin of one of his college notebooks:

> Plato is my friend
> Aristotle is my friend
> but my best friend is truth. (*New Encyclopedia Britannica*, 1979, p. 17)

Through his skepticism and his genius, Newton made contributions to science that were arguably as great as those of any single individual who has ever lived. His major contribution to mechanics was the formulation of the three laws of motion. When today's biomechanists calculate joint moments and forces during running, they rely on the laws formulated by Newton. Einstein may have changed the way in which we view the universe, but our view of the human body is still essentially Newtonian.

Newton was but one of many individuals in the 17th century who were trying to formulate underlying rules to explain natural phenomena. Biological processes, including gait, were under as much scrutiny as physical processes, and a contemporary of Newton, Giovanni Borelli (1608–1679) was among the first to view the human and animal musculoskeletal system in terms of the structural engineering employed by contemporary builders of bridges and designers of pulley systems. There are two excellent accounts of the history of the study of human gait starting at the time of Borelli (chapter 1 in Braune & Fischer, 1987; articles 156 through 162 in Weber & Weber, 1836), and some of the subsequent material has been drawn from these sources.

Borelli's classical volume *De Motu Animalium* ("Animal Motion") (Borelli, 1685) was an echo of Aristotle's treatise on locomotion mentioned earlier, but Borelli's approach was much more mechanistic. A recent symposium on primate locomotion was dedicated to Borelli (Jouffroy, Ishida, & Jungers, 1983, p. 62), and the papers from that symposium each began with a quotation from *De Motu Animalium*, indicating that Borelli had clearly considered all of the various topics that were studied. His bridging of the ancient and modern is emphasized by Jouffroy, who points out that Borelli referred to Aristotle and Galenus as well as to Galileo and Vesalius. Many of Borelli's propositions concerned human and animal locomotion, and it is clear that his work represents the starting point for the biomechanical study of locomotion. The ''compass gait'' illustration shown in Figure 1.5, which was used to further his discussion about ''examining how the mass of the human body is moved forward while walking'' (Jouffroy, 1983, p. 62), is reminiscent of the mechanical analogs of gait shown in Figure 1.6 presented in this century by Inman, Ralston, and Todd (1981).

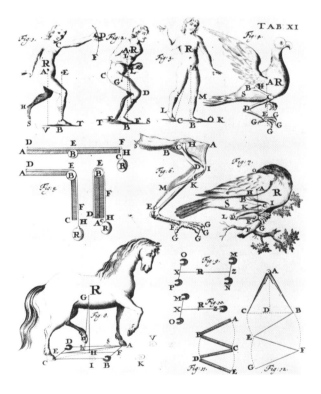

Figure 1.5. A page from Borelli's manuscript (Borelli, 1685) illustrating mechanical principles applied to human and animal locomotion. *Note*. From ''Primate Locomotor Systems: Homage to Giovanni A. Borelli'' by F.-K. Jouffroy, H. Ishida, and W.L. Jungers, 1983, *Annales des Sciences Naturelles, Zoologie* (13th Series), **5**, pp. 53-65. Copyright 1983 by Masson, Paris. Reprinted by permission.

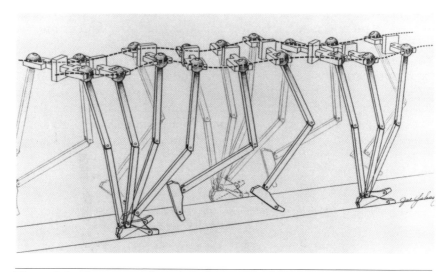

Figure 1.6. A recent analog of the lower extremity during gait, which is a descendent of Borelli's "compass gait." *Note.* From *Human Walking* (p. 13) by V.T. Inman, H.J. Ralston, and F. Todd, 1981, Baltimore: Williams and Wilkins. Copyright 1981. Reprinted by permission.

The Era of Observation— The Weber Brothers and the Pendulum Theory

In 1836, brothers Wilhelm and Eduard Weber published the most detailed and lengthy treatise on human gait—both walking and running—that had appeared thus far (Weber & Weber, 1836). Their work was an excellent example of a thoughtful study that did not depend on complex instrumentation, because, as Marey (1895/1972) was to later remark:

> A level piece of ground of known length, a watch provided with a second hand, and a performer was all they had to work with; they could therefore only obtain a small number of relationships between the frequency and length of stride, of the extent of the vertical head displacements and of the various inclinations of the body, and even these measurements had to be corrected. (p. 127)

Marey's words, written with the benefit of 59 years of hindsight, seem a somewhat harsh review as one peruses the 164 "articles" on 421 pages written by the Weber brothers. Although the Webers were reputed to be mathematicians, the treatise contains a great deal of anatomy in addition to many postulates on walking and running. The Webers are best remembered for their suggestion that the limb can act as a pendulum— which some have interpreted to suggest that they thought muscular action was not needed during the swing phase. Otto Fischer, author of the next major work on gait, assessed the Webers' contribution to the field (Braune & Fischer, 1895–1904), and discounted this interpretation of their work in his introduction: "To assume that Weber and Weber deny any accompanying muscular contraction is to seize too literally the concept of pendulum movement of the leg" (Braune & Fischer, 1987, p. 2).

As part of an attempt to set forth the principles that governed human gait, the Webers presented a number of postulates on running. They separated their remarks according to the speed of running, identifying *Eillauf*, slow or distance running; and *Sprunglauf*, fast running or sprinting. A selection of their statements (Weber & Weber, 1836; translated rather liberally by me) is given here:

§. 26. The difference between walking and running is that a period exists in walking when both feet are in contact with the ground, whereas in running there is a period when neither foot is in contact with the ground. (p. 57)

§. 33. During running, the duration of the stride is shorter, but the length of the stride is greater than during walking. (p. 69)

§. 34. The variability during normal running is less than that which occurs during normal walking. (p. 70)

§. 36. Sprinting is different from slower running in that the stride is longer than would be possible during walking. (p. 74)

§. 112. When the stride during slow running becomes as long as it is during the fastest walking possible, then the duration is the same, that is the duration of one half a pendulum swing of the legs. (p. 279)

§. 117. During running the trunk makes smaller vertical oscillations than during walking. (p. 290)

§. 119. The fastest speed of running is approximately 6.5 meters per second, that is one German mile in 20 minutes. (p. 292)

§. 142. (In slow running) the swinging leg comes to a vertical position but no further. (p. 352)

These postulates are an interesting mixture of "right, wrong, and perhaps." For example, their observations of stride times in walking and running were correct, their observations on maximum speed and the position of the swinging limb were incorrect, and their statements concerning variability still remain to be adequately verified. But the significance of the Webers' work is not what they said but the fact that they said it. By putting their theories—many of which were pure speculation—into print, in meticulous detail, they established an agenda for research, listing almost 150 hypotheses that simply needed the appropriate tools—both mathematical techniques and instrumentation—to be tested.

The Age of Instrumentation—
The Marey School and Vierordt

Etienne Jules Marey (1830–1904) was one of the most prolific scientists that biomechanics has ever known, and his vast output on human locomotion deserves special attention. His regular communications to the *Académie des Sciences* were notable for both their creativity and their frequency. Perhaps the most lasting of Marey's contributions was in the area of instrumentation. Marey was a pioneer of chronography, the automatic recording of the timing of events, and he used such ingenious devices as shoes with air chambers and telegraphic keys triggered by a passing runner to measure the frequency of steps and the speed over a fixed distance, respectively. One of the more famous historical studies of running is shown in Figure 1.7, where Marey's capacity for instrumentation is shown at its fullest. There are (a) chambers in the shoes to show the timing of the foot contacts; (b) a primitive accelerometer attached to the head; and (c) a complete pneumatic recording system, including an on-board chart recorder with rotating drum, to allow the runner freedom from the confines of a laboratory.

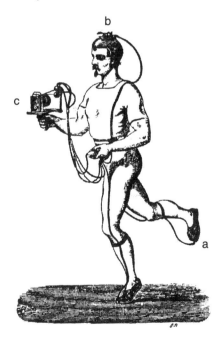

Figure 1.7. A pneumatic experimental arrangement used by Marey (1895) to record swing and support times during locomotion. *Note.* From *Movement* (p. 7) by E.J. Marey, 1895.

Marey was responsible for designing and building what may be the most extensive facility ever devoted exclusively to the study of the biomechanics of locomotion. He describes his equipment and the motivation for building it in the following extract from *Movement* (1895/1972):

> Being fully persuaded that, if one wanted to estimate the exact length of stride, it was necessary to carry out the experiments on a long track, we laid out at the Physiological Station a circular perfectly horizontal course, five hundred meters in circumference; a telegraph wire ran all the way round the track and posts were placed at intervals of fifty meters. Each post was provided with a little contrivance for breaking the circuit the moment the performer came abreast of the post. (pp. 128-130)

The facility is shown in the engraving from Marey's book in Figure 1.8. The debt that A.V. Hill and his contemporaries owed to Marey in their linear arrangement for the study of the velocity curve in sprinting (Best & Partridge, 1928; Furusawa, Hill, & Parkinson, 1927a, 1927b) is apparent in the photograph of their arrangement shown in Figure 1.9. To measure the extent of movement in space, Marey and his associates were among the first to turn the new techniques of photography into true photogrammetric tools. In the study of kinematics Marey used systems with both passive and active targets and devised photographic methods with

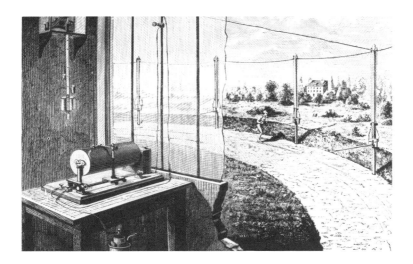

Figure 1.8. Marey's physiological station, which included a 500-m circumference circular track equipped with a variety of monitoring equipment. *Note*. From *Movement* (p. 129) by E.J. Marey, 1895.

Figure 1.9.   A 70-m runway with contact switches and an induction apparatus used by Best and Partridge (1928). *Note.* From ''The Equation of Motion
of a Runner Exerting Maximal Effort'' by C.H. Best and R.C. Partridge, 1928,
Proceedings of the Royal Society, B, **103**, p. 220. Copyright 1928 by The Royal
Society. Reprinted by permission.

multiple images on single fixed plates and others with single images on
moving plates. Stereoscopic chronophotography was also used to better
visualize human and animal movement. The disk camera that Kodak
reintroduced in the 1980s bears a remarkable resemblance to Marey's disk
gun camera of the 1880s. Marey has achieved much more recognition as
a pioneer of the cinema than as an early biomechanist. For this reason,
we have easy access to an excellent account of his endeavors in a contemporary translation of his 1895 book *Le Mouvement*, which has been
reprinted in the *New York Times* Literature of the Cinema series.

Marey also devised and built the first serious force platform and applied
calibration procedures to account for nonlinearities. The platform consisted of spirals of india rubber tubes mounted on an oak frame (Figure
1.10a). A number of spirals in different regions of the platform all com-
municated with a common outlet tube so that the final pressure in the
tube was most probably reasonably independent of the point of application. A multichannel recording tambour (Figure 1.10b) allowed Marey to
record displacement and force simultaneously, and he was also the first
to combine and synchronize photographic and force measurement as
shown in Figure 1.10c. In using the force platform, Marey formulated a

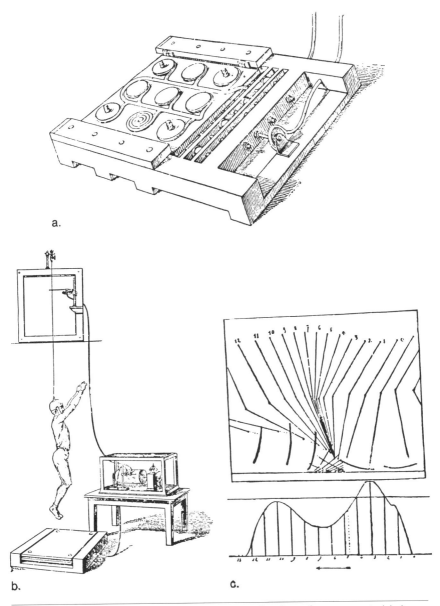

a.

b.

c.

Figure 1.10. Early devices used in the quantification of movement. (a) A pneumatic force platform (with the top removed) used by Marey to record ground reaction force. (b) The device in use to record the simultaneous measurement of force and displacement during a jump. (c) A synchronized record of ground reaction force and lower extremity kinematics during walking. *Note.* From *Movement* (pp. 149, 160, 154, respectively) by E.J. Marey, 1895.

law that governed the variations in "foot-pressure" (more correctly, ground reaction force) and center of gravity motion.

All muscular actions which alter the center of gravity of the body in such a manner as to raise it augment the foot-pressure on the ground.

All actions tending to lower the center of gravity diminish the foot-pressure. (Marey, 1895/1972, p. 150)

There is no doubt that Marey and his pupils were as much—or perhaps more—interested in new ways of collecting data as they were in the meaning of the data itself. They were biologists, and, unlike the Renaissance scientists, they had little background in physics or engineering. During this time, knowledge had started to be compartmentalized and Marey's approach was distinctly biological. It should also be remembered that in almost every direction Marey turned, there were scientists in other fields who were eager to apply his new methods of measurement and recording to their own work. Although human and animal movement was clearly Marey's central scientific interest, he was drawn into areas such as cardiology, astronomy, mechanics, music, civil engineering, hydrodynamics, microscopy, and many more. It is hard to imagine how Marey accomplished any of his own research goals when he was so much in demand.

Nevertheless, concurrent with a flurry of creative instrumentation development, Marey and his associates Demeny and Carlet used their ingenious devices to collect a wealth of new data on the mechanics of running (Marey & Demeny, 1885; Carlet, 1872). Figure 1.11 shows the results of an experiment that to my knowledge has not yet been replicated. The vertical oscillations of the trunk and the step length are plotted as a function of cadence for both walking and running. Marey was deeply involved in the paradox over the apparent lack of external work done during level walking and running despite the clear expenditure of metabolic energy. He divided the work done by the subject into three categories (Figure 1.12):

A. Vertical work

B. Horizontal work

C. Work expended on keeping up the oscillations of the legs during their period of suspension

(Marey, 1895/1972, pp. 157-158)

To calculate vertical work per stride, Marey multiplied the weight of the subject by the amplitude of vertical oscillation, then multiplied the result by 2 for two oscillations in one complete stride, and finally multiplied

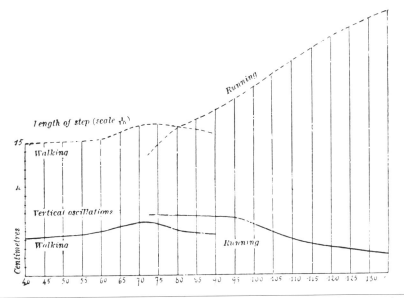

Figure 1.11.   Vertical oscillation of the trunk and step length a function of cadence for walking and running. *Note.* From *Movement* (p. 164) by E.J. Marey, 1895.

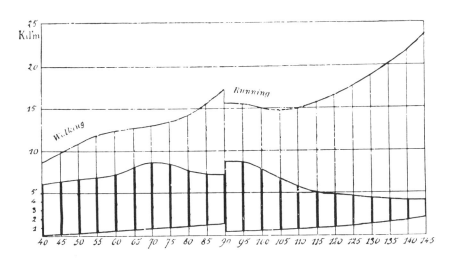

Figure 1.12.   The three components of work calculated by Marey in walking and running at different cadences. *Note.* From *Movement* (p. 165) by E.J. Marey, 1895.

that result by 2 to account for both the rise and the fall. This last step is most interesting as it demonstrates Marey's awareness of the cost of negative work. Horizontal work per stride was calculated from the change in velocity of the center of gravity, and work done on the limbs was determined from a consideration of moment of inertia and angular velocity. This estimate, which considered only rotational effects for the limbs, was correctly found to be "practically a negligible quantity, and only very slightly influences the total amount of energy expended each step" (Marey, 1895/1972, p. 160). The final estimates presented for work done per stride in running are as follows:

Translation of lower limb	3.4 kilogram meters
Vertical oscillations of the body	2.3 kilogram meters
Acceleration and remissions in velocity in the horizontal direction	18.4 kilogram meters
Total	24.1 kilogram meters

(Marey, 1895/1972, p. 161)

What is most astonishing is that Marey was acutely aware of a major error in such a calculation, one that has only recently begun to be reexamined by biomechanists. In his own words,

This estimate . . . is an inexact valuation, which does not really give the expenditure of muscular force: a certain portion of the energy seems to store itself up in the muscles during each period of descent and to be liberated in the next phase of ascent.

Veterinary experts have made a special study of the energy lost by hoofs striking the ground when a horse is travelling at a rapid pace. They maintain that the flexor of the solitary toe which constitutes the foot of the horse, is made up to a great extent of elastic tissue. It possesses in consequence a physical property by means of which a more or less important part of the vital energy lost in falling on the feet is to some extent returned in the form of energy.

This subject deserves re-investigation. It would be interesting to discover whether the tendons in man possess this valuable property to any noticeable degree, and, if so, whether it is retained throughout life. (Marey, 1895/1972, pp. 163-164)

The reinvestigation that Marey suggested is only now being conducted, almost 100 years later. These examples show the tremendous insight that Marey had into what are still unsolved problems in the mechanics of running. He was aware of the dependence of physiological cost on the movement characteristics chosen, he suggested that minimal energy was an operating principle for human locomotion, and he suspected that storage and reutilization of elastic energy was a process that operated to conserve metabolic energy. He was definitely a man well in advance of his time.

Vierordt (1881) was another investigator who constructed ingenious devices to pursue the study of running and was also among the group that was openly skeptical about the Webers' theory of a pendulum swing:

> I must confess that the theory of pure swinging, which excludes any participation by the muscles during swinging of the leg does not seem very plausible to me. Could so precise a movement depend on the result of gravity? (in Braune & Fischer, 1895–1904/1987, p. 4)

The approach taken by Vierordt was an extremely novel one involving the spraying of ink from small nozzles attached to the body. By putting such nozzles on the shoes (Figure 1.13a), he was able to record the details of foot placement, and by attaching them to the trunk and the upper and lower limbs, a trace was left on a vertical roll of paper either alongside or underneath the runner (Figure 1.13b).

Vierordt's methods were described somewhat disdainfully some 20 years later by Otto Fischer, who could have been talking about the modern dilemma about the conflicts of research and clinical gait analysis when he wrote:

> The method of Vierordt cannot be denied a certain historical value. It remains the first experiment to record simultaneously the trajectories of different parts of the body. It could, even today, provide the medical practitioner with a means of obtaining an approximate overall view of abnormalities of walking and running occasioned by particular pathological conditions, though photography is an incomparably better method. The research worker in the field of physiological mechanics would, however, find the projection methods of little use. Quick and easily obtained results are not so essential here, and the research worker can spend time and trouble to achieve as accurate a knowledge as possible. (Braune & Fischer, 1895–1904/1987, p. 4)

According to Fischer, when Vierordt turned to electrical methods to record the timing of the step, his results were very accurate indeed. His principal contribution appears to have been the realization that there is considerable variation in the stride parameters of normal locomotion (time of swing, stride length, left and right step lengths, etc.). This finding has certainly not been exploited by subsequent generations of researchers.

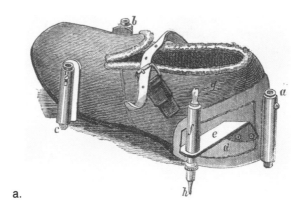

a.

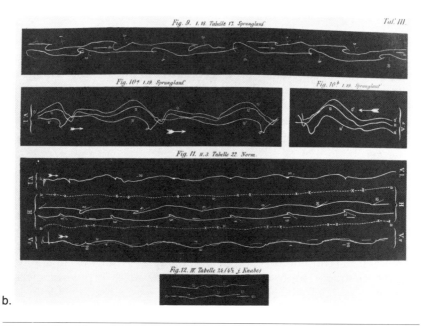

b.

Figure 1.13.   Vierordt's method of ink nozzles attached to the shoe (a), which left a tracing on sheets of paper (b) to reveal movement. Similar devices were applied to various parts of the body. From Vierordt, 1881, p. 7 and Table 3.

Mathematical Methods—Braune and Fischer

The acknowledgment for the initiation of the modern mathematical approach to the study of locomotion belongs to Braune and Fischer, and thanks to a recent translation their major work, *Der Gang des Menschen*, is now available in its entirety to English-speaking readers (Braune & Fischer, 1895–1904/1987). Both the scope and the vision of this research was monumental. In the preface, Fischer describes the difficulties encountered with the data collection, which involved attaching electrical discharge tubes to the bodies of the subjects (Figure 1.14):

> The experiments themselves were very time-consuming and fatiguing. From ten to twelve hours of uninterrupted activity were often necessary since preparing the subject required the utmost care, as did the accurate arranging and insulating of the Geissler tubes. Decisive experiments had to be carried out at night because there was no means of darkening the room in which we performed studies. (p. vii)

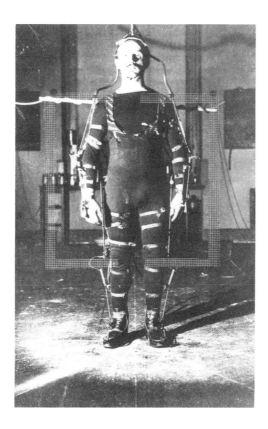

Figure 1.14. A plate from *Der gang des Menschen* showing the subject with electrical discharge tubes attached to the limbs. If the expression on the subject's face is any indication, the subjects were probably at some risk of electric shock! *Note.* From *The Human Gait* (p. 13) by W. Braune and O. Fischer (P. Maquet and R. Furlong, Trans.), 1987, Berlin: Springer-Verlag. Copyright 1987. Reprinted by permission.

Braune died shortly after the experiments were completed, and the enormous task of measuring the photographic plates and making the many calculations was performed under Fischer's direction. Apparently, no studies of running were made, and the entire 437 pages (in translation) are occupied with the description and reporting of just three transits of the experimental area "obtained on the night of 24–25 July 1891." Two of these were "free" walking and the third was walking with "an army regulation knapsack, three full cartridge pouches and an 88 rifle in the 'shoulder arms' position" (p. 18). Four cameras without shutters were used, and the Geissler tubes were pulsed with current from a Ruhmkorff coil at a frequency of 26.09 exposures per second. The coordinate geometry needed to extract three-dimensional coordinates from the film plates is set out in detail, and the equations needed to calculate resultant forces and moments at the joints of a 12-segment rigid body model are formulated. The film plates were digitized using a precision optical device, and once the three-dimensional coordinates of the landmarks were calculated, various graphical and physical model reconstructions were made. By combining their data previously collected on the masses and moments of inertia of the body segments (Braune & Fischer, 1889/1985) with the coordinate data (double differentiated by hand where necessary) these authors presented the first available account of joint moments during the swing phase of walking. On the very last page of the text, the authors conclude, "The much discussed pendulum theory of Weber and Weber is thus erroneous" (Braune & Fischer, 1895–1904/1987, p. 437). This sentence clearly defines the end of an era and the beginning of a new one. The many centuries of speculation concerning the nature of limb movement in gait that had been ongoing since the time of Aristotle was finally in this simple statement beginning to be replaced by quantitative methods. With minor changes, the methods formalized by Braune and Fischer are almost exactly those used by today's investigators in the biomechanics of gait.

Photographic Frenzy—Eadweard Muybridge

Eadweard Muybridge (1830–1894) was a character only the 19th century could have produced. He was not a scientist but a photographer and adventurer who, having been born in England under modest circumstances, emigrated to America to seek his fortune. In doing so he changed his original name of Edward Muggeridge several times, became a successful businessman in San Francisco, and survived such escapades as a near-fatal stagecoach accident and a trial for the murder of his wife's lover. He became a landscape photographer following his accident, and after he showed some of his well-regarded prints to Governor Leyland Stanford of California, the famous association between Muybridge and Stanford began. Muybridge is said to have shown in 1872 that all four feet of a trotting horse left the ground at the same time, and there is no doubt that this assignment was the catalytic event of his professional life. He read Marey's *Animal Mechanism*, and with the patronage of Stanford and later the University of Pennsylvania embarked on an ambitious series of photographs of humans and animals in motion. His enduring 11-volume legacy, *Animal Locomotion* (Muybridge, 1887), has been reproduced in its entirety in a 3-volume set (Muybridge, 1979) that contains all 781 original plates—in total, over 20,000 individual photographs. The two or three orthogonal views were taken by multiple cameras, usually oriented perpendicular to the cardinal planes, although there are occasional oblique views. The plate of a "man running at a half mile gait" (Muybridge, 1955, plate 18) shown in Figure 1.15 is one of many examples of running from Muybridge's work. It is interesting to note that the functional varus of the limb about to land and the initial contact with the outside border of the foot are both visible in the second picture in the frontal series.

Figure 1.15. A plate from Muybridge's *Animal Motion* showing a man running at a distance pace. *Note.* From *The Human Figure in Motion* (plate 18) by E. Muybridge, 1955, New York: Dover. Copyright 1955. Reprinted by permission.

Other than defining the sequence of foot contacts for various animal gaits, there is no record of any scientific analysis being done from Muybridge's photographs; their impact was entirely at the visual level. Scientists and artists who had an interest in motion could, for the first time, see good-quality sequential images, and this allowed them to correct the many misinterpretations that had arisen due to the inadequacy of the human eye. Muybridge toured the world with an illustrated lecture that was received by academic and lay audiences alike with rave reviews. He believed that zoology and art departments in universities around the world would want a copy of his complete works and thought that this would finally make him a rich man. He died in impoverished circumstances in his native England, and ironically, only in the last decade with the publication of the Dover edition has his dream of widespread ownership of his work been realized.

The Muscle Men Discover Running Mechanics— Hill and Fenn

The first quarter of the 20th century was a time of relative neglect for scientific studies of running. Graham-Brown (1912) published one of the few papers on gait that have ever appeared in the *British Medical Journal*. He advanced the theory that locomotion consisted of losing and regaining balance and considered, in a somewhat philosophical manner, the question of how kinetic energy was lost and regained during contact with the ground and how potential energy might be transformed into kinetic energy. Amar (1920) continued the tradition of Marey in the analysis of human motion and ground reaction forces, but his efforts were primarily directed toward an analysis of injured veterans from World War I. Walking was the principal locomotor activity that Amar examined, but there is one example of a force-time curve from running at an unknown speed in his book.

Du Bois-Reymond (1925) conducted some early experiments on the drag experienced by the human body due to air resistance. Though not specifically designed to apply to running, the results of these experiments certainly influenced those investigators who were beginning to initiate calculations concerning the distribution of the metabolic energy expended against the various forms of resistance encountered in running. In particular, the study of running was about to benefit from the minds of A.V. Hill and his colleagues (Best & Partridge, 1928; Furusawa, Hill, & Parkinson, 1927a, 1927b; and Hill 1927, 1928). Although these individuals were primarily engaged in the formulation of mechanical and structural theories for muscle action based on the study of the frog gastrocnemius, they were drawn toward the study of running as an expression of maximal muscular effort in the intact human. A.V. Hill was a master of explaining complex ideas in a straightforward manner, as a review of his lectures on human performance will show (Hill, 1928). In many respects, he made the scientific study of running respectable among his highbrow colleagues and fellow members of the Royal Society. If a Nobel prize winner such as Hill thought running was important, then surely lesser mortals could venture into this arena.

Hill's specific contributions include the development of a velocity curve for sprint running, a consideration of the additional external work done in uphill running (Furusawa, Hill, & Parkinson, 1927a), a discussion of the efficiency of running (Furusawa, Hill, & Parkinson, 1927b), and calculations of the energy expended against air resistance at both sprint and distance running speeds. His estimates for the latter quantity were that about 3% of a distance runner's energy is expended against drag in still air. Subsequent measurements have at least doubled this value (Pugh, 1971). In a more general sense, Hill's contribution was to show that a combined physiological and mechanical approach to running could provide considerably more insight than either approach in isolation, and he gave encouragement to a number of his colleagues to pursue their own studies of running.

Wallace Fenn was one such individual who benefited from A.V. Hill's tutelage. Fenn had performed experiments on isolated muscle that led Hill to name the additional heat of shortening in an active muscle the "Fenn effect." Clearly influenced by Hill's studies of running, Fenn began his own experiments in the spring of 1928 at the University of Rochester. Acknowledging his debt to Marey and Demeny, and Otto Fischer in the methods section of his first paper, Fenn describes his association with the Eastman Kodak Company, which made it possible for him to use the "moving picture" technique (Fenn, 1929). The camera was hand cranked, and thus the unknown and irregular speed (of about 120 frames per second) was determined by dropping croquet balls in the field of view. The film was projected onto a screen with a lantern and analyzed by direct measurement with a protractor and subsequent graphical calculation of the center of gravity position. Although he mentions that one or two of his experimental runs were done at less than sprinting speed, there is no published record of any analysis that he may have conducted on distance running.

Fenn attempted to link the estimates of energy expenditure based on a kinematic analysis of the work done with metabolic estimates for the cost of sprint running. He performed a segment-by-segment calculation of kinetic and potential energies, setting the stage for the widespread use of this technique that has occurred recently (Winter, 1978). His discussion is a remarkably broad treatment of the problem and includes a consideration of errors due to exchange of kinetic and potential energy, storage of elastic energy in tendons and muscles, and intersegmental transfer of energy, which he realized might have resulted in "double counting" of certain components (see Williams, this volume).

In his second paper (Fenn, 1930) the work done against gravity was estimated from film records, and a force platform was used to study the retarding and propulsive shear forces (Figure 1.16). The migration of the center of gravity with respect to a fixed point on the body was also studied, and there is a long discussion of the action of the knee joint in the early braking phase of contact, including the suggestion that more economical runners may "waste less energy on forward pressure with the ground" (p. 459). This statement, which makes excellent mechanical sense, still remains to be proven.

Taken together, Fenn's two papers represented the most complete biomechanical analysis of running that had ever been conducted. They are important in that they form a bridge between the investigators of the 19th century and those of the present day. Fenn had thought about many of the problems that we are still attempting to solve, and he made a first

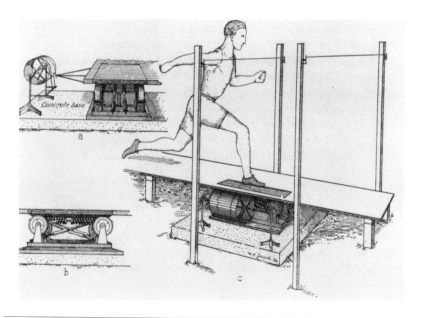

Figure 1.16.   A force platform used by Fenn (1930) to study the energy changes due to the anteroposterior forces during sprinting. *Note.* From "Work Against Gravity and Work Due to Velocity Changes in Running" by W.O. Fenn, 1930, *American Journal of Physiology,* **93,** p. 448. Copyright 1930 by The American Physiological Society. Reprinted by permission.

attempt at a method of solution that has, with some refinement, stood the test of 50 years of hindsight.

Soviet Science—Bernstein

The Pavlovian tradition of biological science in the Soviet Union fostered an individual who made considerable contributions to the theory of the motor control aspects of running. Nicholas Bernstein (1896–1966) was the initiator in the Soviet Union of the study of human activity. His group appeared to work in isolation from much of the biomechanical and physiological activities in the rest of Europe during the period from 1923 to 1944 as their work rarely appeared in languages other than Russian and was rarely quoted by other investigators. Bernstein was an experimentalist only to the extent that it served his very considerable abilities to develop over-arching theories of motor coordination and control; he developed the cyclographic method for the study of motion and pioneered the determination of body segment parameters in living subjects (Bernstein, 1967).

His major work on the biomechanics of running and walking (Bernstein, 1927) has not appeared in translation, but the details given in *The Coordination and Regulation of Movements* (Bernstein, 1967) indicate that Bernstein made many kinematic studies of both children and adults running in an attempt to analyze those factors common to running and walking that would give insight into the "biomechanical structures," as he called them, and reflect neurophysiological principles. That he attempted to relate developmental processes to biomechanical and physiological factors is apparent from the following quote:

> For a given value of the velocity in running in various children, such of them as have at the time of observation a more differentiated biodynamic structure give as a rule smaller amplitudes of acceleration, that is to say, a smaller range of dynamic forces. In order to arrive at the same final result the child with qualitatively less differentiated dynamic picture must expend more energy. (Bernstein, 1967, p. 57)

Why children are less economical in the energy cost of their locomotion is still a problem without solution today, yet few investigators have followed Bernstein's lead in viewing running from a developmental perspective. Bernstein's book also contains what may be the earliest record of a kinematic study of an elite runner. The data shown in Figure 1.17 are from "the world famous runner, Jules Ladoumeg" (Bernstein, 1967, p. 124), and the careful annotation of the force curves for the legs are characteristic of Bernstein's approach of identifying many small features for the purpose of relating different movement patterns.

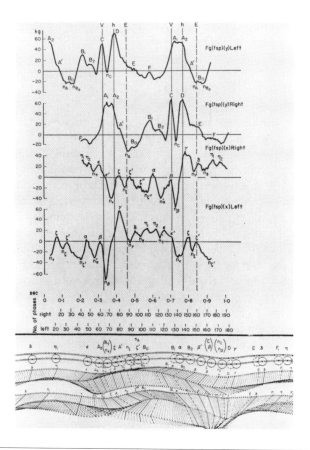

Figure 1.17. Data from the analysis of an elite runner, Jules Ladoumeg, collected by Bernstein in the 1920s. *Note.* From *The Coordination and Regulation of Movements* (p. 124) by N.A. Bernstein, 1967, Oxford, Great Britain: Pergamon. Copyright 1967. Reprinted by permission.

Insight Into Muscle Action— Elftman and Hubbard

The work of Herbert Elftman added a further dimension to biomechanical studies of running (Elftman, 1939a, 1939b, 1940, 1941). Elftman developed methods to study the action of muscles in the body based on kinematic data and realized through these calculations that muscles act, "regulating energy exchange, by transmitting, absorbing, releasing and dissipating energy" (Elftman, 1939b, p. 365). He further realized that two joint muscles are strategically located and activated during locomotion such that they are more efficient than their single joint counterparts would be.

In an interesting and creative discussion, Elftman speculated that three joint muscles would actually be preferable. He was among the first to measure the path of the center of pressure during locomotion from a force platform (Elftman, 1939a) and to present the ground reaction forces as vectors acting under the foot at the center of pressure (Figure 1.18) in a work that was basically an extension of Braune and Fischer's approach to the calculation of forces and moments with the advantage of a device to measure ground reaction forces. However, his understanding of muscular action made this paper a true early example of biomechanics in which mechanical methods were used to gain considerable biological insight.

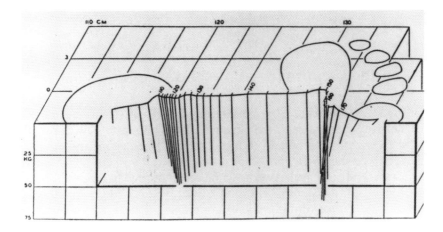

Figure 1.18. Elftman's calculations of the ground reaction forces shown as vectors acting at the center of pressure during walking. *Note.* From "Force and Energy Changes in the Leg During Walking" by H. Elftman, 1939, *American Journal of Physiology,* **125**, p. 339. Copyright 1939 by The American Physiological Society. Reprinted by permission.

To study running without the need to collect his own data, Elftman (1940) obtained permission to reexamine the films used in Fenn's earlier papers (Fenn, 1929, 1930). Elftman determined the net joint moments and the rates at which the muscles did work during one cycle of sprinting. He again theorized on the value of two joint muscles and postulated an almost 50% reduction in mechanical power due to their action.

At about the same time, Hubbard (1939) was engaged in a study of muscle action during running by more direct means. His experiment is important in a number of ways: First, it uses both ancient and modern techniques to study muscle activity—the pneumatic method originally devised by Marey, and the first recording of electromyograms during running that is available in the literature. Also, Hubbard was interested in distinguishing between the characteristics of trained and untrained runners, a topic that has occupied many biomechanists in more recent years. He identified that the muscles act very briefly to initiate and retard limb movements and that "the ballistic strokes have long periods during which the limb moves by momentum" (p. 36). The trained runners had longer strides than the untrained runners (probably due to faster speeds), but the differences between the two groups were not striking. Nevertheless, Hubbard concluded:

> Distance runners should light on their heel with the foot limp and leave the foot flat on the ground as the body passes over. Nurmi was a famous example of this. There is ample photographic evidence of it, and it was evident, in the laboratory study. (p. 36)

A Rounded Approach—Boje and Margaria

The final trend to bring us to the present-day era of scientific studies of distance running was beginning in both Scandinavia and Italy during the 1940s. This was the simultaneous measurement of metabolic and mechanical factors to derive relationships between the two sets of measures. Furusawa, Hill, and Parkinson (1927a) had attempted to forge some relationship in sprint running, but the work of Boje (1944) typifies this approach for distance running. He examined step length and a variety of anthropometric variables as well as determining steady-state oxygen uptake during treadmill running at speeds from 8 to 18 km/hour. In this important early study of running economy, Boje found that the untrained runners were less economical than the trained runners (a group that included the Danish marathon champion). He also discovered that the stride frequency was fairly independent of the speed of running (see Cavanagh & Kram, this volume).

One of the references in Boje's paper is to the work of Rodolfo Margaria, who was at the beginning of what was to be a distinguished career of inquiry into the energetics and mechanics of locomotion, particularly running. Margaria (1938) had earlier made the observation that the energy cost of running, when expressed per kilogram body weight per kilometer, was relatively independent of speed. A review of the various contributions of Margaria and his associates (Margaria, 1976) indicates the considerable extent to which their work has influenced our present thinking on the topic.

Concluding Remarks

This brief review of the history of the study of the mechanics of distance running has covered almost 2,500 years and has traced the progress of knowledge from Aristotle's speculation concerning vertical oscillation during running to measurements and calculations comparing mechanical and metabolic estimates of power output. Much of the remainder of the story is told in the subsequent chapters of this book.

When reviewing the history presented here, it is hard to escape the verdict that despite the tremendous advances in computing power and instrumentation that have occurred in the last 20 years, we are basically still trying to answer many of the questions posed by Fischer and Fenn 90 and 50 years ago respectively. However, there is every indication that progress in the remaining 10 or so years of the 20th century may be so rapid that the findings in the next decade alone could at least double our present knowledge. It should be a decade worth watching, working, and running through!

References

Amar, J. (1920). *The human motor: Or the scientific foundations of labor and industry*. New York: E.P. Dutton.

Bernstein, N.A. (1927). Studies on the biomechanics of walking and running. In *Questions of the dynamics of bridges*. Research of the Science Commission of the Peoples Commisariat of Transport, No. 63. (in Russian)

Bernstein, N.A. (1967). *The coordination and regulation of movements*. Oxford, Great Britain: Pergammon.

Best, C.H., & Partridge, R.C. (1928). The equation of motion of a runner exerting maximal effort. *Proceedings of the Royal Society, B*, **103**, 218-225.

Boje, O. (1944). Energy production, pulmonary ventilation and length of steps in well-trained runners working on a treadmill. *Acta Physiologica Scandinavica*, **7**, 362-375.

Borelli, G.A. (1685). *De motu animalium*. Batavis: Lugduni.

Braune, W., & Fischer, O. (1889) *Ueber den Schwerpunkt des menschlichen Koerpers mit Ruecksich auf de Ausruestung des Infanteristen* [On the center of gravity of the human body with a rucksack and infanty equipment]. (*Abhandlungen der mathematisch-physischen Klasse der Koeniglich Saechisischen Gesellschaft der Wissenschaften*, Vol. 15, No. 7). (English translation 1985). Berlin: Springer.

Braune, W., & Fischer, O. (1895-1904). *Der Gang des Menschen* [The human gait]. Leipzig: B.G. Teubner.

Braune, W., & Fischer, O. (1987). *The human gait* (P. Maquet & R. Furlong, Trans.). Berlin: Springer-Verlag. (Original work published 1895-1904)

Carlet, M. (1872). *Essai experimental sur la locomotion de l'homme*. Annales des Sciences Naturelles,

Cavanagh, P.R. (1981). *The running shoe book*. Mountain View, CA: World.

Clement, D.B., Taunton, J.E., Smart, G.W., & McNicol, K.L. (1981). Survey of overuse running injuries. *The Physician and Sportsmedicine, 9*, 47-58.

Dick, R.W., & Cavanagh, P.R. (1987). A comparison of ground reaction forces (grf) during level and downhill running at similar speeds. *Medicine and Science in Sports and Exercise, 19*(2), 67.

Du Bois-Reymond, R. (1925). Der Luftwiderstand des menschlichen Koerpers [The air resistance of the human body]. *Pfluegers Archives, 208*, 445-453.

Elftman, H. (1939a). Force and energy changes in the leg during walking. *American Journal of Physiology, 125*, 339-356.

Elftman, H. (1939b). The function of muscles in locomotion. *American Journal of Physiology, 125*, 357-366.

Elftman, H. (1940). The work done by muscles in running. *American Journal of Physiology, 129*, 673-684.

Elftman, H. (1941). The action of muscles in the body. *Biological Symposia, 3*, 191-209.

Encyclopedia Britannica: Great books of the Western World. (1952). (Vol. 11, Book 9, pp. 243-252). Chicago: Encyclopedia Britannica.

Fenn, W.O. (1929). Frictional and kinetic factors in the work of sprint running. *American Journal of Physiology, 92*, 583-611.

Fenn, W.O. (1930). Work against gravity and work due to velocity changes in running. *American Journal of Physiology, 93*, 433-462.

Furusawa, K., Hill, A.V., & Parkinson, J.L. (1927a). The dynamics of sprint running. *Proceedings of the Royal Society, B, 102*, 29-42.

Furusawa, K., Hill, A.V., & Parkinson, J.L. (1927b). The energy used in sprint running. *Proceedings of the Royal Society, B, 102*, 43-50.

— Gardner, E.N. (1967). *Athletics in the ancient world*. Oxford, Great Britain: Clarendon.

Graham-Brown, T. (1912, September 28). Note on some dynamic principles involved in progression. *British Medical Journal*, pp. 785-786.

Hill, A.V. (1927). *Living machinery, six lectures*. New York: Harcourt.

Hill, A.V. (1928). The air resistance to a runner. *Proceedings of the Royal Society, B, 102*, 380-385.

Hubbard, A.W. (1939). An analysis of running and of certain fundamental differences between trained and untrained runners. *Research Quar-*

terly of the American Association of Health and Physical Education, **10**(3), 28-38.

Inman, V.T., Ralston, H.J., & Todd, F. (1981). *Human walking*. Baltimore: Williams and Wilkins.

Jouffroy, F.-K., Ishida, H., & Jungers, W.L. (1983). Les systèmes loco-moteurs chez les primates: Hommage à Giovanni A. Borelli (1608-1679) [Primate locomotor systems: Homage to Giovanni A. Borelli]. *Annales des Sciences Naturelles, Zoologie* (Paris) (13th series), **5**, 53-65.

Keele, K.D. (1983). *Leonardo da Vinci's elements of the science of man*. New York: Academic Press.

Klopsteg, P.E., & Wilson, P.D. (1954). *Human limbs and their substitutes*. New York: McGraw Hill.

Marey, E.J. (1972). *Movement*. New York: Arno. (Original work published 1895)

Marey, E.J., & Demeny, G. (1885, November 9). Mesure du travail mécanique effetué dans la locomotion de l'homme. Talk given at the meetings of l'Academie des Sciences.

Margaria, R. (1938). Sulla fisiologia e specialmente sul consumo energetico della marcia e della corsa a varie velocita ed inclinazioni del terreno [Physiology, particularly energy consumption during walking and running at various speeds and grades]. *Atti. Acad. Naz. Lincei Memorie*, (series 6), **7**, 299-368.

Margaria, R. (1976). *Biomechanics and energetics of muscular exercise*. Oxford, Great Britain: Clarendon.

McMahon, P. (Trans.) (1956). *Leonardo da Vinci's treatise on painting*. Princeton, NJ: Princeton University Press.

Muybridge, E. (1887). *Animal locomotion* (Vols. 1-11). Philadelphia: University of Pennsylvania.

Muybridge, E. (1955). *The human figure in motion*. New York: Dover.

Muybridge E. (1979). *Human and animal locomotion*, (Vols. 1-3). New York: Dover. (Original work published 1887)

New Encyclopedia Britannica: Knowledge in depth. (1979). (Vol. 13, pp. 16-21, 15th Ed.). Chicago: Encyclopedia Britannica.

Nussbaum, M.C. (1978). *Aristotle's De moto animalium* (Animal motion). Princeton, NJ: Princeton University Press.

Pugh, L.G.C.E. (1971). The influence of wind resistance in running and walking and the mechanical efficiency of work against horizontal or vertical forces. *Journal of Physiology*, **213**, 255-276.

Vierordt, (1881). *Ueber das Gehen des Menschen in Gesunden und kranken Zustaenden nach Selbstregistrirenden Methoden* [On human gait in health and disease using a self-recording method]. Tuebigen, Germany: Verlag der H. Laupp'schen Buchhandlung.

Weber, W., & Weber, E. (1836). *Mechanik der menschlichen Gehwerkzeuge* [The mechanics of human locomotion]. Gottingen, Germany: Dieterichschen Buchhandlung.

Winter, D.A. (1978). Calculation and interpretation of mechanical energy of movement. In R.S. Hutton, (Ed.), *Exercise and sport science reviews*, (Vol. 6), pp. 183-202.

Chapter 2

Stride Length in Distance Running: Velocity, Body Dimensions, and Added Mass Effects[1]

Peter R. Cavanagh
Rodger Kram

At a given running velocity, an individual chooses, usually subconsciously, a particular combination of stride length (SL) and stride frequency (SF). This choice is made from a wide continuum of possible SL and SF combinations. A number of factors have been shown, or are suspected, to influence the choice of stride variables from that continuum. These include velocity (Nilsson, Thorstensson, & Halbertsma, 1985), grade (Davies, Sargeant, & Smith, 1974), footwear (Clarke, Frederick, & Cooper, 1983), surface properties (McMahon & Greene, 1979), anthropometric dimensions (van der Walt & Wyndham, 1972), developmental status (Amano, Hoshikawa, Toyoshima, & Matsui, 1987), muscle fiber composition (Armstrong, Costill, & Gehlsen, 1984), longitudinal influences (Nelson & Gregor, 1976), state of fatigue (Elliott & Roberts, 1980), and injury history. The exact changes caused by each of these specific factors are not in every case well known, but a knowledge of precisely how they affect the selection of SF and SL would add to our understanding of the general principles that govern locomotion.

Although distance running speeds may be said to range from 2.5 m • s^{-1} to 6 m •s^{-1}, the present series of studies has focused on steady-state, submaximal velocities, between about 3 and 4 m • s^{-1}. Our rationale for selecting this range was twofold. Because SL has been shown to affect the economy of distance running (Cavanagh & Williams, 1982; Hogberg, 1952), any further insight into factors affecting SL in the typical distance running speed range may shed light on the important topic of running

[1]This chapter previously appeared in P.R. Cavanagh and R. Kram (1990) and is adapted with permission of Williams and Wilkins, Baltimore, MD.

economy (Daniels, 1985; Elliott & Roberts, 1980; Fredericks, 1985a). Additionally, most recreational runners use speeds within the present experimental velocity range, so the results from the present studies have the best chance of having some practical as well as theoretical value.

The goal of this series of experiments was to investigate what factors contribute to a given individual's selection of SL at steady-state running velocities under a variety of conditions. The framework against which these experiments should be viewed is that evidence in the literature suggests that stride length is possibly selected to minimize metabolic energy cost, mechanical power output, the potential for injury, or some combination of these (Cavanagh & Williams, 1982; Clarke, Cooper, Hamill, & Clark, 1985; Hogberg, 1952; Kaneko, Matsumoto, Ito, & Fuchimoto, 1987; Plyley, Tiidus, & Pierrynowski, 1985). These experiments were designed to provide a data base against which these competing hypotheses could be discussed. This chapter describes experiments that examine the effects of (a) running velocity, (b) physical dimensions (stature, leg length, leg mass, moment of inertia), and (c) weight added to the lower extremity on preferred SL and SF.

Subjects

All of the subjects were male, well-conditioned, experienced recreational distance runners with considerable experience with treadmill running. Their ages ranged from 18 to 40 years and their competitive 10-km performances from 31 to 39 minutes. Although not measured directly, based on the prediction equations of Daniels and Gilbert (1979, pp. 97-99), their competitive times infer a $\dot{V}O_2$max of at least 54 ml $\cdot$ kg^{-1} $\cdot$ min^{-1}. Thus, at 3.83 m $\cdot$ s^{-1} the least fit subject was probably running at about 75% of his VO_2max. Written informed consent was obtained from all subjects in the study.

Methods

Treadmill Velocity

A Quinton 18-60 treadmill was used for the experiments. The treadmill velocity was measured electronically by attaching a piece of reflective tape to the treadmill belt and detecting revolutions with a photocell unit. The photocell output was conditioned and connected to an electronic timer. Using this arrangement, the velocity could be quickly and precisely adjusted while the subject ran on the treadmill. This was critical for accurate detection of small changes in stride variables. At each footstrike, the output from the tachometer on the treadmill varied, and these small transients were used to detect the time between successive foot contacts using the arrangement described in Appendix B. The accuracy of step times derived from this instrumentation was verified by a comparison with stopwatch-derived estimates of over 100 foot contacts.

The treadmill itself was securely bolted to the concrete floor in an attempt to standardize the mechanical stiffness of the running surface. No attempt was made to have the subjects run at the same place on the treadmill. It should be noted that Zacks (1973) found that running economy could differ by as much as 10% between two treadmills of varying bed and belt materials. This suggests that the mechanics of running may also be dependent on the mechanical properties of the treadmill used.

Stride Variables

The basic unit of running is the stride, the cycle from foot contact to the next contact of that same foot. The terms *stride* and *step* have been used by different groups of investigators to mean the same thing in the past (for example, Cavanagh & Kram, 1985; Cavanagh & Williams, 1982). In this series of papers, stride length is defined as the distance between successive contacts of the same foot, a definition that is consistent with the terminology for forms of locomotion other than bipedal running (Muybridge, 1957). Implicit in this definition is that of a step—half a stride, or the unit consisting of two successive foot contacts of opposite feet. The term *step length* has previously been used to denote the distance the body moves forward during a ground contact period (Cavagna, Thys, & Zamboni, 1976; McMahon & Greene, 1979). Perhaps "support distance" or "support length" as used by Nilsson et al. (1985) would be a less confusing term for this quantity. A reference of considerable antiquity that lays some claim to precedence in the definition of stride and step is from Shakespeare's *The Merchant of Venice*:

I'll . . . turn two mincing steps
Into a manly stride.   (3.4.67-68)

All of the variables studied and their abbreviations are given in Appendix A. Step time (sT) was continuously collected on-line with a microcomputer by electronically detecting the brief fluctuations in belt speed caused by each footstrike. The voltage across the standard Quinton tachometer was connected to the filter and comparator circuit described in detail in Appendix B. The comparator output was used as an interrupt for an ADALAB peripheral card in an Apple II microcomputer. Stride length was calculated from step time and treadmill velocity (V) as follows:

$$SL = 2 \cdot sT \cdot V$$

The similarity between treadmill running and overground running remains a controversial topic, recently discussed by Bassett, Giese, Nagle, Ward, Raab, and Balke (1985). We shall circumnavigate the controversy by reporting the results for treadmill running and assume that the general trends and principles apply to overground running. No previous studies have reported significant SL differences between overground and treadmill running.

Part 1: The Relationship Between Running Velocity, SL, and SF

Review of Literature

It has been established that humans increase both SL and SF as running velocity increases (see reviews by Dillman [1975] and Williams [1985]). The relationship between SL and V has been reported to be curvilinear over the entire velocity range with SL tending to reach a plateau at the higher velocities. SF, therefore, tends to increase relatively more at the higher velocities. It must be emphasized that the plateau in SL begins at velocities much greater than those typically encountered in distance running.

Studies that seek to fit equations relating SL and V over the entire range of human running velocities (Alexander, 1976; Vilensky & Gehlsen, 1984) tend to overlook basic differences between sprinting and distance running. Distance running places a premium on the minimization of metabolic energy expenditure. Sprinting is a process of maximizing power output. Functions that are applicable to both sprinting and distance running velocities may actually obscure important phenomena that occur in the middle of the velocity spectrum. The purpose of the first part of the investigation was to measure the preferred SF and SL over the typical distance running velocity range, a range where economy considerations are thought to dominate.

Subjects and Protocol

Twelve male subjects completed 5 days of testing within a 10-day period. The mean physical characteristics for the group were height (H), 1.796 m (sd 0.054); trochanteric height (bare feet), 0.96 m (sd 0.034); and total body mass, 70.2 kg (sd 7.37). The continuous protocol consisted of 30 minutes of running each day—5 minutes of warm-up and then 5 minutes at each of five velocities (order randomized for each subject on each day). The velocities examined were 3.15, 3.35, 3.57, 3.83, and 4.12 m $\cdot$ s^{-1}. The average SL was obtained by taking the mean SL during the last 2 minutes of each run on each of the five experimental days. The variation about the mean values will be described in more detail in a subsequent paper.

Results

Stride Length.   All subjects exhibited a linear relationship between mean SL and V with correlation coefficients >0.99. The individual best fit lines are shown in Figure 2.1. It should be recalled that the values used to calculate the correlations were means from a 2-minute measurement period. The range of SL was similar at both the lowest and highest velocities— the standard deviation increased from 0.13 m to 0.16 m whereas the CV remained approximately the same. There was some variation in the gradients of the regression lines relating SL and V, with values ranging from 0.528 to 0.708 meters per meter per second. A summary of and descriptive statistics for the individual data are given in Table 2.1.

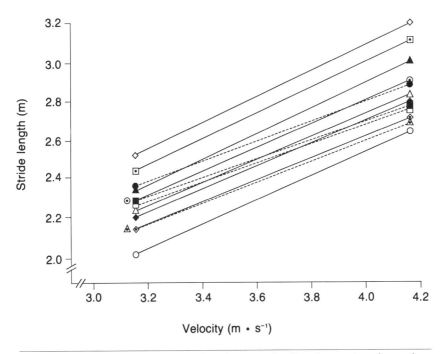

Figure 2.1.   The individual linear best lines to the five data points for each subject relating SL and V. Note the overall similarity in the gradient of these lines between most subjects (each subject is a different symbol).

Table 2.1 The Mean and Standard Deviation for Stride Length (in Meters), Stride Time (in Seconds), and Stride Frequency (in Hertz) at the Five Experimental Velocities

	Velocity (m · s⁻¹)				
	3.15	3.35	3.57	3.83	4.12
Stride length (m)	2.27	2.39	2.53	2.68	2.85
SD (m)	0.13	0.14	0.15	0.15	0.16
CV (%)	5.7	5.9	5.9	5.6	5.6
Stride length/stature	1.26	1.33	1.41	1.49	1.59
Stride length/leg length	2.36	2.48	2.64	2.79	2.97
Stride time (s)	0.723	0.716	0.710	0.703	0.694
SD (s)	0.043	0.042	0.042	0.042	0.041
CV (%)	6.0	5.9	5.9	6.0	5.9
Stride frequency (Hz)	1.38	1.40	1.41	1.42	1.44
SD	0.08	0.08	0.08	0.09	0.09

Note. Mean leg length = 0.96 m. Coefficients of variation (CV) are in percents. The stride lengths are also shown in dimensionless forms after division by the average height of the group and the average leg length.

The equation of the regression line for all the data shown in Figure 2.1 is

$$SL = 0.599 \ V + 0.387 \tag{2.1}$$

where SL is stride length in meters and V is running velocity in m · s⁻¹.

Substituting the appropriate values into Equation 2.1 demonstrates that a 33% increase in velocity between 3 and 4 m · s⁻¹ results in a concomitant change in SL of 28%.

Stride Time and Stride Frequency. The same data shown in Figure 2.1 were recalculated to give stride time (ST) and SF as functions of V, and the results are shown in Figure 2.2. The individual correlation coefficients between ST and V ranged from 0.87 to 0.99. It is apparent from Figure 2.2a that there was a large variation between individuals in the slopes of the regression lines. The gradients of the lines vary from −.0004 to −.0531 seconds per meter per second. The units for the gradient express the change in stride time (in seconds) for each 1 m · s⁻¹ change in running velocity. The stride time data from all subjects are summarized in Figure 2.2b, which shows the linear regression line for the data from all subjects.

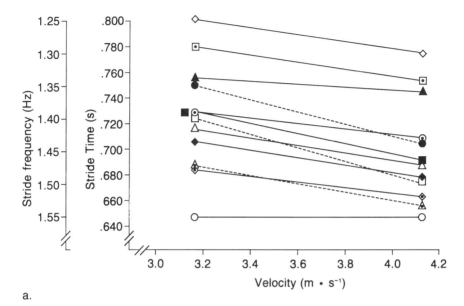

a.

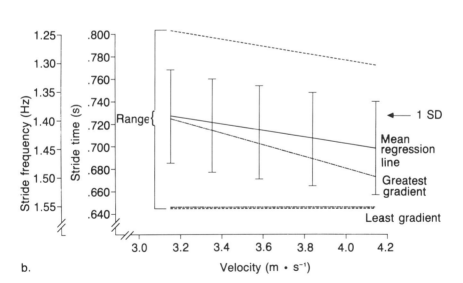

b.

Figure 2.2. The individual linear best fit lines (a) relating ST, SF, and V. Note that the gradients vary much more between subjects than the SL versus V lines (each subject is a different symbol). (b) The mean regression line relating ST, SF, and V for all subjects.

The equation for this regression line is

$$ST = -.029\ V + 0.815 \qquad (2.2)$$

where ST is in seconds and V is in m · s⁻¹.

This can also be expressed as stride frequency (SF) in Hz:

$$SF = 1.203 + 0.0575\ V. \qquad (2.3)$$

The gradient of the SF regression line has the units of Hz per meter per second. The ranges encountered for all subjects are also shown in Figure 2.2b together with the lines for the two subjects with the greatest and least gradients. This range was from 1.25 Hz to 1.55 Hz and the mean values were 1.38 Hz at 3.15 m · s⁻¹ and 1.44 Hz at 4.12 m · s⁻¹. Using the mean regression line, over the range of velocities between 3.0 and 4.0 meters, SF increased by 0.06 Hz, which is only a 4% increase. This is in contrast to the 28% change in SL already noted for the same velocity range. SL might therefore be considered a strong function of velocity and SF a rather weak function of velocity. Note from Figure 2.2a that two individuals in this sample show almost no change in SF with V.

Discussion

Theoretically, one could maintain the SL selected at 3.15 m · s⁻¹ (mean value 2.27 m) while increasing the velocity to 4.12 m · s⁻¹. This would require an SF of 1.8 Hz. Nilsson and Thorstensson (1985) have suggested that a stride frequency of 1.7 Hz may represent the maximum stride rate achievable at any speed. The alternative, keeping the stride time the same and increasing only SL, requires a 32% increase in the SL from 2.27 to 2.98 m. In general, our subjects chose this second alternative, as they increased SF only 4% while lengthening SL by 28%. Although different individuals had different preferred SFs, they appeared to maintain their chosen value across speeds.

Such relatively constant stride frequencies with increasing velocities have been observed in quadrupedal animals galloping and bipedal hopping animals (Heglund, Taylor, & McMahon, 1974). SF in quadrupedal trotting, however, increases substantially with speed; there is an approximate 50% increase in SF for a doubling in speed (Heglund & Taylor, 1988). Bipedal running is usually considered analogous to quadrupedal trotting (Alexander, 1977) because it is bilaterally symmetric, and there are two aerial periods per stride. However, in terms of stride frequency, human running appears to be more similar to quadrupedal galloping. There is no complete source of SF data for other species running bipedally that can be used for comparison.

Running economy may be the reason that one preferred SF is used over a range of speeds. As described by Cavanagh and Williams (1982), most individuals choose a combination of SF and SL that minimizes metabolic cost during level running. Although Cavanagh and Williams studied only one speed (3.83 m • s^{-1}), there is no reason to suspect that this is a "special" speed. In fact, Hogberg (1952) showed, for a single runner, that the most economical stride frequency changed by only 1.4% over a range of speeds from 3.9 to 5 m • s^{-1}. It is likely that at all submaximal speeds the most economical SF and SL combination is usually selected. Viewing our data with this perspective indicates that for each individual there may be a characteristic preferred SF that must be utilized to run most economically at distance running speeds. Though previous studies have suggested that there is a most economical stride length at a *given speed*, our data suggests that there may be a most economical stride frequency at *all speeds* used in distance running.

It has been established that ground reaction impact forces increase with running speed (Munro, Miller, & Fuglevand, 1987), and it is generally assumed that high-impact forces play a causative role in the etiology of running injuries. Although there are no data concerning the effect on ground reaction forces of changing SL at a given running speed, Clarke et al. (1985) have shown that, at any one running speed, tibial accelerations are increased by taking longer strides, providing circumstantial evidence that increased SL results in increased impact force. Because stride length has been shown to increase preferentially over stride frequency with increasing speed, it appears that economy (rather than injury avoidance) is dominating the choice of SL. If a runner chose to progressively understride at higher speeds, it would suggest that maintaining low forces was an objective that determined the SF/SL combination that was chosen.

The increase in SL with increasing speed also relates to the topic of mechanical power output. Kaneko et al. (1987) have reported that at a particular speed, SL is chosen so as to minimize the sum of the external and internal mechanical power outputs. They found increased internal power for shorter SLs and higher external power output for longer SLs. The sum of these mechanical powers was minimized at an SF of approximately 1.4 Hz for running at 2.5, 3.5, and 4.5 m • s^{-1}. This corresponds very closely with our data reported here for preferred SF, suggesting that either mechanical power output, metabolic cost, or both, influence the selection of SL.

Before concluding this section, some practical applications should be noted. Using an individual's SL or SF versus V profile, a new form of all-velocity pedometer can be accurately calibrated. A conventional pedometer estimates distance traveled by simply counting footstrikes and assuming a constant value for SL. Because SL varies directly with velocity, the assumption of constant SL therefore assumes constant velocity. This assumption is unrealistic, and such devices therefore have poor accuracy (Washburne, Chin, & Montoye, 1980). A more accurate pedometer can be constructed based on the individual's previously known ST or SL versus velocity equation. The simpler quantity to measure is clearly ST, as a switch on the shoe can be arranged to detect each footstrike. Once the number of footstrikes is known, the addition of a timing device allows average ST to be calculated. This is substituted in the stride time versus running velocity relationship to determine average running velocity, and the result is multiplied by running time to obtain distance.

The accuracy of such a device depends on a number of factors. The most important of these is the gradient of the individual's stride time–running velocity relationship. Figure 2.2a shows that the prediction for certain subjects would be less accurate than for others tested. Gradient and fatigue may also be confounding factors for the accuracy of any pedometer device.

Individuals who do have flat SF versus V profiles, though not able to use such a pedometer with great accuracy, could receive feedback while training, which could help them to run more economically. Although it is rare, we have occasionally found consistent overstriders or understriders; that is, individuals who choose an uneconomical SF. Such persons, given sufficient laboratory calibration, could train (on the level) with a metronome set to their optimal frequency and presumably learn to run more economically. Such a device, if used during overground running, would not be so simple for an individual with a steeper SF versus V profile, because the device would have to sense overground speed and present the optimal SF for that speed. It is important to realize that only accurate physiological testing can determine the optimal SF to which the metronome should be set. Thus, existing devices that purport to train the runner to use an "appropriate" SF cannot be calibrated without access to sophisticated metabolic measurement apparatus, which is unavailable to most runners.

Part 2: The Relationship Between Anthropometric Variables and SL

Review of Literature

Although it is generally thought that taller individuals inevitably take longer strides at a given velocity of running, there is little evidence to support this idea. Some of the previously published work is confounded by the subject sampling problems discussed in this section. In this portion of the investigation, we measured numerous anthropometric variables (APV) and examined the correlation of these with SL. The robustness of such correlation techniques depends greatly on the nature and size of the sample selected. For example, if the sample has two clusters of subjects (very short and very tall) then the correlation coefficient will be higher than if the calculation was done on one of the clusters. Table 2.2 summarizes previous correlational studies that have used velocities within the distance running range.

SL is not nearly as well correlated with either H or LL as is generally believed. The highest correlation ($r = 0.69$) from any of the distance running studies (van der Walt & Wyndham, 1972) reviewed indicates that leg length accounted for only 48% of the variance in SL in this sample. In the worst case, a slight inverse correlation between SL and LL has been found (Cavanagh, Pollock, & Landa, 1977).

Previous studies (Elliott & Blanksby, 1979) have attempted to determine general equations for prescribing SL at a certain velocity, for a person of a certain size. Such equations are typically of the following form:

$$SL/L = f(V/L);$$

that is, preferred SL—divided by some linear anthropometric measure (usually stature or leg length)—is equal to some function of the running velocity divided by the previously used linear dimension.

Alexander (1976, 1984) has argued that this type of scaling is dimensionally incorrect since V/L is not dimensionless; rather it has the dimension (time^{-1}). Based on considerations of physical and dynamic similarity, he recommended that dimensionless SL should be calculated as a function of the Froude number, V^2/gL, or its square root, $V/SQRT(gL)$, where g is the acceleration due to gravity on the earth and L is an appropriate

Table 2.2 Correlations From the Literature Between SL and Various Anthropometric Measurements

Authors	Velocity (m · s⁻¹)	Number of subjects	SL vs. H		SL vs. LL	
			r	r²	r	r²
van der Walt and Wyndham (1972)	2.23-3.57	6	—	—	.67	.45
Cavanagh and Williams (1982)	3.83	10	.09	.008	—	—
Cavanagh, Pollock, and Landa (1977)	4.97	14 elite	—	—	.67	.45
		8 good	—	—	−.10	.01
Elliott and Blanksby (1979)	3.5	10	.42	.18	.69	.48
	4.5	10	.64	.41	.68	.46

Note. All terms are defined in text.

linear body dimension. This type of normalization is common in engineering and in the analysis of many physical systems (Massey, 1971). Alexander (1976) suggested an equation of the form

$$SL/L = f(V^2/gL),$$

specifically

$$SL/L = a(V^2/gL)b, \qquad (2.4)$$

where a and b are constants, and L is trochanteric height. Note that SL is no longer a simple linear function of some body dimension; the scaling is now allometric rather than isometric. Allometric scaling is a technique frequently used by comparative biologists, and those unfamiliar with this methodology are referred to Schmidt-Nielsen's text (Schmidt-Nielsen, 1984).

In walking, plots of SL/L versus $V/SQRT(gL)$ were shown to better describe data for a diverse group of male and female children and adults than plots of SL/L versus V/L (Alexander, 1984). The improvement in fit was most pronounced at slow walking velocities. Values of 2.3 and 0.3 for a and b, respectively, have been calculated (Alexander, 1976) and used with Equation 2.4 (with hip height as the linear dimension) to fit experimental data from a wide variety of animal species walking and running over a broad velocity range. A deviation from best fit in SL of, say, 25% of leg length is not a serious obstacle for a general biological application, but it does represent a major discrepancy in the prescription of an SL for an individual distance runner.

Subjects and Protocol

In an effort to overcome some of the sampling problems previously discussed, we selected two groups of subjects on the basis of H and LL to form rectangular distributions of these two variables, centered near the mean stature and leg length for U.S. adult males. Intuitively, these two measures were expected to be the best predictors of SL. The first sample, based on stature, had a total of 21 subjects, filling seven "bins" at 2-cm intervals ranging from 1.69 m to 1.83 m. The second sample, based on leg length (greater trochanteric height), contained 16 subjects, filling eight bins at 1.5-cm intervals ranging from 0.83 m to 0.95 m. Many of the subjects were included in both groups.

All subjects ran at the single velocity of 3.83 m • s^{-1} for a period of 5 minutes after a 5-minute warm-up. Measurements of average SL during the last 2 minutes of the run were obtained as described earlier.

Stature and body mass were measured before the run together with a number of anthropometric dimensions. These dimensions were needed to predict masses and moments of inertia, which are the primary mechanical determinants of limb dynamics.

The 11 values required for the estimation of thigh, shank, and foot mass according to Clauser, McConville, and Young (1969) and for the moments of inertia of the leg segments about their proximal joints according to Dempster (1955) were measured. The lengths, masses, and moments of inertia used for correlation are shown in Figure 2.3. The individual body dimensions and SLs for these two groups of subjects are shown in Tables 2.3 and 2.4 respectively.

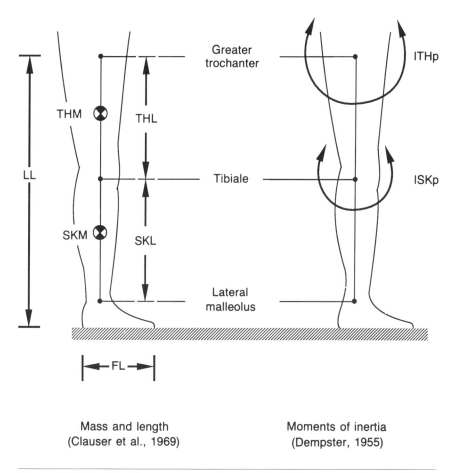

Mass and length
(Clauser et al., 1969)

Moments of inertia
(Dempster, 1955)

Figure 2.3. The lengths, masses, and moments of inertia used for correlation with stride length. All symbols are defined in Appendix A. *Note.* Data from Clauser, McConville, and Young, 1969; Dempster, 1955.

Table 2.3 Anthropometric Dimensions and SL at 3.83 m · s⁻¹ for the Subjects in the Group Based on a Rectangular Distribution of Stature

Table 2.3 Anthropometric Dimensions and SL at 3.83 m · s^{-1} for the Subjects in the Group Based on a Rectangular Distribution of Stature

	Lengths				Masses					Inertia		
H	LL	THL	SKL	FL	Mb	TLM	THM	SKM	FM	ITHp	ISKp	SL
1.695	.866	.430	.364	.254	63.2	11.47	7.13	3.30	1.04	.336	.108	2.67
1.708	.901	.496	.334	.271	70.8	12.72	8.52	3.17	1.05	.504	.102	2.58
1.710	.846	.417	.360	.245	61.7	11.40	7.00	3.35	1.06	.311	.103	2.67
1.715	.832	.406	.357	.268	63.8	11.41	7.08	3.28	1.05	.304	.105	2.69
1.715	.871	.439	.354	.271	72.7	13.07	8.09	3.81	1.17	.404	.113	2.81
1.727	.833	.406	.353	.274	70.2	12.92	8.06	3.70	1.16	.334	.113	2.52
1.742	.880	.473	.339	.269	66.3	11.94	7.93	3.46	1.09	.429	.098	2.67
1.743	.897	.463	.364	.271	64.9	11.91	7.38	3.43	1.09	.400	.109	2.91
1.745	.875	.463	.332	.270	75.6	13.08	8.36	3.60	1.12	.466	.107	2.84
1.751	.885	.468	.349	.281	74.7	13.50	8.33	3.97	1.21	.474	.116	2.58
1.752	.915	.475	.369	.278	65.0	11.79	7.32	3.40	1.07	.424	.114	2.90
1.768	.894	.469	.357	.266	75.8	13.78	8.55	4.01	1.22	.481	.125	2.54
1.771	.875	.475	.328	.250	67.6	12.08	7.96	3.12	1.06	.441	.094	2.55
1.771	.942	.513	.359	.264	69.7	12.97	8.22	3.62	1.13	.529	.115	2.93
1.780	.924	.502	.353	.270	63.6	12.02	7.42	3.50	1.11	.462	.102	2.79
1.797	.873	.443	.361	.280	79.5	15.68	10.17	4.24	1.27	.451	.133	3.01
1.798	.944	.441	.429	.249	71.7	13.13	8.27	3.73	1.13	.401	.169	2.78
1.810	.924	.467	.389	.259	66.5	11.99	7.27	3.61	1.12	.417	.128	2.47
1.811	.984	.485	.431	.29	68.4	12.52	7.92	3.50	1.10	.465	.163	2.75
1.820	.946	.423	.455	.268	78.0	13.90	8.34	4.28	1.28	.403	.207	2.81
1.832	.947	.472	.406	.277	73.5	13.15	8.07	3.90	1.18	.472	.156	2.88

Note. Abbreviations are defined in Appendix A. Lengths are in meters, masses in kilograms, moments of inertia in kilogram meters squared.

Table 2.4 Anthropometric Dimensions and SL at 3.83 m · s⁻¹ for the Subjects in the Group Based on a Rectangular Distribution of Leg Length

Table 2.4 Anthropometric Dimensions and SL at 3.83 m · s^{-1} for the Subjects in the Group Based on a Rectangular Distribution of Leg Length

	Lengths				Masses					Inertia		
H	LL	THL	SKL	FL	Mb	TLM	THM	SKM	FM	ITHp	ISKp	SL
1.727	.833	.406	.353	.274	70.2	12.92	8.06	3.70	1.16	.334	.113	2.52
1.715	.832	.406	.357	.268	63.8	11.41	7.08	3.28	1.05	.304	.105	2.69
1.710	.846	.417	.360	.245	61.7	11.40	7.00	3.35	1.06	.311	.103	2.67
1.753	.860	.423	.364	.270	70.4	12.47	8.01	3.39	1.07	.363	.120	3.04
1.695	.866	.430	.364	.254	63.2	11.47	7.13	3.30	1.04	.336	.108	2.67
1.745	.875	.463	.332	.270	75.6	13.08	8.36	3.60	1.12	.466	.107	2.84

	Lengths					Masses				Inertia		
H	LL	THL	SKL	FL	Mb	TLM	THM	SKM	FM	ITHp	ISKp	SL
1.717	.884	.471	.346	.255	63.6	11.61	6.89	3.61	1.12	.407	.098	2.49
1.751	.885	.468	.349	.281	74.7	13.50	8.33	3.97	1.21	.474	.116	2.58
1.748	.893	.469	.359	.257	63.1	11.17	6.98	3.15	1.03	.401	.104	2.59
1.708	.901	.496	.334	.271	70.8	12.72	8.52	3.17	1.05	.504	.102	2.58
1.742	.915	.415	.423	.268	63.6	11.93	6.86	3.88	1.19	.317	.146	2.66
1.752	.915	.475	.369	.278	65.0	11.79	7.32	3.40	1.07	.424	.114	2.90
1.780	.924	.502	.353	.270	63.6	12.02	7.42	3.50	1.11	.462	.102	2.79
1.810	.924	.467	.389	.259	66.5	11.99	7.27	3.61	1.12	.417	.128	2.47
1.820	.946	.423	.455	.268	78.0	13.90	8.34	4.28	1.28	.403	.207	2.81
1.832	.947	.472	.406	.277	73.5	13.15	8.07	3.90	1.18	.472	.156	2.88

Note. Abbreviations are defined in Appendix A. Lengths are in meters, masses in kilograms, moments of inertia in kilogram meters squared.

Results

Correlation of SL With Anthropometric Variables.   The anthropometric data were then correlated with each other and with SL (see Table 2.5). Considering first the data for the stature-based group, the expected correlations between body dimensions were as seen typically in such studies (Dempster, 1955). For example, stature and leg length were correlated with a coefficient of 0.76. However, correlations between all of the anthropometric measures and SL were remarkably low and none differed significantly from zero ($p > 0.05$). The highest correlation ($r = 0.31$) was between SL and LL. H and SL had a correlation coefficient of 0.26. Thus LL and H explained only 9.6% and 6.8% of the variance in SL respectively.

Table 2.5   Correlation Coefficients Between Anthropometric Dimensions and SL at 3.83 m · s⁻¹ for the Stature and Leg Length Groups

	H	LL	THL	SKL	FL	Mb	TLM	THM	SKM	FM	ITHp	ISKp
Stature[a]	.257	.310	.137	.233	.308	.204	.308	.272	.282	.223	.204	.255
LL[b]		.182	.315	−.043	.209	.364	.291	.233	.279	.065	.016	.088	.287

[a]Based on a rectangular distribution of stature.
[b]Based on a rectangular distribution of leg length.

It should be recalled that this group was structured to exhibit a rectangular distribution of statures, and thus the low correlation cannot be explained as an artifact of the sampling distribution.

Turning to data for the leg length–based group, Table 2.5 shows that the correlations between stride length and the anthropometric variables were again consistently weak ($r \leqslant 0.36$) and no correlation coefficient differed significantly from zero. Correlation coefficients in this group, with a rectangular distribution of leg lengths, were 0.18 between SL and LL and 0.32 between SL and stature.

Scaling. From the data of the 12 subjects used in the running velocity experiment (who were not selected based on particular anthropometric values), Figure 2.4 was constructed by plotting SL/LL versus V (Figure 2.4a) and SL/LL versus V/LL (Figure 2.4b) over the velocity range from 3.15 to 4.12 m • s^{-1}. From the data collected, best fit equations of the form described by Alexander (1984) were calculated and the coefficients a and b were determined to be 2.31 and 0.43 respectively. Dimensionless plots were then constructed where SL/LL is plotted against V/SQRT (g • LL) (Figure 2.4c) and against 2.31 [V^2/(g • LL)]$^{0.43}$ (Figure 2.4d).

Figure 2.4 essentially contains different manipulations of Figure 2.1, to which the reader may wish to refer. Based on our data, the equation that best relates V, LL, and SL using the allometric scaling technique is

$$\text{SL/LL} = 2.31 \times [\text{V}^2/(g \cdot \text{LL})]^{0.43}. \tag{2.5}$$

Discussion

Although we have emphasized how low the correlations were between the physical dimensions and the chosen SL, it should be noted that the correlations were, with only one exception, all positive. Taller, longer

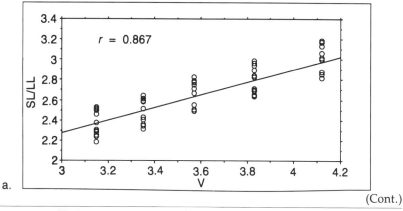

a.

(Cont.)

Figure 2.4. Various presentations of the stride length versus velocity relationship: (a) plot of SL/LL vs. V, (b) plot of SL/LL vs. V/LL, (c) plot of SL/LL vs. V/SQRT (g • LL), and (d) plot of SL/LL vs. 2.31V^2/(g • LL)$^{0.43}$.

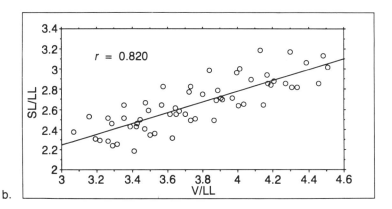

b.

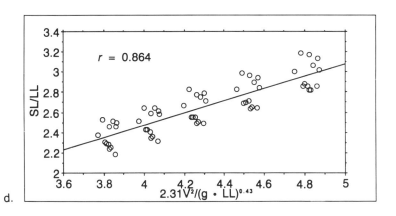

c.

d.

Figure 2.4. Continued.

legged, heavier, and heavier legged individuals and those with greater moments of inertia tended to take longer strides. These variables are, of course, not independent. Though our results predict that an 80-kg person with 1.0 m LL will take longer strides than a 70-kg person with the same LL, they do not predict that an 80-kg person with 0.8 m LL will take longer strides than the 70-kg person with 1.0 m LL. Each of the factors influences the selection of SL to a different degree, which is reflected in the correlation coefficient.

Our values for the scaling coefficients (a = 2.31 and b = 0.43), which relate dimensionless SL with dimensionless velocity, are quite similar to the values of 2.3 and 0.3 previously reported (Alexander, 1976). Table 2.6 was constructed using the previously reported coefficients to predict the SL for the average subject in our study. The agreement between predictions based on the allometric equation derived from studies of multiple species and multiple gaits (including humans, horses, ostriches, and elephants walking, trotting, and galloping) with the present human data is quite impressive. The velocities in the present experiment are toward the fast extreme of the range over which Alexander (1976) fitted his equation, and thus the gradual departure from accuracy is understandable.

Table 2.6   Stride Length at Five Different Velocities Predicted Using Alexander's Coefficients[a]

	Velocity (m · s⁻¹)				
	3.15	3.35	3.57	3.83	4.12
Predicted SL (m)	2.23	2.32	2.41	2.51	2.63
Present SL data (m)	2.27	2.39	2.53	2.68	2.86
Difference (m)	−0.04	−0.07	−0.12	−0.17	−0.23
Difference (%)	−1.08	−2.9	−4.7	−6.7	−8.0

Note. Adapted from Alexander, 1976.
[a]Subject's leg length is equal to the average in our study. The actual observed mean stride lengths are also shown.

Appropriate scaling should reduce the variability of the data—that is, improve the correlation between the two variables. Table 2.7 shows the correlations between the various forms of the stride length versus velocity relationship shown in Figure 2.4. It is notable that the frequently used relationship in which dimensionless SL is plotted against V/LL provides a correlation coefficient similar to that between absolute SL and V. It is also apparent that neither of the dimensionless forms of the relationship provides an improvement over the plot in which dimensionless SL is plotted against absolute V. Presumably the narrow speed range, the single species, and the relatively similar body dimensions of the present subjects (when compared to previous studies [Alexander, 1976, 1984]) have prevented the dramatic rationalization of the data by the nondimensional approach. The mathematical rigor of the latter approach is, however, appealing.

Table 2.7 Correlation Coefficients Between Various Forms of the Stride Length (SL) Versus Velocity (V) Relationship

Stride length form	Velocity form	Correlation coefficient
SL	V	0.815
SL/LL	V	0.867
SL/LL	V/LL	0.820
SL/LL	$V/\mathrm{SQRT}\,(g \cdot LL)$	0.855
SL/LL	$2.31V^2/(g \cdot LL)^{0.43}$	0.864

Note. All terms are defined in Appendix A.

Although scaling equations indicate the overall effect of external variables, it is clear that internal variables also exert substantial influence. Two of our subjects with almost identical leg lengths and heights but greatly different preferred SL serve to illustrate the effect of these internal variables at a running speed of 3.83 m $\cdot$ s^{-1}.

Subject A: height 1.791 m, LL 0.959 m, weight 79.5 kg, SL 2.84 m

Subject B: height 1.791 m, LL 0.957 m, weight 71.8 kg, SL 2.52 m

Further insight into influences on SL can be found in the comparison of identical twins conducted by Frederick, Robinson, and Hamil (1987). Their subjects ($n = 8$) were all monozygotic twins who were competitive distance runners of similar caliber (national championship participants). The mean intrapair difference in SL at the test velocity of 3.83 m $\cdot$ s^{-1} was 0.10 m, only 4%. Thus, when external and internal structures are the same, essentially the same stride length appears to be used.

Conversely, consider two other subjects with substantially different body dimensions but nearly identical SL, again running at a speed of 3.83 m $\cdot$ s^{-1}.

Subject C: height 1.811 m, LL 0.924 m, weight 66.5 kg, SL 2.47 m

Subject D: height 1.597 m, LL 0.838 m, weight 58.4 kg, SL 2.44 m

Though the relevant "internal" variables that determine the choice of SL are difficult to identify, runners are clearly able to integrate all of the factors and arrive at the appropriate SL that minimizes energy cost (Cavanagh & Williams, 1982; Hogberg, 1952). It is clear, however, from the foregoing analysis that anthropometric variables are of little use in prescribing an SL at a given speed for a given individual.

The practical implication of these findings is that distance running coaches would be unwise to attempt any intervention involving alterations in SL based on body dimensions. Only physiological evidence of a discrepancy between preferred and most economical SL should be accepted as evidence that a change in SL is desirable.

Part 3: The Effects of Added Mass

The poor predictability of SL that we have found based on anthropometric dimensions is somewhat counterintuitive; after all, small animals use higher stride frequencies than large animals (Hill, 1950). One might suspect that the correlational approach has introduced too much "noise"— that is, some tall people have light long legs, and some short people have proportionally heavier limbs. This makes it more difficult to separate the individual effects of mass and length. To isolate limb mass rather than limb length as a variable, perturbation experiments were conducted that added mass to the limbs and used the subject's own normal gait as the control.

Review of Literature

Experiments in running with loads carried on the back or on the legs have taken both biomechanical and metabolic approaches. Cureton, Sparling, Evans, Johnson, Kong, and Purvis (1978), who looked at back-carried loads, remarked that "running form and efficiency were not hampered by carrying the added weight" (p. 197). Taylor, Heglund, McMahon, and Looney (1980) measured SF in various animals (including humans) carrying back loads and reported that after training, SF was not significantly different between loaded and unloaded running. Martin (1985) reported, in human runners, only small changes in SL (1% increase) and swing time (2% increase) with 0.5 kg of additional load carried on each foot. Kram, McMahon, and Taylor (1987) also found constant SF with added back loads of up to 30% of body mass in humans but reported substantially increased contact times for loaded running. This adjustment allowed the subjects to limit the peak vertical ground reaction force.

These perturbation studies are consistent with results of our correlational studies shown earlier in Table 2.5. Body mass (Mb) and foot mass (FM) were both found to be rather weak predictors of SL in our human subjects. However, in a broader context, smaller (and therefore lighter) animals have been shown to exhibit higher SF than larger (heavier) animals (Heglund, Taylor, & McMahon, 1974).

Investigators who have measured the increased metabolic cost of carrying additional loads on the back during running (Cureton et al., 1978; Taylor et al., 1980) have found that oxygen consumption increased in direct proportion to the mass carried. Human studies of limb loads (Catlin & Dressendorfer, 1979; Frederick, 1985b; Myers & Steudel, 1985) all report that adding 100 g (50 g/leg) to the feet increased oxygen uptake by about 1% over a fairly wide range of running velocities. Another rule of thumb is that it is metabolically about six times as expensive to carry a given mass on the feet or ankles as it is on the back (Catlin & Dressendorfer, 1979; Frederick, 1985b; Myers & Steudel, 1985). The increase in cost is more extreme as the weights are placed more distally on the leg (Martin, 1985; Myers & Steudel, 1985).

Summarizing the literature, it is apparent that even large back loads do not have an appreciable effect on SL, and their metabolic effect is far less than that of equal loads on the legs. Even small loads on the feet dramatically increase metabolic cost but appear to have little effect on SF or SL.

Subjects and Protocol

Ten subjects running at a test velocity of 3.65 m · s^{-1} were used to study the effects of moderate added leg weights. On a familiarization day the subjects ran for 5 minutes under normal conditions, 5 minutes with 0.18 kg added to each shoe, and finally for another 5 minutes with 0.335 kg added to each shoe. The added weights (metal disks, 0.1 m in diameter) were strapped tightly to the throat of each shoe.

The two days of actual testing used the following protocol: a 5-minute warm-up in the test shoes, then in randomized sequence the three conditions during which SL data were collected (no added weight, 0.18 kg added to each shoe, and 0.335 kg added to each shoe). Each test period lasted 5 minutes with data collected during the last 2 minutes. Five minutes of rest, although not needed, was enforced between conditions. A different randomized sequence was used on the second day.

After the results from this experiment were examined, it was decided to test the effect of even greater added weight at the ankle. A different group of 10 subjects first completed a 5-minute warm-up at the test velocity of 3.83 m · s^{-1}, during which time their preferred stride lengths (PSL) were determined. Ankle weights of 1.1 kg per leg were then attached. Subjects ran for 5 minutes with ankle weights on both legs, and SL data were collected during the last 2 minutes of the run. These weights seem to be close to the limit that can be reasonably studied, because greater loads at this running speed would increase $\dot{V}O_2$ above steady state and cause localized fatigue. Heavier loads may also have increased the risks of lower extremity, particularly knee, injuries in our subjects.

Results

No significant change in SL or SF was brought about by any of the added weights. Mean SLs were as follows: no load, 2.63 m; +0.18 kg/foot, 2.616 m; and +0.336 kg/foot, 2.616 m. Mean SL for the group of subjects with the 1.1 kg/foot weights was an insignificant 0.022 m longer than the unloaded condition (2.737 m vs. 2.760 m).

Discussion

The addition of loads of up to 1.1 kg at the ankle approximately doubled the mass of the foot and dramatically increased the moment of inertia of the leg about the hip. Surprisingly, these additions did not alter the PSL. It would be worthwhile to examine whether the support time/swing time ratio changes under these conditions. Thus, in addition to the lack of dependence of SL on stature or leg length shown earlier, these findings strongly suggest that factors other than limb segment mass or inertia are the primary determinants of SL. Individuals who appear externally to be physically identical may have internal differences in, for example, their muscle fiber type populations, recruitment patterns, or mechanical advantage at important muscle insertions, as well as more subtle differences.

There is indirect evidence to suggest that these important "internal" differences include fiber type and/or training specificity. Armstrong, Costill, and Gehlsen (1984) compared the SL of sprinters and marathon runners running at the same distance running speed. The two groups of runners were essentially the same height and leg length, but the sprinters were substantially heavier and presumably had heavier limbs than the marathoners. Although muscle biopsy data were not available, it is fair to assume that the sprinters had a higher proportion of fast twitch muscle fibers than the distance runners. The results indicated that at 3.9 m • s^{-1}, the sprinters had dramatically longer strides (2.92 vs. 2.52 m) and were less economical (approximately 47 vs. 40 ml • kg^{-1} • min^{-1}). Based on our findings with added limb weights it seems fair to dismiss greater limb mass as the reason for the sprinters' longer SLs. This suggests that such factors as different fiber types, other physiological characteristics, training background, or a combination of these may be causative in the observed differences in SL.

Concluding Comments

In this chapter we have provided a basic set of data on stride length in distance running. The way in which SL changes with velocity and the effects on SL of classical anthropometric variables such as limb length, mass, and inertia have been reported and discussed. The evidence suggests that models based only on externally measurable variables are not adequate to predict or prescribe an individual's SL. Although these variables place constraints on the SL chosen at a given speed, it is clear that anthropometric variables are not the primary determinant of SL. Within the speed range that we have studied the mechanism for increasing speed appears to be one that maintains SF nearly constant, necessitating an increase in SL. Though different individuals have different preferred SFs, they appear to maintain their chosen value of SF across speeds and in response to mass added to the limbs.

References

Alexander, R.McN. (1976). Estimates of the speed of dinosaurs. *Nature*, **261**, 129-130.

Alexander, R.McN. (1977). Terrestrial locomotion. In R.McN. Alexander & G. Goldspink (Eds.), *Mechanics and energetics of locomotion* (pp. 168-203). London: Chapman and Hall.

Alexander, R.McN. (1984). Stride length and speed in adults, children and fossil hominids. *American Journal of Physical Anthropology*, **63**, 23-27.

Amano, Y., Hoshikawa, T., Toyoshima S., & Matsui, H. (1987). Longitudinal study of running in children over an 8-year period. In B. Jonsson (Ed.), *Biomechanics X-B* (pp. 819-824). Champaign, IL: Human Kinetics.

Armstrong, L.E., Costill, D.L., & Gehlsen, G. (1984). A biomechanical comparison of university sprinters and marathon runners. *Track Technique*, **87**, 2781-2782.

Bassett, D.R., Giese, M.D., Nagle, F.J., Ward, A., Raab, D.M., & Balke, B. (1985). Aerobic requirements of overground versus treadmill running. *Medicine and Science in Sports and Exercise*, **17**, 477-481.

Catlin, M.E., & Dressendorfer, R.H. (1979). Effect of shoe weight on the energy cost of running. (Abstract) *Medicine and Science in Sports and Exercise*, **11**, 80.

Cavagna, G.A., Thys, H., & Zamboni, A. (1976). The sources of external work in level walking and running. *Journal of Physiology (London)*, **262**, 639-657.

Cavanagh, P.R., & Kram, R. (1985). The efficiency of human movement: A statement of the problem. *Medicine and Science in Sports and Exercise,* **17,** 304-308.

Cavanagh, P.R., & Kram, R. (1990). Stride length in distance running: Velocity, body dimensions, and added mass effects. *Medicine and Science in Sports and Exercise,* **21**(4), 467-479.

Cavanagh, P.R., Pollock, M.L., & Landa, J. (1977). A biomechanical comparison of elite and good distance runners. *Annals of the New York Academy of Sciences,* **301,** 328-345.

Cavanagh, P.R., & Williams, K.R. (1982). The effect of stride length variation on oxygen uptake during distance running. *Medicine and Science in Sports and Exercise,* **14,** 30-35.

Clarke, T.E., Cooper, L.B., Hamill, C.L., & Clark, D.E. (1985). The effect of varied stride rate upon shank deceleration in running. *Journal of Sports Science,* **3,** 41-49.

Clarke, T.E., Frederick, E.C., & Cooper, L.B. (1983). Biomechanical measurement of running shoe cushioning properties. In B.M. Nigg & B.A. Kerr, (Eds.), *Biomechanical aspects of sports shoes and playing surfaces* (pp. 25-33). Calgary, AB: University of Calgary.

Clauser, C.E., McConville, J.T., & Young, J.W. (1969). *Weight, volume and center of mass of segments of the human body.* AMRL (Aerospace Medical Research Laboratories) (Tech. Rep. No. AMRL-TR-69-70). Ohio: Wright-Patterson Air Force Base.

Cureton, K.J., Sparling, P.B., Evans, B.W., Johnson, S.M., Kong, U.D., & Purvis, J.W. (1978). Effect of experimental alterations in excess weight on aerobic capacity and distance running performance. *Medicine and Science in Sports and Exercise,* **10,** 194-199.

Daniels, J.T. (1985). A physiologist's view of running economy. *Medicine and Science in Sports and Exercise,* **17,** 332-338.

Daniels, J.T., & Gilbert, J. (1979). *Oxygen power.* Tempe, AZ: Oxygen Power.

Davies, C.T.M., Sargeant, A.J., & Smith, B. (1974). The physiological responses to running downhill. *European Journal of Applied Physiology,* **32,** 187-194.

Dempster, W.T. (1955). *The space requirements of the seated operator.* WADC (Wright Air Development Center) (Tech. Rep. No.-TR-55-159). Ohio: Wright-Patterson Air Force Base.

Dillman, C.J. (1975). Kinematic analyses of running. *Exercise and Sport Science Reviews,* **3,** 193-218.

Elliott, B.C., & Blanksby, B.A. (1979). Optimal stride length considerations for male and female recreational runners. *British Journal of Sports Medicine,* **13,** 15-18.

Elliott, B.C., & Roberts, A.D. (1980). A biomechanical evaluation of the role of fatigue in middle-distance running. *Canadian Journal of Applied Sport Science,* **5,** 203-207.

Frederick, E.C. (1985a). Synthesis, experimentation, and the biomechanics of economical movement. *Medicine and Science in Sports and Exercise,* **17,** 44-47.

Frederick, E.C. (1985b). The energy cost of load carriage on the feet during running. In D.A. Winter, R.W. Norman, R.P. Wells, K.C. Hayes, & A.E. Patla (Eds.), *Biomechanics IX-B* (pp. 809-812). Champaign, IL: Human Kinetics.

Frederick, E.C., Robinson, J.R., and Hamil, C.L. (1987). Rearfoot kinematics and ground reaction forces in elite caliber identical twins. In B. Jonsson (Ed.), *Biomechanics X-B* (pp. 295-300). Champaign, IL: Human Kinetics.

Heglund, N.C., & Taylor, C.R. (1988). Speed, stride frequency and energy cost per stride: How do they change with body size and gait? *Journal of Experimental Biology,* **138,** 301-318.

Heglund, N.C., Taylor, C.R., & McMahon, T.A. (1974). Scaling stride frequency and gait to animal size: Mice to horses. *Science,* **186,** 1112-1113.

Hill, A.V. (1950). The dimensions of animals and their muscular dynamics. *Science Progress,* **38,** 209-230.

Hogberg, P. (1952). How do stride length and stride frequency influence the energy-output during running? *Arbeitsphysiologie,* **14,** 437-441.

Kaneko, M., Matsumoto, M., Ito, A., & Fuchimoto, T. (1987). Optimum step frequency in constant speed running. In B. Jonsson (Ed.), *Biomechanics X-B* (pp. 803-807). Champaign, IL: Human Kinetics.

Kram, R., McMahon, T.A., & Taylor, C.R. (1987). Load carriage with compliant poles—physiological and/or biomechanical advantages? (Abstract) *Journal of Biomechanics,* **20,** 893.

Martin, P.E. (1985). Mechanical and physiological responses to lower extremity loading during running. *Medicine and Science in Sports and Exercise,* **17,** 427-433.

Massey, B.S. (1971). *Units, dimensional analysis and physical similarity.* London: Van Nostrand Reinhold.

McMahon, T.A., & Greene, P.R. (1979). The influence of track compliance on running. *Journal of Biomechanics,* **12,** 893-904.

Munro, C.F., Miller, D.I., & Fuglevand, A.J. (1987). Ground reaction forces in running: A reexamination. *Journal of Biomechanics,* **20,** 147-155.

Muybridge, E. (1957). *Animals in motion.* New York: Dover.

Myers, M.J., & Steudel, K. (1985). Effect of limb mass and its distribution on the energetic cost of running. *Journal of Experimental Biology,* **116,** 363-373.

Nelson, R.C., & Gregor, R.J. (1976). Biomechanics of long distance running: A longitudinal study. *Research Quarterly*, **47**, 417-428.

Nilsson, J., & Thorstensson, A. (1985). Adaptability in frequency and amplitude of leg movements during locomotion at different speeds. *10th International Congress of Biomechanics Abstract Book*, **20**, 194. Solna, Sweden: Arbetar-Skydd Sverket.

Nilsson, J., Thorstensson, A., & Halbertsma, J. (1985). Changes in leg movements and muscle activity with speed of locomotion and mode of progression in humans. *Acta Physiologica Scandinavica*, **123**, 457-475.

Plyley, M.J., Tiidus, P.M., & Pierrynowski, M.R. (1985). The effect of stride frequency variation on oxygen uptake during downhill running. (Abstract) *Canadian Journal of Applied Sport Sciences*, **10**, 25P.

Schmidt-Nielsen, K.S. (1984). *Scaling: Why is animal size so important?* Cambridge, UK: Cambridge University.

Taylor, C.R., Heglund, N.C., McMahon, T.A., & Looney, T.R. (1980). Energetic cost of generating muscular force during running. *Journal of Experimental Biology*, **86**, 9-18.

van der Walt, W.H., & Wyndham, C.H. (1972). An equation for the prediction of energy expenditure of walking and running. *Journal of Applied Physiology*, **34**, 559-563.

Vilensky, J.A., & Gehlsen, G. (1984). Temporal gait parameters in humans and quadrupeds: How do they change with speed? *Journal of Human Movement Studies*, **10**, 175-188.

Washburne, R., Chin, M.K., & Montoye, H.J. (1980). Accuracy of pedometer in walking and running. *Research Quarterly*, **51**, 695-702.

Williams, K.R. (1985). Biomechanics of running. *Exercise and Sports Science Reviews*, **13**, 389-441.

Zacks, R.M. (1973). The mechanical efficiency of running and bicycling against a horizontal impeding force. *International Zeitschrift Angewande Physiologie*, **31**, 249-258.

Appendix A

Abbreviations Used

APV —Anthropometric variables (such as stature, leg length, and limb segment mass)

CV —Coefficient of variation

FL —Foot length

FM —Foot mass

g —Acceleration due to gravity

H —Stature (height without shoes)

ISKP—Moment of inertia of shank about knee joint in sagittal plane

ITHP—Moment of inertia of thigh about hip joint in sagittal plane

L —An appropriate linear body dimension that is used for nondimensionalized scaling; usually stature or leg length

LL —Leg length (trochanteric height from floor in barefoot stance)

LMH—Lateral malleolus height

Mb —Body mass

OSL —Optimal stride length

PSL —Preferred stride length

sF —Step frequency

SF —Stride frequency

SKL —Shank length (TH − LMH)

SKM—Shank mass

sT —Step time

ST —Stride time

sL —Step length

SL —Stride length

TH —Tibiale height

THL —Thigh length (LL − TH)

THM—Thigh mass

V —Running velocity

$\dot{V}O_2$ —"Steady" oxygen uptake rate

Appendix B

The Treadmill Foot Strike Detector

The tachometer, which is the standard velocity measuring device supplied with the treadmill, generates a DC voltage that is proportional to treadmill velocity (0 to 5 volts full scale). When a footstrike occurs the velocity is momentarily reduced, causing a slight dip in the tachometer voltage. The tachometer signal passes through a 1.2 uF series capacitor to eliminate the DC component (see appendix Figure B.1). It then passes through a series of four bandpass filter/amplifiers to accentuate the voltage dip. The signal then passes through a regenerative comparator device (Schmitt trigger) to produce a TTL-compatible pulse. The time between such pulses can easily be measured by a counter timer or computer.

Note. The footstrike detector was designed by Dennis Dunn and Ewald Hennig.

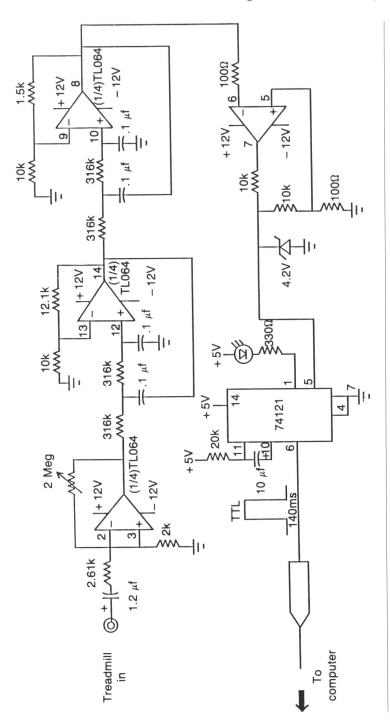

Figure B.1.   Footstrike detector circuit.

Chapter 3

Sagittal Plane Kinematics of the Lower Extremity During Distance Running

Monica J. Milliron
Peter R. Cavanagh

Kinematics is defined as that branch of mechanics dealing with the motion of points or bodies without regard to the forces that create that motion. One can therefore look at the kinematics of running simply as a description of the way the parts of the body move in space. Such a view, however, tells only part of the story. Kinematics is the starting point for a number of other analyses that can give considerable insight into both the biological and mechanical aspects of running. For example, a kinematic description of lower extremity motion throughout the running cycle provides a good definition of the way the major joints act in a coordinated manner to produce the desired motion. The kinematics of individual landmarks on the body may be interesting in their own right—for example, the velocity of the heel at the footstrike may provide insight into differential injury patterns between two runners. It is also possible, from kinematic information, to calculate total center of gravity location, certain muscle lengths and rates of change of length, lower extremity net moments of inertia, angular velocities of joints, and the displacement/velocity profiles for "extensible struts" that could, theoretically, replace the lower extremity. The addition of body segment parameter information opens the doors for energy analysis and, in addition, kinematic values are necessary as input to all kinetic models. Thus kinematics deals with the foundation for much of what biomechanics is about.

A number of review articles have dealt with this topic in the past. Atwater (1973) briefly mentioned a number of relevant studies, and Dillman (1975) reviewed data on running kinematics over the whole continuum of running speeds, as did Vaughan (1984). Williams (1985) presented an exhaustive review of literature in this area. The objective of this chapter is to synthesize the data available in the literature on the sagittal plane kinematics of the lower extremity during unfatigued distance running and to show how such data can be extended to give insight into the mechanics of running. We shall also present a coherent set of new data on the lower extremity kinematics of a small group of subjects running on a treadmill over a range of speeds and gradients typically encountered. Because this book contains separate chapters on arm action and rearfoot motion, these topics will not be covered here.

After a brief discussion of marker placement, the first section of this chapter is a review focusing on the results of studies of level running at approximately 3.83 m • s^{-1} (7 minutes per mile). In the second section, speeds other than 3.83 m • s^{-1} will be considered and the within-subject effects of changes in speed will be discussed. The few studies available concerning grade running will be reviewed and the new data set will be presented. Some of the various quantities that can be derived from kinematic data will also be presented. A set of numerical values for the coordinates of the hip, knee, lateral malleolus, heel, and fifth metatarsal head during level running at 3.83 m • s^{-1} are presented in the appendix to give the reader access to kinematic data without the need for data collection.

Any discussion of kinematics must face the problem of *marker placement*: Values derived from skin markers are only an approximation of the actual movements of the bones that comprise a joint of interest. At some joints this is more problematic than at others. Determining the endpoints for the trunk segment, for example, is particularly difficult. In running, the trunk can be approximated only as a single segment or rigid body because there is little motion at the intervertebral joints. Clearly in activities such as gymnastics or diving such an approximation would not be valid. Even when the rigid body assumption is made, the proximal marker presents a problem. Any marker on the shoulder girdle is likely to give erroneous data because the shoulder is one of the most mobile joints in the body, and during the sometimes vigorous arm swing in running, considerable motion occurs at the shoulder. A marker on the neck (such as the chin-neck intersect) is probably the best solution although it can be affected by excessive head motion.

Both knee and hip joint markers present problems of their own. At the hip the superior border of the greater trochanter, a landmark frequently used to approximate the center of the hip joint, is difficult to locate in an individual with much subcutaneous fat. Andriacchi and Strickland (1985) conducted comparative X-ray and external measurements to locate the hip joint center in a sagittal plane. They found an average error of 0.8 cm in a group of 20 subjects. They further studied the implications of this error for the estimation of joint moments, but this will not be explored here. The instantaneous center of rotation of the knee joint migrates as a function of joint angle, and thus a single marker will be wrong most of the time.

The review of available literature presented here will show some considerable differences in values reported for various joint angles by different investigators. In addition to simple variability between subjects, part of the variation encountered in different investigations is likely to be due to landmark location and conventions. The hip and knee landmarks, for example, could be located over a fairly broad range. Figure 3.1 is a tracing from film of a subject standing on a stationary treadmill. The knee angle as measured from the landmarks is 9 degrees, but taking the worst case—when the knee marker is moved 1 cm anteriorly and the hip marker is moved 1 cm posteriorly—the resulting angle is 16 degrees.

The markers placed on the lateral malleolus and on the base or head of the fifth metatarsal are less of a problem even when shoes are worn, but landmarks in the heel region are less obvious. Typically the heel marker is arbitrarily placed, and a standing calibration is taken before data collection. The angles during stance are measured from this view and all dynamic values are referenced as changes from standing. Though this

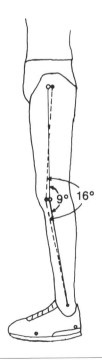

Figure 3.1.   Tracing of a subject standing erect. The change in angle results from a 1-cm displacement of the knee and hip markers.

approach has the merit of allowing for some variability in marker placement, it does not easily allow between-subject comparisons to be made. Suppose, for example, that a runner habitually stands with hyperextended knees; such a situation could not be determined from arbitrarily placed markers. Because the choices of landmark locations for the measurement of the ankle joint angle are usually the lateral malleolus and the head of the fifth metatarsal, it is usual to locate a heel marker so that the standing ankle angle is 90 degrees. The conventions used in this article for the major joints are shown in Figure 3.2. The results from the various authors presented below have been transformed to these conventions to allow for comparison of results.

A further problem in kinematic studies on the treadmill is the accurate determination of cycle landmarks such as footstrike and toe-off. During overground running, the best method of determining these events is by

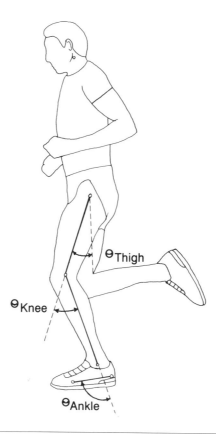

Figure 3.2. Conventions used for the thigh, knee, and ankle joint angles.

measuring the ground reaction forces simultaneously with the motion and using some criterion of vertical force value. Thus footstrike might be said to occur, for example, when the vertical component of ground reaction force is greater than 30 N, and similarly toe-off might be defined as when the force falls below this value. On the treadmill this process is not possible, and simply observing the foot in its approach to the treadmill can be deceptive, particularly with an individual who strikes the ground with a fairly flat foot rather than a clear rearfoot or forefoot strike.

A Review of the Literature

Kinematics During Level Running at 3.83 m · s⁻¹

This section is limited to studies that have presented data for level running at approximately 3.83 meters per second. This has been used throughout the book as a typical distance running speed; the comparison of data within a fairly narrow range of speeds should avoid the confusion that multiple speeds in the literature present. The range of speeds that has been included is between 3.57 and 4.08 m · s⁻¹. Each section will begin with a general statement about motion at that joint, and then the individual research reports will be cited.

Hip Joint and Thigh Angle. In considering the kinematics of the hip joint, one must distinguish between studies that have measured the orientation of the thigh with respect to the vertical and those that have measured a hip joint angle using some markers on the trunk or head to define the upper segment of the joint. As there is generally a forward lean in running, inclusion of the trunk orientation will generally tend to increase the measured flexion angles and decrease the measured extension angles compared to the measurement of thigh orientation alone. Where possible we will report hip flexion and extension, but because most studies have not measured true hip joint angle, a general description based on thigh position, according to the conventions shown in Figure 3.2, will be given.

At footstrike, the thigh has an inclination with respect to the vertical of approximately 25 degrees because the hip is flexed. Some difference of opinion exists concerning the changes in thigh position and hip joint angle during the early phases of support. Miller's data (1978) shows a thigh segment that is moving backwards immediately after footstrike. Williams (1980) shows that the initial posture of the thigh is maintained almost to the point of maximum support phase knee flexion. Our own data at 3.83 m · s⁻¹ (presented later in this chapter) show the thigh to be actually moving forward for the first portion of support at this speed, but this pattern disappears at faster speeds. Thus the discrepancy in the literature may simply be a speed effect.

After maximum support phase knee flexion has been reached, thigh extension and knee extension occur simultaneously and at approximately the same rate. Toe-off occurs shortly before or at the point of maximum backswing of the thigh, coincident with maximum knee extension during late support or early swing. During swing, the thigh motion reverses direction almost immediately, and the hip flexes at a slower average rate than the simultaneous knee flexion. The thigh angle changes by about 45 degrees during the same time period that the knee flexes through a range of approximately 86 degrees. The thigh continues to flex approximately 13 degrees after maximum knee flexion. This next phase of the gait cycle can be characterized as one in which knee extension occurs around a thigh that is relatively fixed in space. Actually, the thigh moves forward at first and then begins to move slightly backward well before footstrike as the hip flexes during the majority of the knee flexion phase. This motion is continued right through swing into footstrike. These changes are most clearly seen in the thigh-knee angle diagram (Figure 3.6a) discussed later in this chapter.

A number of studies have reported specific values for thigh position and hip joint angle at 3.83 m $\cdot$ s^{-1} (see Table 3.1). Teeple (1968) found a mean value of 31 degrees for the thigh position at maximum hip flexion in a group of 16 college women runners. Miller (1978) showed extreme thigh positions of 35.2 and 26.1 degrees during hip flexion and extension respectively in a single female subject running at 3.8 m $\cdot$ s^{-1}. Williams

Table 3.1   Hip Joint Angles and Thigh Positions (Degrees) During Running at 3.83 m $\cdot$ s^{-1} (±0.5) From Various Studies in the Literature

Authors	Hip joint			Thigh Position		
	Extension	Flexion	Range	Extension	Flexion	Range
Teeple (1968)	—	—	—	—	31.0	—
Williams (1985)	—	—	—	−25.8	33.5	59.3
Nilsson et al. (1985)	−11.0	23.8	34.8	—	—	—
Sinning (1970)	−8.0	26.0	34.0	—	—	
Miller (1978)	—	—	—	−26.1	35.2	61.3
Milliron and Cavanagh (this chapter)	—	—	—	−23.0	39.3	62.3
Means	−9.5	24.9	34.4	−25.0	34.7	60.9

(1980), in studies of 31 subjects running overground, found mean extremes for thigh position of 33.5 degrees during hip flexion and 25.8 degrees during hip extension. His data also indicated that maximum thigh position in flexion occurred well before footstrike, when the knee angle was still 70 degrees of flexion. Likewise, maximum thigh position during hip extension occurred after toe-off when the knee had already begun to flex through about 8.5 degrees. The heterogeneity of subjects used as well as possible differences between overground and treadmill running must be considered when comparing the results of various investigators.

Nilsson, Thorstensson, and Halbertsma (1985) included trunk position in their definition of hip angle, but their angle was defined, somewhat unusually, as that between the long axis of the femur and a line joining the anterior superior iliac spine and the hip joint center. Their values at 4.0 m $\cdot$ s^{-1} indicated a range of hip motion of 34.8 degrees, with 11.0 degrees of extension and 23.8 degrees of flexion. Sinning and Forsyth (1970) reported on seven male subjects running on a treadmill at various speeds. Maximum hip angles of 26 degrees of flexion and 8 degrees of extension—a total range of 34 degrees—can be derived from their data by interpolating for values at 3.83 m $\cdot$ s^{-1}.

The almost 100% discrepancy between total range of hip joint motion and total range of thigh segment motion shown in Table 3.1 is puzzling. It does not appear to be due simply to variations in the trunk angle, as Elliot and Ackland (1981) have indicated that the amplitude of trunk movement during running at 5.52 m $\cdot$ s^{-1} is only 13 degrees. Such variation would not completely reconcile the differences found, although it would tend to reduce them.

Knee Joint. At footstrike, the knee is not fully extended but exhibits a flexion angle of 10 to 20 degrees. Following footstrike, flexion occurs during what is commonly called the cushioning phase and continues through midstance, followed by an extension phase that lasts up to or until slightly after toe-off. During level running, the maximum value of knee extension that occurs during late stance is approximately equal to the value at footstrike. The knee then begins to flex rapidly at about the same angular velocity that occurred in the cushioning phase, reaching maximum flexion of about 110 degrees near midswing. During the extension phase, the angular velocity is approximately the same as the earlier swing phase flexion velocity. The endpoint of the extension movement is approximately 10 to 15 degrees short of full extension, and approximately 5 degrees of flexion occurs before footstrike.

Though the overall description given above is generally supported, the exact values for the joint angle at different stages of the running cycle vary between investigators as shown in Table 3.2. Interpolation from the data of Sinning and Forsyth (1970) to 3.83 m $\cdot$ s^{-1} indicated maximum

Table 3.2   Maximum Knee Extension and Knee Flexion (Degrees) During Running at 3.83 m · s⁻¹ (²0.5) From Various Studies in the Literature

	Support		Swing	
	Extension at footstrike	Max flexion	Max flexion	N
Sinning and Forsyth (1970)	12	27	82	7
Miller (1978)	19.2	42	104	1
Nilsson et al. (1985)	14	37	106	10
Williams (1980)	9.4	42.2	101	31
Weighted means (excluding Sinning and Forsyth)	10.7	41.0	102.3	

stance phase values of 12 degrees and 27 degrees for extension and flexion respectively. It is unclear exactly which phases of flexion and extension these values represent. During swing phase, Sinning and Forsyth reported a maximum flexion angle of 82 degrees. All other investigators who have studied the knee joint in running report greater values at most of the cycle landmarks. Maximum swing phase knee flexion values of 104, 106, and 101 degrees (Miller, 1978; Nilsson et al., 1985; and Williams, 1980, respectively) agree closely between these investigators.

Thigh-Knee Diagrams.   Miller (1978) presented a thigh-knee diagram for one subject running at 3.8 m · s⁻¹ and Williams (1980) presented the mean thigh-knee diagram for 31 male distance runners running overground at 3.57 m ·s⁻¹. These curves (Figure 3.3) indicate that the initial knee flexion after footstrike is accompanied by little thigh motion, although the disagreement mentioned suggests that there are likely to be individual differences in the exact pattern. As support phase knee extension begins, the rate of thigh extension is almost the same as the rate of knee extension, shown by this part of the curve having an angle of approximately 45 degrees. Toe-off occurs just before the maximum position of the thigh and maximum knee extension. During the swing phase, the knee is maximally flexed when the thigh has already moved through 75% of its range of motion. The manner in which the knee extends about an almost stationary thigh segment—and in the later part of the swing while slight thigh flexion is underway—is clearly shown in the diagrams.

Ankle Joint.   At footstrike, the ankle is at approximately 90 degrees (according to the convention that the standing angle is 90 degrees), and

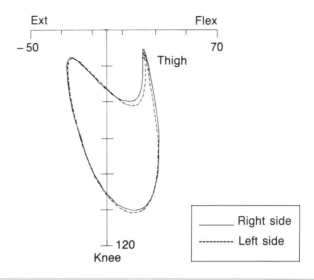

Figure 3.3. Mean thigh-knee diagrams for 31 subjects running at 3.57 m • s⁻¹. *Note.* From *A Biomechanical Evaluation of Distance Running Efficiency* (p. 157) by K.R. Williams, 1980. Unpublished PhD dissertation, Pennsylvania State University. Copyright Pennsylvania State University, University Park. Adapted by permission.

there is generally only a slight amount of plantar flexion (approximately 5 degrees) immediately following footstrike. The commonly held belief that a major amount of plantar flexion occurs following footstrike is not correct. Depending on the type of footstrike, the foot is, in fact, almost flat on the ground at footstrike and, because the knee is flexing, the shank is rotating forward relative to the foot segment. This motion results in about 20 degrees of dorsiflexion at the ankle that continues to a maximum near midstance—approximately coincident with maximum knee flexion. Individual available ranges of motion most likely have an important effect on how much ankle motion occurs. For example, individuals with tighter calf musculature will probably exhibit heel-off earlier in support than those who are more flexible. "Tighter" individuals may therefore use a smaller range of ankle dorsiflexion than those who are more flexible.

Most reports have indicated that a maximum plantar flexion angle of approximately 70 degrees (20 degrees of plantar flexion from neutral) occurs shortly after toe-off, and then a steady return to dorsiflexion (or to a neutral 90-degree position) occurs during the majority of the swing phase. There is sometimes a very slight overshoot when more dorsiflexion than is needed at footstrike is attained, and slight plantar flexion occurs just prior to footstrike. Values reported in the literature for both plantar and dorsiflexion during the support and swing phases of the cycle are shown in Table 3.3.

Table 3.3   Ankle Plantar Flexion and Dorsiflexion During Swing
and Support Phases of Running at 3.83 m • s⁻¹ (±0.5)
From Various Studies in the Literature

	Support		Swing	
	Plantar	Dorsiflexion	Plantar	Dorsiflexion
Sinning and Forsyth (1970)	—	104	64	—
Nilsson et al. (1985)	84.3	107.5	57	94
Williams (1980)	74	115	68	98
Means	79.2	108.8	63.0	96.0

Ankle-Knee Diagrams.   The ankle-knee diagram shown in Figure 3.4 is drawn from 31 subjects studied by Williams (1980). The ankle dorsiflexes as the initial knee flexion following footstrike is underway. The reversal from dorsiflexion to plantar flexion is almost exactly coincident with the transition from knee joint flexion to extension. Slight plantar flexion continues through toe-off, but 65% of the range of motion toward the most dorsiflexed position has already been achieved by the time of maximum knee flexion. During the remainder of the swing phase, as the knee extends, the ankle dorsiflexes slightly and asymptotically to the value some 3 to 5 degrees greater than the angle at footstrike. This is similar to the movement of the thigh (see Figure 3.3) that overshoots the footstrike value and then returns.

Kinematic Data at Discrete Speeds Other Than 3.8 m • s⁻¹

Although human and animal locomotion is often characterized in terms of speed and the transitions of locomotor form, the distinction between human walking and running is not always entirely clear. For example, in racewalking it is often controversial as to whether a flight phase exists. In describing kinematic adaptions to speed, Grillner, Halbertsma, Nilsson, and Thorstensson (1979) measured the amplitude of knee movements for a range of velocities of walking and running. They found that the curves for running and walking intersected at about 2.5 m • s⁻¹. This speed was close to that at which the transition from walking to running occurred spontaneously in their subjects and, according to the authors, marked the beginning of decreasing hip flexion and increasing knee flexion as speed was increased.

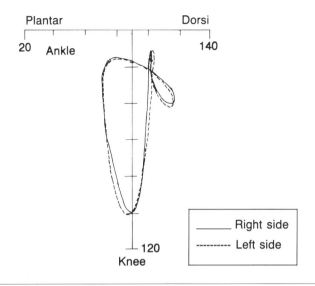

Figure 3.4. Mean ankle-knee diagrams for 31 subjects running at 3.57 m •
s⁻¹. *Note.* From *A Biomechanical Evaluation of Distance Running Efficiency* (p.
158) by K.R. Williams, 1980. Unpublished PhD dissertation, Pennsylvania
State University. Copyright Pennsylvania State University, University Park.
Adapted by permission.

In all reviews of kinematic data pertaining to running, it is critical to
note the speed at which the data were collected. In this section, literature
describing kinematic changes at single speeds other than the 3.83 m •
s⁻¹ standard used in this chapter will be briefly reviewed.

Hip Joint. Giradin and Roy (1974) attempted to identify changes in the
running patterns of 22 sedentary male subjects after a 6-week training
period. Approximately one third of the group maintained a speed of 2.8
m • s⁻¹ for all the tests. Thigh angles at footstrike for this group were
reported as 25 degrees. Osterhoudt (1969) reported a thigh angle at footstrike of 30 degrees in 16 collegiate males running at 3.35 m • s⁻¹. Elliot
and Blanksby (1975) reported thigh angles of 25.7 degrees at maximum
hip flexion and thigh angles of 20.8 degrees at footstrike during running
at 3.5 m • s⁻¹ (Elliot & Blanksby, 1979). Cavanagh, Pollock, and Landa
(1977) found little difference in the thigh-knee diagrams between groups
of elite and good runners on a treadmill at 4.96 m • s ⁻¹. Maximum values
for thigh position during hip flexion and extension were approximately
46 and 23.5 degrees respectively.

Knee Joint. Gollnick and Karpovich (1964) studied the kinematics of
knee joint motion during running at 2.2 m • s⁻¹. They reported 25 degrees
of knee flexion in the cushioning phase after footstrike, followed by 10

degrees of extension prior to toe-off. Their subjects exhibited maximum flexion during swing of approximately 75 degrees. Giradin and Roy (1974) found knee angles at footstrike of 20 degrees in their subjects who ran at 2.8 m • s⁻¹. At a speed of 4.96 m • s⁻¹, the elite male runners studied by Cavanagh, Pollock, and Landa (1977) exhibited knee flexion of 27 degrees at footstrike, with the knee flexing to 50 degrees during the support phase. Maximum flexion during swing was 122 degrees. Mann and Hagy (1980) studied running speeds of 5.4 m • s⁻¹ and reported knee angles at footstrike of 20 degrees, followed by a further 20 degrees of flexion during the cushioning phase. Their subjects flexed the knee to 130 degrees during the swing phase.

Ankle Joint.   In their electrogoniometric study of running at 2.2 m • s⁻¹, Gollnick and Karpovich (1964) divided the ankle motion into four components: initial dorsiflexion, support phase plantar flexion, second dorsiflexion period, and plantar flexion during swing. The values associated with each of these phases were 110, 80, 110, and 95 degrees respectively. Mann and Hagy (1980) reported maximum dorsiflexion of 100 degrees during running at 5.4 m • s⁻¹. The maximum plantar flexion angle occurred slightly after toe-off and measured 60 degrees.

Kinematics for the Same Subjects Over a Range of Speeds

There have been few systematic studies of the change in kinematic patterns over a range of distance running speeds. The three exceptions are the papers of Sinning and Forsyth (1970), Grillner et al. (1979), and Nilsson et al. (1985). Miller (1978) also compared kinematic patterns at 3.8 m • s⁻¹ and 5.6 m • s⁻¹ for a single elite female distance runner. Elliot and Ackland (1981) reported certain kinematic values during a 10,000-m run in which the speed progressively decreased. Because these results were undoubtedly influenced by fatigue, they will not be discussed here. Hoshikawa, Matsui, and Miyashita (1973) measured the kinematic changes for an excellent sprinter at speeds of 3.3, 5.0, 6.6, and 8.3 m • s⁻¹. His style at the slower speeds was said to be markedly different from that exhibited by their poorer runners. At the higher speeds, this individual showed increases in the ranges of motion at the hip and knee but not at the ankle. No quantitative values were provided by the authors.

Hip Joint.   Sinning and Forsyth (1970), studying seven subjects running on the treadmill, found increases of only 3 degrees in the maximum hip extension as the speed of running increased over a wide range from 3.30 to 6.67 m • s⁻¹. This finding is, at first, quite counterintuitive, as one feels that faster running is achieved by increased extensor thrust, which should imply greater hip extension. These investigators found that the maximum hip flexion angle increased by more than 22 degrees over the same speed range.

The trend shown by the results of Sinning and Forsyth was confirmed by Nilsson et al. (1985), who measured hip angles as functions of speed over a wide range of speeds. They showed that over a speed range from 3 to 6 m • s^{-1}, the maximum hip flexion angles increased by 13 degrees whereas the extension angle increased by only 5 degrees. Miller (1978) showed increases in maximum thigh position during hip extension of 5 degrees and increases in thigh position during hip flexion of 25 degrees as the speed of running changed from 3.8 to 5.6 m • s^{-1}.

Knee Joint. Grillner et al. (1979) averaged the knee angle for one subject over 10 consecutive strides at 13 different speeds between 0.5 and 8.0 m • s^{-1}. The resulting graph showed a monotonic increase in the amplitude of knee joint motion as speed increased. For example, the ranges of motion at 4, 5, 6, and 7 m • s^{-1} were 85, 100, 105, and 110 degrees respectively. In another study, Saito, Kobayashi, Miyashita, and Hoshikawa (1974) measured the phasic patterns of knee joint motion in one trained and one untrained runner at five speeds from 4.94 to 9.17 m • s^{-1}. The most interesting feature of the data presented was the absence of knee flexion prior to footstrike in certain trials at higher speeds (> 6.0 m • s^{-1}).

Sinning and Forsyth (1970) found that over the speed range from 3.33 to 6.67 m • s^{-1} the knee angle at footstrike changed by only 3 degrees, tending to be more flexed at the faster speed. The maximum knee flexion during support increased by 3.5 degrees over the same speed range whereas the flexion during swing increased by 13 degrees. Miller (1978) reported a slightly less extended knee at footstrike at the faster speed of 5.6 m •s^{-1} when compared to 3.8 m • s^{-1} and a change of only 6 degrees in the maximum knee flexion during support. She found that the knee was more extended by approximately 4 degrees in late support at the faster speed and also reported an increase in knee flexion during swing of about 41 degrees. In contrast to the above findings, Nilsson et al. (1985) found that as the running speed changed from 3 to 6 m • s^{-1} there was a small decrease (1.5 degrees) in the maximum knee flexion during support. Maximum knee flexion during swing incrased by 28 degrees over the same speed range. This is double the change reported by Sinning and Forsyth (1970), but still less than Miller (1978) through a smaller range of speeds.

Ankle Joint. Compared to the more substantial changes at the hip and knee, the changes reported for ankle angles at faster running speeds are fairly small. Nilsson et al. (1985) reported changes during the support phase of approximately 3 degrees in the maximum values of plantar flexion and dorsiflexion as the speed changed from 3.0 to 6.0 m • s^{-1}. Sinning and Forsyth (1970) reported similar changes in dorsiflexion over an approximately similar speed range but found almost no change in the plantar flexion range. The study of Saito et al. (1974) showed initial dorsiflexion for both runners to be consistent from a temporal point of view. This phase continued until about 10% of cycle time and was followed by plantar flexion for the next 15% of the cycle. The relative lengths of the subsequent plantar flexion and dorsiflexion phases appeared to vary erratically over the speed range.

The Effects of Grade
on Lower Extremity Kinematics

Although distance runners frequently encounter varied slopes during their training and competition, very few studies have been conducted on the kinematics of grade running. Nelson and Osterhoudt (1971) studied 16 experienced runners at three grades: +10%, 0%, and −10%. These investigators filmed the subjects over a range of speeds from 3.38 m • s^{-1} to 6.46 m • s^{-1} at each grade. However, only stride rate, stride length, and periods of support and nonsupport were reported. Osterhoudt (1969) found that with a change in grade from level to a −10% slope, changes in trunk and shank angles occurred. Shank angle relative to the vertical was greater for the downhill grade at footstrike. At toe-off, the shank angle was only slightly greater for the downhill run (relative to vertical). The thigh angle (to vertical) was greater at takeoff for the level condition compared to the −10% grade results. Trunk angle relative to the vertical was greater for the level condition at both footstrike and toe-off; therefore the runner was leaning forward less in the downhill condition.

Further insight into the changes in kinematics over a range of grades was provided by Hamill, Clarke, Frederick, Goodyear, and Howley (1984). In this study 10 male runners were studied at six slopes from −9% downhill to 6% uphill in 3% increments. As the grade changed from −9% to +6%, knee flexion at footstrike increased from 15.3 degrees to 28.3 degrees, representing a more extended knee for the downhill condition. Peak knee angular flexion velocity decreased from 12.4 rad • s^{-1} at the −9% slope to 7.4 rad • s^{-1} for the +6% grade trial.

Ankle angles ranged from 88.3 degrees (−1.7 degrees of plantar flexion) at footstrike during the −9% trial to 96.1 degrees (6.1 degrees of dorsiflexion) for the +6% grade. The authors suggested that the kinematic changes at the knee and ankle joints occur to attenuate the increased shock associated with greater downhill slope.

Buczek and Cavanagh (in press) compared the lower extremity kinematics of seven male runners between level runway and during downhill (−8.3%) running on a wooden ramp. In agreement with Hamill et al.

(1984), greater knee extension was found at footstrike for the downhill trial—17.0 degrees (−8.3% slope) versus 24.6 degrees (level). Peak knee flexion in the cushioning phase following footstrike was also significantly greater ($p < 0.05$) for the downhill condition (48 degrees vs. 44 degrees). The angular velocity of knee flexion during the cushioning phase was greater, although not significantly so, for the downhill slope. The ankle joint tended to show greater dorsiflexion at footstrike during downhill running. These various differences between downhill and level running are much smaller than one might anticipate when considering the muscular consequences of downhill running, such as delayed onset muscle soreness (Abraham, 1977). This is one of the many reminders that only relatively small changes in external kinematics are required to produce what the runner perceives as fairly major physiological changes.

A Coherent Data Set for Several Speeds and Grades

The information presented in this part of the chapter was collected by the authors to provide a coherent data set to document lower extremity kinematic changes during varied grade and varied speed running in the same group of subjects. This was thought desirable because, as the preceding review of literature has demonstrated, discrepancies in methodology between investigators make the comparison of results difficult. The data have also been used here to present some of the quantities that can be derived from such kinematic information.

Methods

The subjects, four experienced male treadmill runners, ran nine trials each on a Quinton 18-60 motorized treadmill. Kinematic data were collected over a range of grades and typical distance running speeds using a NAC high-speed camera at 200 fields per second. The grade was varied in 10% increments from +20% to −20%, including a level run. All grade trials were collected at a speed of 3.4 m • s⁻¹. The five speed trials ranged from 3.4 m • s⁻¹ to 5.0 m • s⁻¹ in 0.4 m • s⁻¹ increments at a 0% gradient. All subjects wore the same model running shoe.

Prior to data collection, markers were placed at five anatomical landmarks: the superior border of the greater trochanter, the lateral femoral epicondyle, the lateral malleolus, on the shoe heel counter overlying the lateral aspect of the calcaneus, and on the shoe over the head of the fifth metatarsal on the left side of the body. A standing calibration was collected for each subject. Three-second periods of data were extracted from the video using standard software on the ExpertVision System from Motion Analysis Corporation. A sample of the raw data is given in the appendix to enable the reader to perform other analyses than those given here.

Results

Path Diagrams. Because all forms of kinematic data analysis start by determining the coordinates of the targets on the body (digitizing the analog movement pattern), a plot of the movement of the parts of the body with respect to a laboratory reference frame is frequently the starting point. Examples from one subject for the extremes of speed and gradient in the present study are shown in Figure 3.5. These path diagrams, which

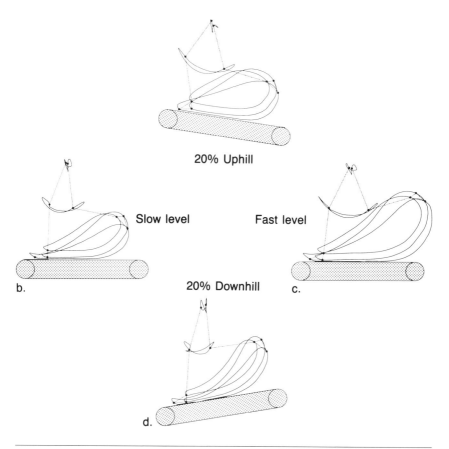

Figure 3.5. Path diagrams for the extremes of grade (−20% and +20%) and speed (3.4 m • s⁻¹ to 5.0 m • s⁻¹) in a single subject. The stick figures represent the posture of the lower extremity at footstrike and at contralateral footstrike.

have been presented previously for running by Hoshikawa, Matsui, and Miyashita (1973), are rather similar to chronocyclograms (Murray, Kory, Clarkson, & Sepic, 1966). The present set in Figure 3.5 has been augmented by the addition of two stick figures representing the posture of the lower extremity at footstrike and at contralateral footstrike. Though these diagrams are devoid of information on coordinated action, they do give a good impression of the range of movement of the parts of the body in space, and in this sense they are more useful than conventional stick diagrams in which successive frames overwrite in parts of the cycle. The narrow loops drawn out by the ankle marker and the restricted motion of the knee target are characteristic of downhill running (Figure 3.5d). Similarly, the higher lift of the foot during running at the faster speed on the level is apparent from a comparison of Figures 3.5b and 3.5c. The distance of the ankle in front of the hip at footstrike can also be appreciated from the path diagrams. Table 3.4 represents a reduction of the numerical data to show the various horizontal and vertical ranges of motion for the diagrams shown in Figure 3.5.

Table 3.4 **Ranges of Motion in Meters for the Hip, Knee, and Ankle Targets of a Subject Running on a Treadmill at Various Speeds and Grades**

Speed (m · s^{-1})	3.4	3.8	4.2	4.6	5.0
Hip					
X range	0.06	0.09	0.10	0.10	0.11
Y range	0.11	0.12	0.12	0.12	0.11
Knee					
X range	0.42	0.05	0.56	0.58	0.64
Y range	0.14	0.16	0.19	0.19	0.22
Ankle					
X range	0.75	0.93	1.00	0.98	1.11
Y range	0.38	0.45	0.51	0.55	0.59
Xha					
at fs[a]	0.08	0.24	0.26	0.16	0.27
Grade (%)	+20	+10	0	-10	-20
Hip					
X range	0.08	0.09	0.06	0.07	0.07
Y range	0.13	0.12	0.11	0.14	0.15

(Cont.)

Table 3.4 (Continued)

Speed (m • s⁻¹)	3.4	3.8	4.2	4.6	5.0
Knee					
X range	0.60	0.57	0.42	0.40	0.32
Y range	0.27	0.22	0.14	0.15	0.13
Ankle					
X range	0.86	0.89	0.75	0.86	0.76
Y range	0.39	0.39	0.38	0.53	0.58
Xha					
at fs	0.23	0.25	0.08	0.17	0.14

Note: The subject was 1.82 m tall with a leg length (superior border of greater trochanter to floor) of 0.995 m.
ªXha at fs = The distance of the ankle in front of the hip at footstrike.

Angle-Angle Diagrams.   Thigh, shank, and foot segment angles were calculated from the coordinate data and used to determine knee and ankle joint angles according to the conventions shown in Figure 3.2. From the standing calibration data, the thigh and knee angles were set to 0 degrees, and the ankle angle to 90 degrees in upright standing. The shank angle was calculated using the knee and lateral malleolus markers, and the foot segment was determined from the calcaneal and fifth metatarsal head markers. The cycles were normalized to 100 points (beginning and ending at footstrike) and average intrasubject segment and joint angles were determined for each subject. Intersubject averages were then calculated for each condition. The mean patterns for the four subjects are shown in the families of thigh-knee diagrams in Figure 3.6 and the corresponding ankle-knee diagrams in Figure 3.7.

As speed increased, the main visual impression of the thigh-knee diagrams (Figure 3.6a) is that they simply get larger in all directions except knee extension (the upper range of the vertical axis). This is confirmed by the values for maximum flexion and extension of the thigh and knee that are shown in Table 3.5. The early- and mid-support phases at all speeds show remarkable similarity except that the slight hip flexion that occurs following footstrike at the slower speeds disappears at the faster speeds. The position of the thigh is maintained or even extends slightly at 5 meters per second.

The ankle-knee diagrams for the range of speeds (Figure 3.7a) also show only small changes in many of the kinematic variables. Maximum stance phase dorsiflexion varied by only 5 degrees whereas the faster speed involved 7 degrees more plantar flexion than the slowest speed. This

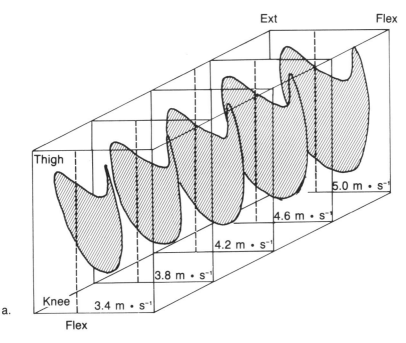

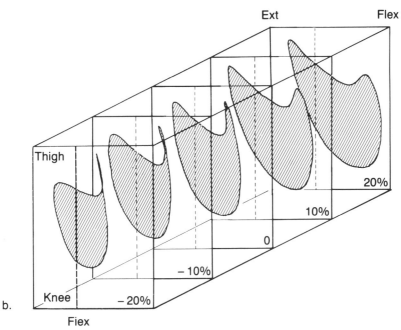

Figure 3.6. Families of thigh-knee diagrams for the range of speeds (a) and grades (b) used in the experiment.

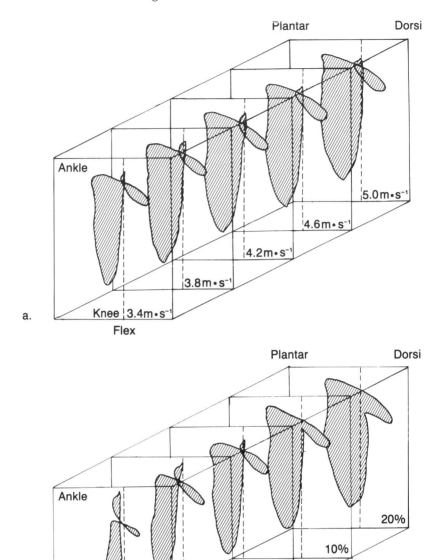

Figure 3.7. Families of ankle-knee diagrams for the range of speeds (a) and grades (b) used in the experiment.

Table 3.5 Maximum Values for Thigh, Knee, and Ankle Angles (in Degrees) During Running at Different Speeds (in m • s⁻¹) on a Level Treadmill

	Speed (m • s⁻¹)				
	3.4	3.8	4.2	4.6	5.0
Thigh					
Flexion	39.0	39.9	41.0	44.0	46.6
Extension	−20.6	−22.0	−23.1	−26.4	−28.0
Total range of motion	59.6	61.9	64.1	70.4	74.6
Knee					
Extension prior to footstrike	13.4	9.9	16.7	15.3	15.5
Cushioning flexion	40.6	38.6	39.4	42.7	42.9
Range of cushioning flexion	27.2	28.7	22.7	27.4	27.4
Propulsive phase extension	15.1	13.5	16.8	14.1	14.2
Swing phase flexion	103.9	109.3	108.1	111.5	117.3
Total range of motion	90.5	99.4	91.4	97.4	103.1
Ankle					
Dorsiflexion prior to footstrike	91.3	90.0	91.8	92.9	92.5
Stance phase dorsiflexion	112.4	108.0	108.7	113.0	110.8
Plantar flexion	64.2	54.9	58.7	58.3	57.1
Total range of motion	48.2	53.1	50.0	54.7	53.7

Note. Mean values are shown in degrees from the four subjects.

difference in the extent of plantar flexion was sufficient to give the faster speed diagram considerably greater area, as the final angle of plantar flexion is maintained for a large part of the knee flexion phase.

In contrast to the fairly small kinematic adaptions involved over the speed range studied, the progression from uphill to downhill running shown by the thigh-knee diagrams in Figure 3.6b involved marked changes in the coordinated patterns of joint movements. Most notable were the addition of a major new lobe to the curve in the upper left-hand region (knee extension and hip extension) during uphill running, and the increased range of knee flexion during the cushioning phase. In agreement with previous work (Buczek & Cavanagh, in press), the maximum knee flexion during the cushioning phase was only slightly less during −20% downhill running compared to +20% uphill (41.2 vs. 46.3 degrees). However, because the knee was much more extended prior to footstrike in the downhill condition (4.0 vs. 37.7 degrees), the total range of motion at the knee was much greater in the downhill condition (37.2 vs. 8.6 degrees). The increased size of the diagram to the right of the vertical axis in uphill running is indicative of the greater range of motion in flexion of the thigh (25.8 vs. 54.5 degrees for −20% downhill and +20% uphill, respectively). The numerical data for the various ranges of motion in grade running are given in Table 3.6.

Table 3.6   Maximum Values for Thigh, Knee, and Ankle Angles (in Degrees) During Treadmill Running at 3.4 m · s⁻¹ at Various Grades

	Grade (%)				
	−20	−10	0	10	20
Thigh					
Flexion	25.8	29.4	29.0	46.9	54.5
Extension	−18.4	−22.4	−20.6	−20.7	−18.7
Total range of motion	44.2	51.8	49.6	67.6	73.2
Knee					
Extension prior to footstrike	4.0	4.6	13.4	25.8	37.7
Cushioning flexion	41.2	39.7	40.6	42.7	46.3
Range of cushioning flexion	37.2	35.1	27.2	16.9	8.6
Propulsive phase extension	32.1	16.9	15.1	11.4	13.0
Swing phase flexion	99.6	100.8	103.9	107.5	113.2
Total range of motion	95.6	96.2	90.5	96.1	100.2
Ankle					
Dorsiflexion prior to footstrike	—	—	91.3	93.3	97.6
Stance phase dorsiflexion	103.5	107.5	112.4	113.4	116.4
Plantar flexion	76.4	62.8	64.2	60.9	59.4
Total range of motion	27.1	44.7	48.2	52.5	57.0

Note. Mean values are shown in degrees from the four subjects.

The ankle-knee diagrams over the range of gradients (Figure 3.7b) show changes not only in extent but also in shape. The diagram for +20% uphill running is a large, open pattern whereas that for −20% downhill is narrow and involuted. This difference is due to the limited knee extension during swing in downhill running mentioned earlier, and the considerably greater range of ankle motion, particularly in plantar flexion, that occurred in uphill running (range of motion of 27 vs. 57 degrees for downhill and uphill, respectively). The maximum dorsiflexion during downhill running was also less than that observed during uphill running (103.5 vs. 116.4 degrees for downhill and uphill, respectively). It is interesting to note that the ankle is held in plantar flexion during most of the swing phase for all conditions but the 20% uphill.

Variability in Thigh Knee Patterns. Though only 3 seconds of data (approximately four cycles) were extracted from each subject to form the group mean data presented here, data were collected from two subjects for over 7.5 seconds. The thigh-knee diagrams shown in Figure 3.8 give some indication of the stride-to-stride variability of the two subjects. Although it is unlikely that each subject had precisely the same amount of prior experience on the treadmill as the other, both subjects were experienced recreational runners and had completed a 15-minute accommodation run prior to data collection. Subject B appears to be more variable than Subject A at all stages of the cycle, but the variability in swing phase knee flexion is the most marked. Optoelectronic techniques make the investigation of within-subject variability possible and can be expected to provide considerable insight into running mechanics in the future.

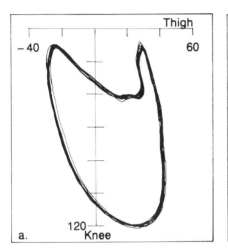
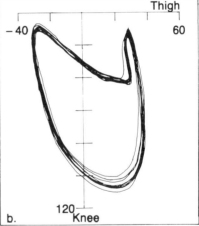

Figure 3.8. Multiple thigh-knee diagrams for two subjects collected from 7.5-s bouts of treadmill running. The greater variability of subject B is apparent at all stages of the running cycle, particularly swing phase knee flexion.

Angular Velocities. The mean thigh segment angle and the knee and ankle joint angle versus time plots were smoothed using a three-point moving average filter and then differentiated using a finite difference algorithm. The resulting velocity curves are shown in Figure 3.9 for the speed range and Figure 3.10 for the range of grades used in the experiment.

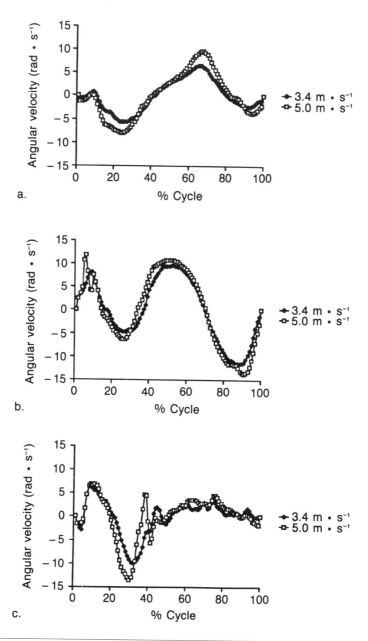

Figure 3.9. Angular velocity versus time curves for (a) the thigh segment, (b) the knee joint, and (c) the ankle joints for the 3.4 m • s⁻¹ and 5.0 m • s⁻¹ conditions.

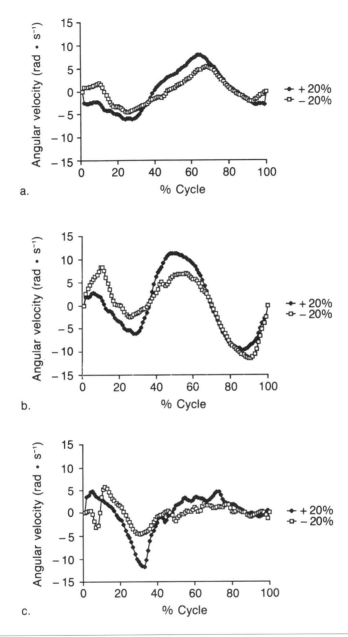

Figure 3.10. Angular velocity versus time curves for (a) the thigh segment, (b) the knee joint, and (c) the ankle joints for the −20% and +20% conditions.

Various peaks have been extracted from the data, and these are presented for speed in Table 3.7 and grade in Table 3.8. The thigh velocity data indicate that the velocity in flexion was always greater than the peak velocity in extension. This is hardly surprising, because hip flexion is performed on the unloaded limb whereas extension occurs during the support phase. It should be recalled that these values are thigh segment velocity and not hip joint velocity, as the motion of the trunk was not accounted for in our data. At slower speeds, there was a brief period when a small initial velocity of flexion occurred following footstrike, but this disappeared at the faster speeds. Peak values of both extension and flexion velocities increased with speed, the values at 5.0 m $\cdot$ s^{-1} being approximately 40% greater than those at 3.4 m $\cdot$ s^{-1}. The knee flexion velocity during the cushioning phase showed an overall trend to increase as running speed increased. This was not, however, consistent, as only the value at 5.0 m $\cdot$ s^{-1} was markedly different from those at other speeds. Trends in other peaks of the knee velocity profile showed only moderate changes with

Table 3.7 Various Maximum Values for Segmental and Joint Velocities (in Radians per Second) During Running at Different Speeds on a Level Treadmill

	Speed (m $\cdot$ s^{-1})				
	3.4	3.8	4.2	4.6	5.0
Thigh					
Cushioning flexion	0.8	0.9	—	—	—
Cushioning extension	—	—	—	0.9	0.3
Stance extension	−5.8	−6.2	−6.3	−7.5	−8.2
Swing flexion	6.8	7.4	7.6	9.1	9.9
Swing extension	−2.4	−2.6	−2.8	−3.4	−3.9
Knee					
Cushioning flexion	8.3	9.5	7.6	9.9	7.9
Stance extension	−4.9	−4.9	−5.2	−5.8	−6.5
Swing flexion	9.9	9.4	9.8	10.5	11.1
Swing extension	−11.8	−12.6	−12.7	−13.2	−14.1
Ankle					
Cushioning plantar flexion	−2.9	−3.3	n/a	−3.1	−1.5
Stance dorsiflexion	6.7	6.5	5.4	7.5	7.1
Plantar flexion	−10.1	−11.1	−10.3	−12.9	−13.9
Swing dorsiflexion	3.1	4.1	4.5	3.9	4.7

Note. The points at which velocity values were extracted from the curves are shown in Figure 3.9.

Table 3.8 Various Maximum Values for Segmental and Joint Velocities (in Radians per Second) During Treadmill Running at 3.4 m · s⁻¹ at Various Grades

	Grade (%)				
	−20	−10	0	10	20
Thigh					
Cushioning flexion	1.8	2.4	0.8	—	—
Cushioning extension	—	—	—	0.6	−2.4
Stance extension	−4.6	−5.5	−5.8	−5.6	−6.1
Swing flexion	5.6	6.5	6.8	7.1	8.2
Swing extension	−2.3	−2.5	−2.4	−2.7	−2.9
Knee					
Cushioning flexion	8.3	10.0	8.3	5.0	2.7
Stance extension	−2.5	−4.6	−4.9	−5.4	−6.1
Swing flexion	6.9	8.2	9.9	10.0	11.3
Swing extension	−11.5	−11.7	−11.8	−9.9	−9.6
Ankle					
Cushioning plantar flexion	−3.2	−3.1	−2.9	−0.7	—
Stance dorsiflexion	5.8	6.3	6.7	4.9	5.0
Plantar flexion	−4.7	−7.5	−10.1	−9.9	−12.1
Swing dorsiflexion	2.0	2.9	3.1	3.5	4.8

speed. Ankle velocity values did not show a single monotonic change with speed (Table 3.7). The maximum plantar flexion velocity, which occurred in late support, increased by approximately 40%, from −10.1 rad · s⁻¹ at 3.4 m · s⁻¹ to −13.9 rad · s⁻¹ at 5.0 m · s⁻¹.

During level and downhill running, there was always a small peak in velocity of hip flexion immediately following footstrike. In uphill running there was consistently a velocity of extension during this phase. The peak extension velocity of the thigh during the support phase showed no trend between −10% downhill and +10% uphill but, at the extreme values of grade, there was a 32% higher value of extension velocity at the +20% uphill grade when compared to −20% downhill. Peak flexion velocity of the thigh during swing showed a monotonic increase with increasing gradient, with the value at +20% uphill being 46% greater than the value at −20%.

Knee flexion velocities during the cushioning phase were markedly less than those at level or downhill, but the subsequent stance phase extension velocities increased progressively with increasing gradient. This latter change was large—the value at +20% uphill was more than 2.4 times greater than that at −20%. Peak knee flexion velocity during the swing phase of +20% uphill running was more than 60% greater than the corresponding value at −20%, and this change was monotonic across the grade range. Peak swing phase extension at the knee was similar for downhill and level conditions but was slightly less for uphill running.

At the ankle, early stance phase values for peak velocities of dorsiflexion and plantar flexion showed only minor trends with change in grade. The peak value for support phase plantar flexion velocity was dramatically greater at +20% uphill when compared to −20% downhill (−12.1 vs. −4.7 rad • s⁻¹). Swing phase dorsiflexion velocity was also greater in uphill running.

Derived Values From Kinematic Data.   A number of interesting derivations can be made from lower extremity kinematic data, and three examples will be considered here: muscle length, "strut length," and moment of inertia of the whole limb about the hip. This is by no means an exhaustive list of the possible derivations, but it should serve to show that kinematic data can be a useful window on parameters of somewhat more far-reaching scope.

Muscle Length.   The prediction of gastrocnemius muscle length from the angular data for the knee and ankle joints was based on the technique described by Grieve, Pheasant, and Cavanagh (1978). These investigators measured the change in length of the gastrocnemius relative to a reference position (90-degree angles at both the knee and ankle) for eight cadaveric limbs as the knee and ankle joints were varied. Second-degree polynomial regression equations were determined for each set of joint motions, and a combination of the length changes obtained from these regression equations yields a net length change for the muscle. The regression coefficients of Grieve et al. (1978) were used on the data set described above to estimate length and rate of change of length values for the gastrocnemius muscle during running. In addition, surface electromyography of the gastrocnemius was conducted for one of the subjects whose data was presented previously to show the benefit of a combined kinematic and electromyographic approach.

Figure 3.11 shows the length changes for all gradients plotted for the complete cycle, and beneath the length plots the periods of EMG activity are indicated. The units used for length are percentage of the shank segment length. It is apparent from the diagram that the periods of muscle activity occurred when the muscle was at relatively long lengths. At all gradients, the muscle was also lengthening at the onset of muscle activity in late swing, and thus eccentric muscle action was occurring. In Figure

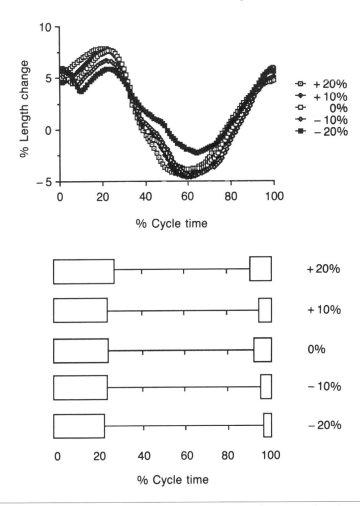

Figure 3.11. Mean gastrocnemius muscle length as a function of cycle time for running at 3.4 m • s⁻¹ on the five grades. The period during which surface EMG activity in the gastrocnemius was greater than baseline for one of the subjects is shown beneath the diagram.

3.12, only the length data for the periods of muscle activity have been plotted. This enables the details of the individual curves to be appreciated more easily. In every case, the termination of muscle activity occurred at or just beyond the point at which the muscle has reached maximum length. Concentric muscle action occurred during only the first 5% to 10% of the cycle in level and downhill running, and this was both preceded and followed by eccentric action. The conclusion that in running the

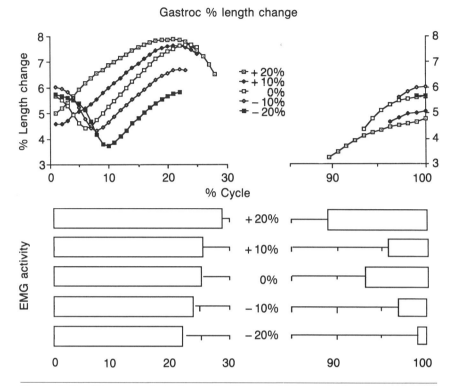

Figure 3.12. The same data shown in Figure 3.11 but only for those periods of muscle activity at the various grades. Note that the muscle is lengthening at the onset of activity at all speeds and that no concentric activity occurs at all during uphill running.

gastrocnemius acts principally to control dorsiflexion rather than to generate active plantar flexion would not likely have been reached without the combined benefits of kinematic and EMG data.

Strut Length. The topic of walking and running robots has received considerable attention in both popular and scientific literature. One of the best known locomoting robots consists of a single extensible strut extending from a gyroscopic guidance assembly. If such a device were to mimic the trajectory of the human hip joint during the support phase of the running cycle, the necessary length changes of such an extensible strut could be calculated rather simply from the raw coordinate data as the distance between the hip and the fifth metatarsal head. Swing phase kinematics could follow a different scheme, but for the purposes of illustration, the calculations have been performed for the entire cycle. The results of such a set of calculations are presented in Figure 3.13, which

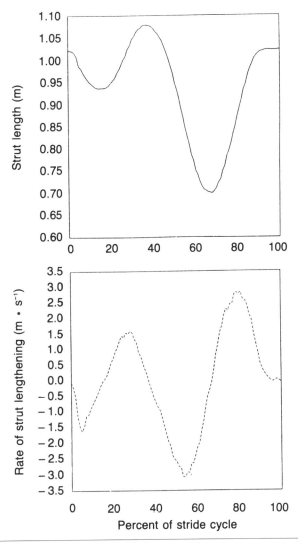

Figure 3.13. The length and rate of change of length for an extensible strut that would mimic the displacement of the hip in space throughout the cycle of running at 3.8 m • s⁻¹.

includes the length, L, and the rate of change of length, L', for level running at 3.8 m • s⁻¹. Peak values for other speeds and grades are given in Table 3.9. The length versus cycle curve bears a striking resemblance to the knee angle versus cycle curve, which is not surprising given the length of the segments proximal and distal to the knee joint. The total length change during support would need to be approximately 10 to 15

Table 3.9   Maximum and Minimum Values for Strut Length and Rate of Change of Strut Length During the Support Phase of Gait

	Strut length (m)				Moment of inertia (kg • m²)	
	L min	L max	L′ min	L′ max	$I_h$min	$I_h$max
Speed						
3.4	0.71	1.07	−2.0	2.2	1.58	2.99
3.8	0.70	1.08	−2.2	2.5	1.56	2.99
4.2	0.65	1.09	−2.3	2.7	1.44	3.04
4.6	0.61	1.08	−2.0	2.8	1.35	3.01
5.0	0.60	1.09	−2.3	3.2	1.30	3.03
Grade						
+20%	0.62	1.07	−1.7	2.4	1.35	2.89
+10%	0.65	1.08	−2.4	2.3	1.41	2.98
0%	0.71	1.07	−2.0	2.2	1.58	2.99
−10%	0.70	1.05	−2.3	2.4	1.54	2.93
−20%	0.66	1.03	−2.0	1.7	1.42	2.86

Note. These values are for a single subject at five different speeds on a level treadmill and five grades at 3.4 m • s⁻¹. The moment of inertia of the entire lower extremity about the hip (I_h) is also shown (in kilograms • square meter); L is in meters; L′ is in meters per second.

cm depending on speed and grade. The strut would be required to shorten at a rate of approximately 2 m • s⁻¹ during the cushioning phase, and surprisingly, this velocity would be fairly independent of grade or speed. However, the required rate of extension during propulsion would vary from 1.7 m • s⁻¹ for −20% downhill running to 3.2 m • s⁻¹ for 5.0 m • s⁻¹ level running.

Total Moment of Inertia About the Hip Joint Center. The hip musculature must overcome both the effects of gravity on the segments of the lower extremity and the inertial effects of these segments. Swing phase in running is sometimes described as involving knee flexion to reduce the moment of inertia of the limb during swing. Because in the rigid body assumption the local moments of inertia of the segments about their own centers of mass are fixed, the total moment of inertia of the lower extremity about the hip is simply a function of the knee and ankle angles. If these angles and the dimensions of the segments are known, the parallel axis theorem can be used to determine the net moment of inertia (I_h) for any given configuration of the lower extremity. These calculations were performed for the mean data sets presented above at all speeds and gradients, and the maximum ($I_h$max) and minimum ($I_h$min) values are presented in Table 3.9. The increased knee flexion at 5.0 m $\cdot$ s^{-1} results in a reduction of approximately 17% below the value at 3.8 m $\cdot$ s^{-1}. It is interesting to note that the moment of inertia during the swing phase is largest during level running and less during both uphill and downhill running. Regardless of speed or grade the maximum value varies by only 5%.

Concluding Remarks

The review of kinematic data on running presented in this chapter provided a quantitative basis for the visual observations on the lower extremity that are frequently made. In some senses, kinematic studies can represent the greatest limitations of sports biomechanics in that they can be purely descriptive and lacking in a mechanistic or hypothesis-driven rationale. Yet as mentioned in the introduction, kinematics are the starting point for many more fundamental and biologically relevant calculations and can, as we have shown, lead to a number of interesting derived quantities. Neither should it be forgotten that the coaching of distance running style is conducted almost exclusively at a kinematic level. Phrases such as "more knee lift," "lengthen your stride," and "stop letting your body roll" are frequently heard exhortations from coach to athlete. Thus, the kinematic data presented here are also important to insure accuracy of coaching information. Future studies will clearly need to escape from the two-dimensional boundaries to which we have limited ourselves in this chapter. In particular, the measurement of axial rotations of the segments of the lower extremity during running remains a distant goal at present. More information is also required on the movement of the trunk, head, and upper extremities. The collection of kinematic data during the later stages of fatigue at racing pace would also be of tremendous interest.

Acknowledgments

The authors are indebted to Brian Davis for assistance with the review of literature and to Bob Snyder for data processing help.

References

Abraham, W.M. (1977). Factors in delayed muscle soreness. *Medicine and Science in Sports and Exercise*, **9**, 11-20.

Andriacchi, T.P., & Strickland, A.B. (1985). Gait analysis as a tool to assess joint kinematics. In N. Berme, A.E. Engin, & K.M. Correia Da Silva (Eds.), *Biomechanics of normal and pathological human articulating joints* (pp. 83-102). NATO ASI series. Dordrecht, The Netherlands: Martinius Nihoff.

Atwater, A.E. (1973). Cinematographic analyses of human movement. *Exercise and Sports Science Reviews*, **1**, 217-258.

Buczek, F.L., & Cavanagh, P.R. (in press). Knee and ankle kinematics and kinetics during level and downhill running. *Medicine and Science in Sports and Exercise*.

Cavanagh, P.R., Pollock, M.L., & Landa, J. (1977). A biomechanical comparison of elite and good distance runners. *Annals of the New York Academy of Sciences*, **301**, 328-345.

Dillman, C.J. (1975). Kinematic analysis of running. *Exercise and Sports Science Reviews*, **3**, 193-218.

Elliot, B., & Ackland, T. (1981). Biomechanical effects of fatigue on 10,000 meter running technique. *Research Quarterly for Exercise and Sport*, **52**, 160-166.

Elliot, B., & Blanksby, B. (1975). A cinematographic analysis of overground and treadmill running by males and females. *Medicine and Science in Sports and Exercise*, **8**(2), 84-87.

Elliot, B., & Blanksby, B. (1979). The synchronization of muscle activity and body segment movements during a running cycle. *Medicine and Science in Sports and Exercise*, **11**(4), 322-327.

Giradin, Y., & Roy, B.G. (1974). Effects of a non-directive type of training program on the running patterns of male sedentary subjects. In R.C. Nelson & C.A. Morehouse (Eds.), *Biomechanics IV* (pp. 112-120). Baltimore: University Park.

Gollnick, P.D., & Karpovich, P.V. (1964). Electrogoniometric study of locomotion and of some athletic movements. *The Research Quarterly*, **35**, 357-369.

Grieve, D.W., Pheasant, S., & Cavanagh, P.R. (1978). Prediction of gastrocnemius length from knee and ankle joint posture. In E. Asmussen & K. Jorgensen (Eds.), *Biomechanics VI-A* (pp. 405-412). Baltimore: University Park.

Grillner, S., Halbertsma, J., Nilsson, J., & Thorstensson, A. (1979). The adaptation to speed in human locomotion. *Brain Research*, **165**, 177-182.

Hamill, C.L., Clarke, T.E., Frederick, E.C., Goodyear, L.J., & Howley, E.T. (1984). Effects of grade running on kinematics and impact force. (Abstract) *Medicine and Science in Sports and Exercise*, **16**(2), 185.

Hoshikawa, T., Matsui, H., & Miyashita, M. (1973). Analysis of running pattern in relation to speed. In S. Cerquiglini, A. Venerando, & J. Wartenweiler (Eds.), *Biomechanics III* (pp. 342-348). Basel, Switzerland: Karger.

Mann, R.A., & Hagy, J. (1980). Biomechanics of walking, running and sprinting. *American Journal of Sports Medicine*, **8**, 345-350.

Miller, D.I. (1978). Biomechanics of running—what should the future hold? *Canadian Journal of Applied Sports Sciences*, **3**, 229-236.

Murray, M.P., Kory, R.C., Clarkson, B.H., & Sepic, S.B. (1966). Comparison of free and fixed speed walking patterns of normal men. *American Journal of Physical Medicine*, **45**(1), 8-24.

Nelson, R.C., & Osterhoudt, R.G. (1971). Effects of altered slope and speed on the biomechanics of running. In J. Vredenbrecht & J. Wartenweiler (Eds.), *Biomechanics II* (pp. 220-224). Basel, Switzerland: Karger.

Nilsson, J., Thorstensson, A., & Halbertsma, J. (1985). Changes in leg movements and muscle activity with speed of locomotion and mode of progression in humans. *Acta Physiologica Scandinavica*, **123**(4), 457-475.

Osterhoudt, R.G. (1969). *A cinematographic analysis of selected aspects of the running stride of experienced track athletes at various speeds and on different slopes*. Unpublished master's thesis, The Pennsylvania State University, University Park.

Saito, M., Kobayashi, K., Miyashita, M., & Hoshikawa, T. (1974). Temporal patterns in running. In R.C. Nelson & C.A. Morehouse (Eds.), *Biomechanics IV* (pp. 106-111). Baltimore: University Park.

Sinning, W.E., & Forsyth, H.L. (1970). Lower limb actions while running at different velocities. *Medicine and Science in Sports and Exercise*, **2**, 28-34.

Teeple, J.B. (1968). *A biomechanical analysis of running patterns of college women*. Unpublished master's thesis, The Pennsylvania State University, University Park.

Vaughan, C.L. (1984). Biomechanics of running gait. *CRC Critical Reviews of Biomedical Engineering*, **12**, 1-48.

Williams, K.R. (1980). *A biomechanical evaluation of distance running efficiency*. Unpublished doctoral thesis, The Pennsylvania State University, University Park.

Williams, K.R. (1985). Biomechanics of running. *Exercise and Sports Science Reviews*, **13**, 389-441.

Appendix

The following is sample raw (unsmoothed) two-dimensional coordinate data at a spacing of 0.005 milliseconds (200 Hz collection) for five points on the left lower extremity during one cycle of level treadmill running at 3.8 m • s⁻¹. The data were collected using the video system described earlier, and the origin of the right-handed coordinate system (x horizontal, y vertical) was arbitrarily placed in the lower left corner in front of the subject (who was facing the left side). A plot of these coordinates, after filtering, would result in a diagram similar to that shown in Figure 3.5. The cycle (from footstrike to the next ipsilateral footstrike—frames 1 and 146) is "padded" with data from 16 frames before (labeled -16 to -1) and 16 frames after footstrike (labelled $+1$ to $+16$) to allow experimentation with different smoothing algorithms.

Note that the subject was moving slightly backward on the treadmill during the cycle so that the x coordinate of the hip at the second footstrike is approximately 10 cm greater than the similar value at the first footstrike. Footstrike was detected by an algorithm that examined the change in the y coordinate of the heel target after the most forward movement of the same target.

Sample Data for One Subject Running at 3.8 m • s⁻¹

Frame	Hip x	Hip y	Knee x	Knee y	Lateral malleolus x	Lateral malleolus y	Heel x	Heel y	5th Metatarsal x	5th Metatarsal y
-16	57.02	104.51	31.81	65.92	43.71	20.31	45.82	13.05	32.99	8.88
-15	57.02	104.47	32.13	65.59	41.86	19.59	43.50	12.17	30.43	8.75
-14	57.02	104.41	32.72	65.15	40.27	18.83	41.45	11.42	27.97	8.66
-13	57.02	104.31	33.09	64.95	38.69	18.11	38.99	10.66	25.93	8.68
-12	56.72	104.06	33.54	64.38	37.31	17.49	37.00	10.13	23.92	8.66
-11	57.02	103.85	33.77	63.95	35.88	17.03	35.36	9.49	21.97	8.80
-10	57.02	103.74	34.80	63.54	34.59	16.52	34.19	9.02	20.72	9.13
-9	57.02	103.70	35.57	62.64	32.68	15.93	31.27	8.43	17.74	9.45
-8	57.02	103.70	35.72	62.45	31.86	15.73	30.31	8.13	16.72	9.62
-7	57.02	103.49	36.05	62.05	31.26	15.29	29.57	8.02	16.05	9.55
-6	56.83	103.08	36.54	61.74	30.91	15.29	28.81	7.85	15.64	9.76
-5	57.15	102.65	37.31	61.08	30.65	14.57	28.86	7.35	15.23	9.45
-4	57.28	102.54	37.05	60.77	31.04	14.34	28.75	7.03	15.69	9.09

Frame	Hip x	Hip y	Knee x	Knee y	Lateral malleolus x	Lateral malleolus y	Heel x	Heel y	5th Metatarsal x	5th Metatarsal y
−3	57.48	102.21	37.19	60.58	31.16	14.16	29.41	6.76	16.21	8.79
−2	57.43	102.15	37.40	60.31	31.78	13.76	29.98	6.25	16.51	8.22
−1	57.68	101.83	37.47	60.00	32.44	13.39	30.80	5.91	17.28	7.81
1	57.49	101.33	37.47	59.88	32.95	12.78	31.68	5.74	18.15	7.15
2	57.84	101.24	37.76	59.36	34.31	12.60	32.85	5.56	19.42	6.63
3	58.01	100.62	37.87	59.03	35.12	12.23	34.49	5.33	20.82	5.61
4	57.89	100.42	37.95	58.61	36.54	11.91	36.18	5.27	22.60	4.63
5	58.11	99.80	38.01	58.39	38.01	11.85	38.58	5.33	24.73	4.17
6	58.31	98.81	38.17	57.90	39.65	11.85	40.35	5.40	26.34	4.08
7	58.25	98.57	38.17	57.65	41.45	11.91	42.05	5.40	28.34	4.06
8	58.50	98.38	37.87	57.44	43.25	12.17	44.13	5.40	30.65	4.17
9	58.74	98.20	37.82	57.43	45.25	12.32	45.96	5.45	32.75	4.06
10	58.98	98.02	37.82	57.25	46.89	12.21	48.01	5.68	34.65	4.17
11	59.23	97.83	38.06	56.97	48.52	12.32	49.89	5.63	36.85	4.06
12	59.46	97.65	38.12	56.81	50.64	12.36	52.16	5.74	38.70	4.17
13	59.71	97.46	38.99	56.55	52.36	12.37	53.80	5.80	40.39	4.17
14	59.96	97.27	39.46	56.20	54.40	12.50	56.26	5.80	42.76	4.22
15	60.19	97.51	40.47	55.71	56.20	12.41	58.31	5.74	44.58	4.17
16	59.70	97.68	41.10	55.26	58.17	12.28	60.11	5.69	46.89	4.22
17	59.75	97.68	42.27	54.91	60.14	12.39	62.05	5.80	48.63	4.17
18	59.94	97.55	43.34	54.35	61.94	12.41	64.40	5.80	50.47	4.13
19	60.00	97.50	44.73	54.09	63.69	12.35	66.04	5.80	52.78	4.17
20	60.30	97.58	45.74	53.53	65.97	12.41	68.31	5.81	54.88	4.13
21	60.52	97.68	47.00	53.22	67.95	12.50	70.32	5.95	56.83	4.17
22	60.82	97.60	48.17	52.87	69.97	12.50	72.40	5.95	58.97	4.13
23	61.12	97.80	49.53	52.61	72.04	12.50	74.48	5.95	61.12	4.13
24	61.39	97.73	50.71	52.66	74.23	12.57	76.40	6.04	63.03	4.26
25	61.78	97.73	52.16	52.39	75.92	12.62	78.58	6.07	65.22	4.13
26	62.10	97.73	53.91	52.22	77.92	12.76	80.60	6.26	67.27	4.13
27	62.27	97.73	55.55	51.84	79.80	12.99	82.43	6.32	69.27	4.08
28	62.92	97.79	56.86	51.84	81.66	13.23	84.70	6.70	71.42	4.17
29	62.70	98.16	58.44	51.84	83.84	13.38	86.52	6.67	73.39	4.10
30	63.00	98.23	60.14	51.89	85.87	13.76	88.93	6.94	75.58	4.22
31	62.95	98.35	61.72	52.05	88.00	14.14	90.68	7.44	77.51	4.13
32	63.23	98.52	63.00	52.40	89.64	14.53	92.73	7.73	79.82	4.26
33	63.17	98.75	64.28	52.66	91.44	15.08	94.53	8.34	81.77	4.26
34	63.17	99.19	65.87	52.87	93.14	15.63	96.58	8.84	83.92	4.26
35	63.17	99.26	67.37	53.35	95.19	16.34	98.65	9.28	86.07	4.54

(Cont.)

Sample Data (Continued)

Frame	Hip x	Hip y	Knee x	Knee y	Lateral malleolus x	Lateral malleolus y	Heel x	Heel y	5th Metatarsal x	5th Metatarsal y
36	63.30	99.32	68.66	53.97	96.83	16.86	100.10	9.95	88.16	4.63
37	62.60	99.88	69.54	54.30	98.33	17.42	102.10	10.77	90.16	4.84
38	62.82	100.15	70.79	54.84	99.96	18.19	103.73	11.59	92.45	4.84
39	62.54	100.33	71.88	55.44	101.33	18.91	105.37	12.41	94.42	5.04
40	62.31	100.78	73.00	56.20	102.91	19.73	107.18	13.32	96.36	5.36
41	62.54	101.15	74.07	56.75	103.79	20.55	108.65	14.21	98.24	5.61
42	61.69	101.28	74.89	57.52	104.80	21.26	110.01	15.12	100.62	5.95
43	62.01	101.78	75.32	58.01	105.84	21.89	111.65	16.16	101.98	6.44
44	61.66	102.26	76.22	58.71	106.82	22.77	112.75	17.27	104.55	6.99
45	61.53	102.47	76.69	59.12	108.02	23.47	114.03	18.16	106.41	7.58
46	61.39	102.81	77.35	59.76	109.00	24.18	115.31	19.59	108.08	8.34
47	61.23	102.97	77.85	60.35	109.66	25.04	116.44	20.84	110.35	9.08
48	61.12	103.36	78.39	60.76	110.54	26.05	117.48	22.27	112.17	10.10
49	60.82	103.79	78.21	61.29	111.36	26.93	118.49	23.42	113.98	11.01
50	60.53	104.02	78.99	61.67	112.20	27.80	119.36	24.70	115.86	12.17
51	60.30	104.18	78.80	61.99	112.99	28.95	120.37	25.99	117.55	13.32
52	60.12	104.43	79.15	62.34	114.04	29.86	121.13	27.31	118.73	14.57
53	59.95	104.51	78.80	62.58	115.04	31.02	121.83	28.44	120.31	15.78
54	59.71	104.84	79.03	62.88	116.03	32.08	122.99	29.90	121.27	17.09
55	59.60	104.84	78.99	62.98	116.68	33.37	123.81	31.04	122.42	18.24
56	59.24	105.31	78.85	63.08	117.71	34.60	124.44	32.38	123.52	19.67
57	58.93	105.27	78.55	63.09	118.53	35.77	125.45	33.77	124.49	20.75
58	58.93	105.27	78.44	63.22	119.14	37.09	126.08	34.84	125.45	22.13
59	58.84	105.27	78.17	63.09	119.96	38.34	126.75	36.43	126.33	23.42
60	58.66	105.43	77.80	63.10	120.29	39.60	127.16	37.80	127.15	24.88
61	58.11	105.27	77.43	62.90	120.59	40.98	127.97	38.99	127.56	26.29
62	58.11	105.27	77.15	62.97	120.94	42.13	127.91	40.63	128.65	27.80
63	57.84	105.43	76.69	62.72	120.94	43.44	128.16	42.07	129.02	29.33
64	57.60	105.31	75.92	62.52	121.05	44.58	128.59	43.54	129.61	30.74
65	57.43	104.92	75.41	62.00	121.00	45.90	128.13	44.90	129.72	32.26
66	57.02	104.56	75.05	61.85	120.64	46.84	127.50	46.43	130.29	33.76
67	56.92	104.61	74.45	61.40	120.64	48.17	127.16	48.01	130.52	35.27
68	56.61	104.31	73.60	60.95	119.82	49.15	126.95	49.45	130.89	36.88
69	56.61	103.74	73.00	60.42	119.53	50.15	126.34	50.56	130.78	38.52
70	56.45	103.74	72.29	59.90	119.00	51.09	125.67	52.00	130.65	39.67
71	56.10	103.27	71.36	59.29	118.12	52.02	125.22	53.37	130.20	41.24
72	55.98	102.79	70.36	58.88	117.36	52.89	124.44	54.43	130.02	42.62
73	55.75	102.48	69.83	58.26	116.44	53.75	123.47	55.34	129.61	43.73
74	55.98	101.97	68.50	57.80	115.57	54.48	122.80	56.30	128.95	45.03
75	55.60	101.60	67.73	57.14	114.43	54.87	121.83	57.06	128.16	46.05

| | Hip | | Knee | | Lateral malleolus | | Heel | | 5th Metatarsal | |
Frame	x	y	x	y	x	y	x	y	x	y
76	55.63	101.06	67.09	56.61	113.93	55.69	120.68	57.84	127.34	46.92
77	55.57	100.31	65.89	55.89	112.70	56.28	119.57	58.50	126.57	47.69
78	55.32	99.99	64.81	55.34	111.27	56.92	118.55	59.38	125.16	48.41
79	55.20	99.17	63.58	54.62	110.24	57.33	117.31	59.94	124.49	49.11
80	55.16	98.16	62.60	53.97	108.84	57.62	116.09	60.20	123.24	49.38
81	55.03	98.05	61.64	53.35	107.63	58.01	114.63	60.85	121.98	49.87
82	55.20	97.48	60.75	52.87	106.33	58.01	113.38	60.85	120.64	50.02
83	55.22	96.86	59.12	52.37	104.79	58.16	111.99	61.02	119.30	50.11
84	55.08	96.01	58.06	51.84	103.64	58.24	110.54	60.97	117.78	50.05
85	55.06	95.52	57.02	51.35	102.14	58.01	109.12	61.00	116.68	50.15
86	55.08	95.14	55.38	51.09	100.46	58.11	107.67	60.85	115.02	49.94
87	54.92	94.93	53.80	50.87	99.18	57.79	106.25	60.70	113.63	49.35
88	54.77	94.71	52.56	50.68	97.90	57.47	105.02	60.35	112.07	48.92
89	54.61	94.49	51.12	50.53	96.40	57.19	103.38	59.94	110.35	48.24
90	54.47	94.27	49.70	50.43	94.91	56.85	102.15	59.47	109.12	47.89
91	54.31	94.05	48.01	50.38	93.60	56.37	100.64	58.89	107.07	47.13
92	54.15	93.84	46.48	50.43	92.12	55.83	99.47	58.12	105.72	46.31
93	54.20	93.66	45.25	50.73	90.76	55.26	98.00	57.48	103.96	45.49
94	54.23	93.54	43.63	51.08	89.32	54.69	96.90	56.75	102.32	44.24
95	54.01	93.69	42.11	51.40	88.16	54.05	95.24	55.79	100.46	43.44
96	53.90	93.91	40.68	51.91	87.01	53.34	93.96	54.91	98.87	42.13
97	53.79	93.91	39.51	52.48	85.61	52.52	92.86	53.86	97.01	41.18
98	53.79	94.02	38.01	53.04	84.38	51.66	91.61	52.66	95.19	39.86
99	53.44	94.57	36.70	53.86	83.29	50.82	89.99	51.52	93.54	38.54
100	53.60	94.84	35.55	54.51	81.83	49.87	88.73	50.20	91.44	37.21
101	53.09	95.27	34.13	55.19	80.48	48.89	87.41	48.93	89.31	36.00
102	52.93	95.55	33.09	55.93	79.34	47.89	85.95	47.74	87.58	34.64
103	52.52	96.07	32.24	56.85	77.51	46.99	84.52	46.08	85.71	33.11
104	51.96	96.48	31.21	57.84	76.49	45.72	83.30	44.78	83.36	31.96
105	51.91	97.06	30.22	58.54	75.05	44.69	81.83	43.37	81.01	30.59
106	51.91	97.64	29.16	59.39	73.25	43.64	80.08	41.92	78.99	29.28
107	51.14	98.12	28.58	60.18	71.88	42.33	79.20	40.23	76.39	27.84
108	51.09	98.72	27.93	60.88	70.30	41.18	77.35	38.83	74.40	26.49
109	50.42	99.27	27.22	61.70	68.95	40.08	75.87	37.36	71.88	25.28
110	50.27	99.80	26.54	62.49	67.27	38.64	73.87	35.77	69.32	23.94
111	49.70	100.42	25.78	63.07	65.27	37.41	72.04	34.24	66.69	22.77
112	49.50	101.02	25.47	63.79	63.99	36.18	70.13	32.81	64.09	21.38
113	49.40	101.49	24.96	64.36	61.99	34.91	68.09	31.17	61.51	20.21
114	48.83	102.11	24.46	64.90	60.14	33.81	66.35	29.53	58.71	18.93
115	48.63	102.65	24.29	65.23	57.95	32.45	64.09	27.94	56.06	17.66

(Cont.)

Sample Data (Continued)

Frame	Hip x	Hip y	Knee x	Knee y	Lateral malleolus x	Lateral malleolus y	Heel x	Heel y	5th Metatarsal x	5th Metatarsal y
116	48.42	103.18	23.94	65.64	56.04	31.35	61.94	26.51	53.38	16.78
117	47.86	103.56	23.89	66.03	54.03	30.05	59.53	25.02	50.37	15.73
118	47.86	104.15	23.88	66.36	52.11	28.80	57.07	23.43	47.70	14.78
119	47.79	104.45	23.64	66.59	50.16	27.53	55.02	22.05	44.73	13.71
120	47.70	104.88	23.47	66.66	48.23	26.43	52.62	20.57	42.22	12.83
121	47.34	105.38	23.42	66.82	46.21	25.34	50.30	19.27	39.28	11.93
122	47.60	105.64	23.76	66.69	43.96	24.30	48.06	17.70	36.54	11.25
123	47.50	105.69	23.88	66.46	42.27	23.23	45.77	16.43	33.81	10.48
124	47.41	105.88	24.08	66.59	40.07	22.41	43.28	15.39	30.96	10.10
125	47.30	106.15	24.24	66.38	38.38	21.49	40.63	14.35	28.09	9.47
126	47.19	106.15	25.06	66.05	36.17	20.52	38.44	13.10	25.68	9.27
127	47.19	106.15	25.11	65.79	34.49	19.75	36.31	12.23	23.18	9.13
128	47.35	106.03	25.33	65.44	32.72	19.01	33.67	11.65	20.92	8.98
129	47.19	106.05	25.93	65.13	30.99	18.39	31.78	11.13	18.51	8.98
130	47.04	105.88	26.60	64.61	29.16	18.06	29.57	10.71	16.14	9.21
131	46.89	105.69	26.94	64.28	27.76	17.52	27.39	10.04	14.41	9.39
132	47.04	105.54	27.22	63.95	26.40	17.24	25.88	9.66	12.52	9.62
133	47.19	105.43	28.19	63.70	25.47	16.88	24.30	9.37	10.88	9.95
134	47.04	105.25	28.58	63.32	24.43	16.61	22.79	9.16	9.73	10.19
135	47.09	105.11	28.80	62.82	23.46	16.35	21.78	9.02	8.48	10.44
136	47.19	104.86	29.00	62.49	22.77	16.27	20.78	8.78	7.69	10.53
137	47.19	104.61	29.27	62.34	22.79	15.93	20.33	8.72	6.89	10.77
138	47.43	104.43	29.82	62.06	21.78	16.01	19.67	8.61	6.45	10.81
139	47.35	104.34	29.82	61.67	21.53	15.55	19.33	8.26	6.07	10.81
140	47.50	104.06	30.25	61.56	21.64	15.44	19.33	8.20	6.56	10.42
141	47.34	103.74	30.39	61.11	21.78	15.19	19.60	7.91	6.50	10.24
142	47.76	103.56	30.63	60.85	22.19	14.73	19.57	7.64	6.67	10.02
143	47.55	103.24	30.80	60.70	22.60	14.42	20.20	7.20	6.93	9.66
144	47.79	102.86	30.50	60.18	22.96	13.96	20.96	6.67	7.44	9.04
145	47.84	102.81	30.80	59.94	23.42	13.60	21.60	6.19	8.31	8.53
146	47.91	102.26	30.45	59.65	24.55	13.09	22.42	5.88	9.03	7.76
+1	48.01	102.01	30.45	59.18	25.06	12.73	23.83	5.63	10.09	6.93
+2	48.01	101.29	30.61	58.96	26.19	12.32	25.58	5.58	11.49	6.13
+3	47.70	100.88	30.58	58.45	27.16	11.96	27.11	5.45	13.29	5.04
+4	47.84	100.25	30.63	58.29	29.00	11.68	28.75	5.27	15.37	4.22
+5	47.82	99.93	30.45	57.84	30.03	11.76	30.39	5.27	16.92	3.87
+6	47.37	99.26	30.25	57.57	32.24	11.63	32.78	5.33	19.13	3.97
+7	48.42	98.98	29.82	57.52	34.03	11.96	34.67	5.34	21.10	3.97

Frame	Hip x	Hip y	Knee x	Knee y	Lateral malleolus x	Lateral malleolus y	Heel x	Heel y	5th Metatarsal x	5th Metatarsal y
+8	48.66	98.74	29.82	57.25	35.72	12.01	36.54	5.50	22.96	4.08
+9	48.90	98.50	29.41	57.19	37.40	12.01	38.40	5.51	24.91	3.97
+10	49.15	98.25	29.63	56.90	38.99	12.27	40.57	5.68	26.84	3.92
+11	49.39	98.01	29.86	56.75	40.90	12.17	42.54	5.72	28.86	3.97
+12	49.65	97.75	30.09	56.61	42.79	12.27	44.57	5.62	30.94	3.97
+13	50.30	97.79	30.45	56.20	44.73	12.35	46.18	5.88	32.85	4.08
+14	49.93	97.74	31.68	55.79	46.25	12.39	48.28	5.72	34.99	3.97
+15	49.81	97.73	32.31	55.43	48.35	12.35	50.28	5.95	36.54	4.02
+16	50.11	97.65	33.37	54.97	50.16	12.32	52.16	5.80	38.75	4.17

Chapter 4

Upper Extremity Function in Distance Running[1]

Richard N. Hinrichs

In searching the literature on the biomechanics of running, one can find only an occasional mention of the arms and the role they play in the running process. It appears that little importance has been placed on the arm swing. In the opinion of Mann (1981) and Mann and Herman (1985), for example, the arms are simply used for balance and do not play a significant role in dictating the quality of running performance.

This lack of interest in arm action may not have always been the case. Figure 4.1 shows an ancient Greek artist's rendition of a sprint running

Figure 4.1. Ancient Greek artist's rendition of sprint running event. *Note.* From *Greek Athletic Sports and Festivals* (p. 283) by E.N. Gardiner, 1910, London: Macmillan. Copyright 1910 by Macmillan and Co.

[1]Portions of this chapter previously appeared in R.N. Hinrichs (1987), and R.N. Hinrichs, P.R. Cavanagh, & K.R. Williams (1987). Adapted by permission of Human Kinetics Publishers, Champaign, IL.

event. The arm swing seems greatly exaggerated in these runners, leading one to believe that the ancient Greeks attributed much importance to the arm swing in running (Gardiner, 1910).

The first reference to arm swing in the literature was by Aristotle (1961). In demonstrating a remarkable understanding of what would come to be known as Newton's laws of motion, Aristotle commented on the importance of swinging the arms in running:

> . . . the animal that moves makes its change of position by pressing against that which is beneath it. . . . Hence athletes jump farther if they have the weights in their hands than if they have not, and runners run faster if they swing their arms; for in the extension of the arms there is a kind of leaning upon the hands and wrists. (p. 498)

The possible propulsive role of the arm swing alluded to by Aristotle seems to have eluded scientists until just recently. In fact, the entire topic of arm swing in running remained dormant for more than 2,000 years.

This chapter summarizes the little that has been done in this area. The bulk of the information presented here comes from the author's PhD dissertation (Hinrichs, 1982). Portions of the work have been published (Hinrichs, 1985, 1987; Hinrichs, Cavanagh, & Williams, 1983, 1987). The reader is referred to one or more of these works for details.

This chapter is organized into three sections. First, descriptions of "ideal form" are presented, which are based on opinions from various parties as to how the arms should be swung in running. Second, an overview of the author's own research is presented. This section, in turn, has three major subsections: (a) electromyographic (EMG) and net joint moment considerations, (b) center of mass and propulsion considerations, and (c) angular momentum considerations. Finally, the author's research is discussed in relation to other studies dealing with the arm swing, and recommendations for future work in this area are provided.

Descriptions of "Ideal Form"

Most discussions about the arm swing have been descriptive in nature with little attention directed toward explaining the motion. Wickstrom (1970), for example, described the arm swing in sprint running as follows (see Figure 4.2):

> Arm action in running is compensatory and synchronous with the action of the legs. Since the leg action in sprinting is forceful and extensive, the arms must move in like manner. The hand swings nearly as high as the chin and slightly toward the midline on the forward

Figure 4.2. Pattern of running at sprinting speed. *Note.* From *Fundamental Motor Patterns* (p. 34) by R.L. Wickstrom, 1970, Philadelphia: Lea & Febiger. Copyright 1970 by Lea & Febiger. Reprinted by permission.

swing, and the elbow reaches almost as high as the shoulder on the back swing. Both shoulders are raised slightly at the ends of the swing because of the forceful and extreme arm movement but this is achieved without producing tension.

During the excursion of the arms in the forward-backward movement, the angle of the elbow changes somewhat in many expert sprinters. The angle increases on the downward movement of the backward swing and decreases to about a right angle on the upward swing of the backward movement. Then this position is retained on the forward swing until just before the end when the angle tends to decrease slightly.[a]

Wickstrom went on to describe the changes in the action of the arms as the pace is slowed from the sprint through middle distance and distance paces down to "jogging." As the pace slows, the arms move through shorter arcs, there is slightly more flexion at the elbows, the arms swing more across the trunk toward the midline, and the shoulders are elevated less at the ends of the arm swing.

[a]From *Fundamental Motor Patterns* (p. 34) by R.L. Wickstrom, 1970, Philadelphia: Lea & Febiger. Copyright 1970 by Lea & Febiger. Reprinted by permission.

Track and field coaches have definite opinions as to how the arms should be used in running. In a *Runner's World* article entitled "Up in Arms" (1980), several famous coaches were interviewed and gave the following recommendations.[a]

1. *The arms should be carried low and relaxed.* They recommended against carrying the arms rigidly and too high (i.e., with an elevated scapula and elbows flexed). In addition, the hands "should be cupped or closed with the fingers pinched together or touching. Running with the hands open or the hand clinched tenses the arms and as a consequence the entire body" (p. 55).

2. *The elbows should swing forward and backward in a straight line.* "Any lateral elbow motion hinders efficiency. Such outward rotation defeats the principle that all of the runner's parts should be moving in the direction he is running" (p. 54). This principle, if taken literally, is rather absurd. It is impossible in a cyclic motion such as running for all body parts to move in a single direction and rather unlikely that the elbows would ever move in straight lines. Its intent is, perhaps, that all body segments should move only in the sagittal plane. Some coaches, however, are realizing that many good runners do not follow this "ideal form." Len Miller, who was head track and field coach at Arizona State, qualified this notion a bit. He was quoted as saying, "The elbow must swing straight through. . . . Some runners because of their anatomical composition, are more comfortable carrying their forearms a little bit across their bodies. Other runners have more of what may be considered a classic style, with the forearms being more aligned with the direction in which the runner is moving. But the important thing is the elbow" (p. 54).

3. *The pattern of movement of the arms is opposite that of the legs.* "When the left leg comes forward so does the right arm. This pattern . . . allows the runner to maintain a squareness in the hips and shoulders in the direction the runner wants to go" (p. 55). Although this rule of opposing action is a statement of "ideal form" as such, it is unlikely any runner would ever violate this rule. It is hard enough to walk and force the same arm and leg to go forward at the same time. Try it running sometime. We will see later in this chapter that the important variable is angular momentum, and for normal airborne phases to occur, one *must* have opposing action of upper and lower body. This point is commonly missed by artists and models in clothing ads that show "runners" with the same arm and leg forward (see Figure 4.3).

Figure 4.3. Example of advertisement for running apparel. Note the same arm and leg forward.

Summary of the Author's Research

The purpose of this section is to summarize my research on upper extremity function in running and to relate various aspects of this research to the few other studies that have been done in this area.

The purpose of the original study (Hinrichs, 1982) was to answer the following questions:

1. What contributions do the arms make to the total-body angular momentum in running?
2. Do the arms aid in the forward and upward propulsion of the body, and if so, to what extent?
3. Does the arm swing tend to reduce the magnitudes of the vertical and horizontal oscillations of the body center of mass, and if so, to what extent?
4. What are the EMG patterns of selected upper extremity and trunk muscles believed to be involved in the motion of the arms?

5. What are the net moments at the shoulder and elbow joints and how do they relate to the EMG activity described above?
6. What is the effect of running speed on the arm swing and on each of the above parameters?

Procedures

Ten male recreational runners ranging in age from 20 to 32 years (mean 22.5 years) were used as subjects. They ranged in height from 1.61 to 1.83 m (mean 1.76 m) and ranged in mass from 51.3 to 75.9 kg (mean 64.8 kg). Three trials were performed by each subject on a treadmill, one at a slow speed of 3.83 m • s⁻¹, one at a medium speed of 4.47 m • s⁻¹, and one at a fast speed of 5.36 m • s⁻¹. These represented running paces of 7, 6, and 5 minutes per mile, respectively. A four-camera, three-dimensional (3D) cinematographic technique was employed using the DLT algorithm (Abdel-Aziz & Karara, 1971). Each camera operated at a nominal rate of 100 frames per second.

Simultaneous electromyograms were obtained from eight upper extremity and trunk muscles believed to be involved in the arm swing. These muscles were the clavicular pectoralis major (PEC), anterior deltoid (AD), middle deltoid (MD), posterior deltoid (PD), upper latissimus dorsi (LAT), long head of the biceps brachii (BIC), brachioradialis (BR), and the long head of the triceps brachii (TRI). The raw EMG data were sampled by an analog to digital (A/D) converter at the rate of 833 Hz, full-wave rectified, integrated with a time-reset period equal to 1/20 of the running cycle, and normalized to percentages of maximal voluntary contractions (%MAX) in each respective muscle. The EMG and film data were synchronized by means of a digital pulse elicited at each left footstrike from the output of an insole footswitch device placed inside each subject's shoe.

A 14-segment mathematical model of the body was used and consisted of the head, the trunk, and both upper arms, forearms, hands, thighs, calves, and feet. The 3D estimates of joint centers digitized from the films represented the endpoints of each of the segments. Several different combinations of body segment parameters (BSPs) were obtained from the literature, and a certain set was deemed most appropriate for a given subject if it minimized the fluctuations in the computed linear and angular momentum values during the airborne phases of the running cycle.

The raw 3D segment endpoint data were smoothed using a second-order Butterworth digital filter (Winter, Sidwall, & Hobson, 1974) with cutoff frequencies ranging from 2 to 8 Hz. Different cutoff frequencies were chosen objectively for each coordinate of each endpoint using an algorithm similar to the one described by Jackson (1979) adapted for use with a digital filter. The vertical (Z) coordinates of segment endpoints were generally found to need higher cutoff frequencies for smoothing than did the X or Y coordinates.

Finite difference differentiation (Miller & Nelson, 1973) was used whenever the analysis called for taking the first or second time derivative of the smoothed displacement data. No further smoothing was done following differentiation.

The smoothed 3D coordinates of the 20 points on the body were entered into a computer program that calculated various kinematic and kinetic quantities of the body and its segments at successive instants of time throughout the running cycle. Specifically, the following quantities were computed:

1. The paths of the centers of mass (CMs) of the arms, the body-minus-arms, and the whole body over the running cycle
2. The contributions of the arms, head-plus-trunk (HT), and legs to the changes in total-body linear momentum over each contact phase—these were called "lift" and "drive," respectively
3. The angular momentum of each body segment and various systems of segments about the body CM
4. The net moments at the shoulder and elbow joints

Results and Discussion

Unless otherwise indicated, the results shown in this section have been averaged over all 10 subjects, providing mean curves and mean values of selected variables.

EMG and Net Moments.   The mean results for joint angles, net moments, and EMG for the shoulder joint at the medium speed are shown in Figure 4.4. In this figure and in subsequent figures, circles have been drawn on stick figures to indicate left wrist, shoulder, and ankle. Shaded areas represent airborne phases. The MD is not shown because its activity closely mirrored that of the PD. Although both sides of the body were measured, results for the left and right sides were similar on the average. Shown here are the results for the left side. The arm action consisted of a forward swing (flexion phase of the shoulder) ending shortly after ipsilateral toe-off (ITO), followed by a backward swing (extension phase of the shoulder) ending shortly after contralateral toe-off (CTO).

The angle β shows the shoulder slightly abducted (10 to 25 degrees) throughout the running cycle. At this speed of running, the angle α never became positive, indicating that the shoulder always remained somewhat extended. If measured relative to the trunk long axis, rather than to the vertical as was done here, the angle α would have been somewhat larger (less negative) because of the slight forward lean of the trunk.

EMG activity in the PEC and AD correlated well with the occurrence of a net flexor moment at the shoulder, which halted the backward swing and initiated the forward swing of the arm. Activity in the MD, PD, and LAT produced a net extensor moment at the shoulder throughout most

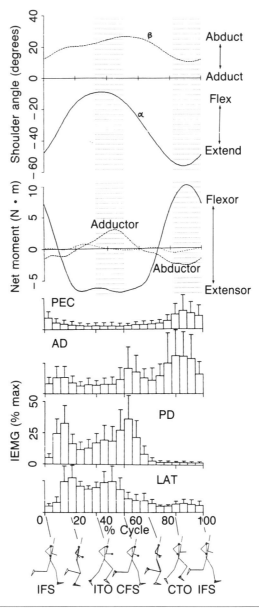

Figure 4.4. Relationship between angles, moments, and EMG activity at the shoulder joint for running at 4.5 m • s⁻¹. *Note.* From ''A Three-Dimensional Analysis of the Net Moments at the Shoulder and Elbow Joints in Running and Their Relationship to Upper Extremity EMG Activity'' by R.N. Hinrichs, 1985, in D.A. Winter, R.W. Norman, R.P. Wells, K.C. Hayes, and A.E. Patla (Eds.), *Biomechanics IX-B*, p. 339, Champaign, IL: Human Kinetics. Copyright 1985 by Human Kinetics Publishers. Reprinted by permission.

of the forward swing and into the first half of the backward swing of the arm. The eccentric activity during the early part of the forward swing appears to have been used to control the forward swing of the arm, which was aided by gravity.

The mean results for joint angle, net moment, and EMG for the elbow joint are shown in Figure 4.5. Instead of a single phase each of flexion and extension per cycle as for the shoulder joint, the elbow showed two phases of flexion and two of extension per cycle. The primary extension phase (PEP) occurred at contralateral footstrike (CFS) followed by a much smaller secondary extension phase (SEP) at ipsilateral footstrike (IFS). The EMG activity in the BIC and the BR was closely associated with flexor moments at the elbow that arrested each extension phase and initiated elbow flexion. The PEP was initiated by a peak extensor moment occurring at the time of peak activity in the TRI. The SEP, absent in some subjects, occurred without either an extensor moment or TRI activity and was probably due to gravity.

The close relationship between EMG and net moments at the elbow is highlighted in Figure 4.6. The first peak in BR EMG activity was slightly larger than the second. The same was found for the two flexor moment peaks, although there was an electromechanical delay of approximately 50 ms between each peak in BR EMG activity and the corresponding peak flexor moment. There was a similar delay between peak TRI EMG activity and the peak extensor moment. These delays were present at all speeds and are consistent with the findings of Cavanagh and Komi (1979).

The results of the EMG and net joint moment analyses showed each of the eight muscles to have bursts of phasic activity at different times throughout the running cycle. The PEC and BIC showed the least activity, with mean peak integrated values ranging from 10 to 24 %MAX. The PD and LAT showed the most activity, with mean peak integrated values ranging from 30 to 60 %MAX. The EMG activity in each muscle, as well as the net moments and ranges of motion at the joints, all increased with running speed.

Center of Mass Excursions and Ranges of Motion.   It was suggested by Murray, Sepic, and Bernard (1967) and confirmed by Hinrichs and Cavanagh (1981) that the arms serve to reduce the magnitude of the vertical oscillations of the body CM during walking and hence may reduce energy expenditure. To see if this might be true for running, the body was divided into two parts—(a) the arms and (b) the body-minus-arms (BMA). The CM of each part was calculated and its respective path observed over the running cycle. These were compared with each other and with the motion of the CM of the whole body.

The results in Table 4.1 show the mean values ($\pm$ SD) over all subjects for the vertical range of motion of the CMs of the body, BMA, and the

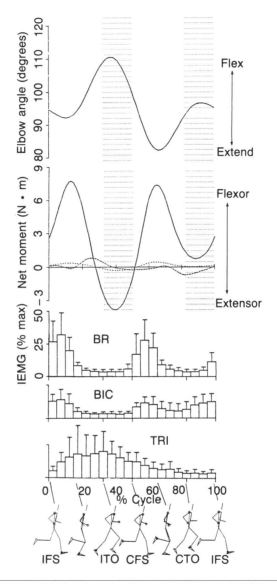

Figure 4.5. Relationship between angles, moments, and EMG activity at the elbow joint for running at 4.5 m • s⁻¹. *Note.* From ''A Three-Dimensional Analysis of the Net Moments at the Shoulder and Elbow Joints in Running and Their Relationship to Upper Extremity EMG Activity'' by R.N. Hinrichs, 1985, in D.A. Winter, R.W. Norman, R.P. Wells, K.C. Hayes, and A.E. Patla (Eds.), *Biomechanics IX-B*, p. 340, Champaign, IL: Human Kinetics. Copyright 1985 by Human Kinetics Publishers. Reprinted by permission.

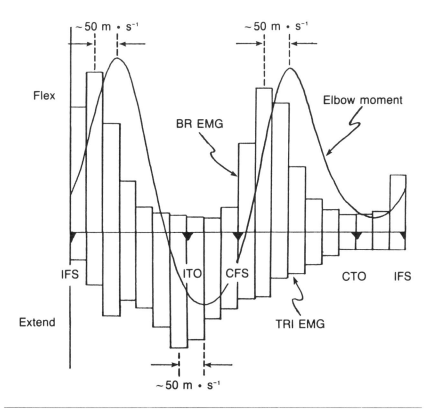

Figure 4.6. Relationship between brachioradialis (BR) and triceps (TRI) EMG activity and the net elbow moment for running at 4.5 m • s⁻¹ *Note.* From "A Three-Dimensional Analysis of the Net Moments at the Shoulder and Elbow Joints in Running and Their Relationship to Upper Extremity EMG Activity" by R.N. Hinrichs, 1985, in D.A. Winter, R.W. Norman, R.P. Wells, K.C. Hayes, and A.E. Patla (Eds.), *Biomechanics IX-B*, p. 341, Champaign, IL: Human Kinetics. Copyright 1985 by Human Kinetics Publishers. Adapted by permission.

arms at each running speed. The vertical excursion of the CM of the body was larger than that of the BMA, indicating that the addition of the arms actually increased the magnitude of the vertical oscillations of the body CM rather than decreasing it as in walking. In addition, the vertical range of motion of the body CM was found to decrease as running speed increased. This result is consistent with the findings of Cavanagh, Pollock, and Landa (1977).

The paths of the CMs in the side-to-side (X) and forward (Y) directions depended to some extent on the manner in which the subjects ran on

Table 4.1   Mean Values for Center of Mass Vertical
Ranges of Motion Across Subjects

	Running speed		
System	Slow	Medium	Fast
Arms	13.84 ± 2.08	13.62 ± 1.62	13.61 ± 1.43
Body-minus-arms	8.25 ± 0.95	8.08 ± 1.14	7.22 ± 1.55
Body	8.69 ± 0.96	8.56 ± 1.13	7.73 ± 1.43

Note. Mean range of motion ± standard deviation in centimeters.

the treadmill. Although the amounts were small, most subjects achieved a net gain or loss in horizontal displacement (both X and Y) during the particular cycle analyzed. This rendered the range of motion comparison described for the vertical direction less useful because the starting and ending points were not the same. Therefore, a different procedure was devised for comparing CM paths in the two horizontal directions. For each direction separately, the total *excursion* of each CM over the running cycle was calculated by computing the absolute value of the CM displacement between two frames and summing over all the frames in the cycle.

The results of the excursion analysis are presented in Table 4.2. Included are the mean X and Y excursions (±SD) of the CMs of the arms, BMA,

Table 4.2   Mean Values for Horizontal Center
of Mass Excursions Across Subjects

	Running speed		
System	Slow	Medium	Fast
X direction			
Arms	6.85 ± 2.25	6.44 ± 2.37	6.20 ± 2.78
Body-minus-arms	2.73 ± 0.94	2.43 ± 0.91	2.64 ± 0.46
Body	2.56 ± 1.15	2.30 ± 1.11	2.43 ± 0.50
Y direction			
Arms	6.47 ± 1.62	7.20 ± 2.48	8.29 ± 2.60
Body-minus-arms	5.63 ± 0.85	5.88 ± 0.65	6.10 ± 0.84
Body	5.31 ± 0.86	5.60 ± 0.69	5.75 ± 0.95

Note. Mean excursion ± standard deviation in centimeters.

and whole body for each running speed. The results showed that at each speed, the mean excursions of the body CM in both X and Y were less than the corresponding excursions of the CM of the BMA. Eight out of 10 subjects showed these results. In those who did not, the body and BMA excursions were very close in value. This indicates that the use of the arms tended to *reduce* the total excursion of the body CM both side to side (X) and in the direction of progression (Y).

To illustrate how this occurs, examples of the CM paths in the X and Y directions are shown in Figures 4.7 and 4.8, respectively. In Figure 4.7,

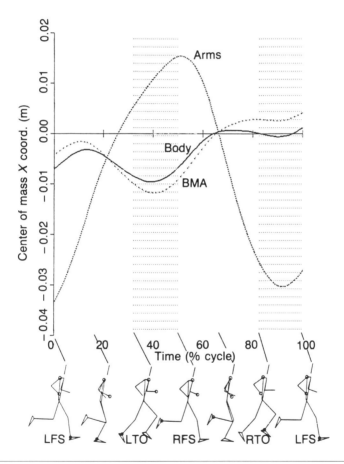

Figure 4.7. X (mediolateral) component of the centers of mass of the arms, body-minus-arms, and whole body of subject 3 running at 3.8 m • s⁻¹. *Note.* From "Upper Extremity Function in Running. I: Center of Mass and Propulsion Considerations" by R.N. Hinrichs, P.R. Cavanagh, and K.R. Williams, 1987, *International Journal of Sport Biomechanics*, **3**, p. 231. Copyright 1987 by Human Kinetics Publishers. Reprinted by permission.

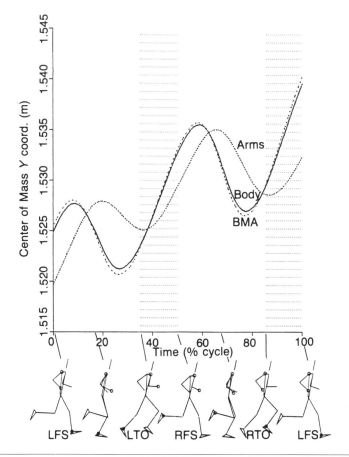

Figure 4.8. Mean curves for the Y (AP) component of the centers of mass of the arms, body-minus-arms, and whole body for running at 3.8 m • s⁻¹. *Note.* From ''Upper Extremity Function in Running. I: Center of Mass and Propulsion Considerations'' by R.N. Hinrichs, P.R. Cavanagh, and K.R. Williams, 1987, *International Journal of Sport Biomechanics, 3*, p. 232. Copyright 1987 by Human Kinetics Publishers. Reprinted by permission.

the curves come from an individual subject. The effect described previously did not show up in the mean curves due to rather large variations between subjects in the CM paths in the X directions. From side to side, the arms covered a relatively large excursion. A portion of this came from the crossover of the forearm and hand in front of the body at the end of the forward swing. This side-to-side motion, however, was coordinated with an opposite motion of the rest of the body. The result was a smaller lateral deviation of the whole-body CM.

A similar phenomenon can be seen in Figure 4.8 for the forward direction. Because the run took place on a treadmill, one would expect the CMs to fluctuate around a fairly constant position. The general pattern was that the body CM moved forward during each airborne phase, changing to a backward motion during each contact phase. (The body CM path in Figure 4.8 also shows the average subject "gaining" slightly on the treadmill.) By moving somewhat out of phase with the rest of the body, the arms tended to reduce these forward and backward fluctuations in the position of the body CM.

The CM results indicate that the arms tend to reduce fluctuations in the horizontal velocity of the runner, in both the side-to-side and forward directions. This could result in a reduction in energy cost. The arm swing does not, however, reduce the *vertical* range of motion of the body CM, but rather tends to increase it. We will see in the next section that this is a beneficial effect rather than a detrimental one.

Lift and Drive.   The terms *lift* and *drive* were adopted to refer to segmental contributions to the propulsive impulses (i.e., changes in linear momentum) of the runner over the contact phase of the running cycle. Lift refers to impulses in the vertical direction; drive to impulses in the forward or anteroposterior (AP) direction. Using the relative momentum approach of Ae and Shibukawa (1980), the *relative momentum* of the arms was defined as the mass of the arms multiplied by the velocity of the CM of the arms relative to a point fixed in the upper trunk. Thus, any passive motion of the arms due to the motion of the trunk is assigned to the trunk's contribution to the total-body momentum. Similarly, any passive motion of the arms and/or trunk resulting from the motion of the hips is assigned to the legs' contribution to the total-body momentum.

The mean values (along with standard deviations and ranges) of the segmental contributions to the total lift are listed in Table 4.3. At the slow speed, the arms contributed roughly 5% to the total lift and the trunk contributed roughly -3%, leaving the legs to contribute 98%. As running speed increased, the arms' contribution to lift increased and the legs' contribution decreased. It is evident that the arms do add a small but potentially important part to the total propulsive lift in running and tend to contribute even more at faster speeds. What was unexpected was the negative lift contribution by the trunk. In effect, the trunk "undoes" half the lift that the arms provide, making the legs do that much more.

The propulsive potential of the arms in the AP direction turned out to be quite different than in the vertical direction. Table 4.4 lists the mean drive values for the arms, legs, and trunk. Also listed are the standard deviations and ranges, which, it can be seen, were rather large. Because the total AP impulse is approximately zero over the contact phase in "constant speed" running, total drive (100%) has been defined here as the AP impulse over the latter (propulsive) portion of the contact phase.

Table 4.3 Mean Values of Segmental Contributions to Lift (Across Subjects), Expressed as Percentages of the Total Lift

Segment	Running speed		
	Slow	Medium	Fast
Arms			
Mean ± SD	4.5 ± 1.7	5.2 ± 1.5	7.1 ± 2.8
Range	(2.0)-(7.6)	(3.2)-(7.9)	(3.1)-(12.9)
Head-plus-trunk			
Mean ± SD	−2.9 ± 1.7	−2.6 ± 1.8	−3.4 ± 2.0
Range	(−5.4)-(0.1)	(−5.3)-(−0.1)	(−7.0)-(−0.4)
Legs			
Mean ± SD	98.3 ± 1.6	97.4 ± 2.2	96.4 ± 2.2
Range	(94.7)-(100.2)	(94.1)-(100.7)	(92.8)-(99.0)

Table 4.4 Mean Values of Segmental Contributions to Drive (Across Subjects), Expressed as Percentages of the Total Drive

Segment	Running speed		
	Slow	Medium	Fast
Arms			
Mean ± SD	−2.6 ± 4.9	−3.3 ± 5.2	−4.8 ± 6.0
Range	(−9.1)-(5.7)	(−10.1)-(5.0)	(−11.5)-(3.7)
Head-plus-trunk			
Mean ± SD	−45.9 ± 44.0	−58.8 ± 45.6	−64.4 ± 44.8
Range	(−147.0)-(−6.3)	(−157.4)-(−12.0)	(−126.3)-(−26.7)
Legs			
Mean ± SD	49.7 ± 43.7	60.4 ± 40.6	66.2 ± 41.0
Range	(19.2)-(153.6)	(19.4)-(159.5)	(14.2)-(135.1)

The results showed that, in general, the arms do not contribute to drive, at least in the manner defined here. The reasons for this can be seen in Figures 4.9 and 4.10. Figure 4.9 shows the mean curves for the AP component of the relative momentum of the arms at the slow speed. Figure 4.10 shows the corresponding momentum curves for the whole body, including contributions of the arms, legs, and trunk.

It is clear in these figures that the arms oppose each other in the AP direction. The forward relative momentum of one arm is nearly canceled out by the backward relative momentum of the other arm. A net decrease can be seen in the total relative momentum of the arms over each contact phase followed by a net increase over each airborne phase. The mean results listed in Table 4.4 thus show small negative values for drive of the arms over the contact phase. The variability was large, however, as some subjects were able to generate a small amount of positive drive from the arms over the contact phase. Most, however, produced close to zero or slightly negative drive from the arms.

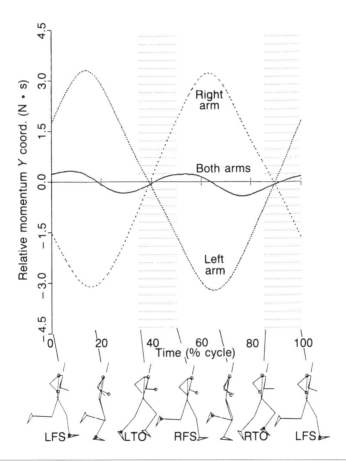

Figure 4.9.   Mean curves of the AP component of relative momentum of the arms for running at 3.8 m • s⁻¹. *Note.* From "Upper Extremity Function in Running. I: Center of Mass and Propulsion Considerations" by R.N. Hinrichs, P.R. Cavanagh, and K.R. Williams, 1987, *International Journal of Sport Biomechanics,* **3,** p. 237. Copyright 1987 by Human Kinetics Publishers. Reprinted by permission.

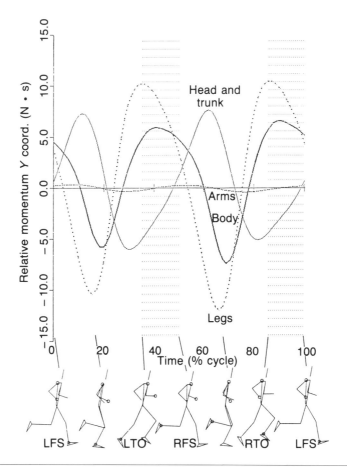

Figure 4.10. Mean curves of the AP component of relative momentum of the arms, head-plus-trunk, legs, and whole body for running at 3.8 m • s⁻¹. *Note.* From "Upper Extremity Function in Running. I: Center of Mass and Propulsion Considerations" by R.N. Hinrichs, P.R. Cavanagh, and K.R. Williams, 1987, *International Journal of Sport Biomechanics*, **3**, p. 238. Copyright 1987 by Human Kinetics Publishers. Reprinted by permission.

Contrary to its minimal contribution to lift, the trunk played a major role in the AP momentum of the whole body. As seen in Figure 4.10, the trunk initially possessed a phase of positive momentum relative to the hips, thus indicating flexion. Later in the contact phase the trunk extended, producing negative relative momentum. Flexion occurred again during the airborne phase and into the next contact phase. Overall the trunk achieved a net decrease in its momentum over the contact phase of the cycle. Numerically, therefore, the trunk was assigned negative values for drive. The legs were the only part of the body consistently producing positive drive over the contact phase.

The numerical results by themselves seem rather puzzling. Do the arms and trunk actually hinder the runner by tending to slow him or her down? Close examination of the curves in Figures 4.9 and 4.10 reveals that they do not. On the contrary, the flexion and extension of the trunk seem to have a beneficial effect on the forward momentum of the runner. The momentum oscillations of the trunk tended to be out of phase with those of the legs, reducing the amount that the whole-body forward momentum decreased during the braking phase and thus requiring less increase over the propulsive phase. If there had been no relative motion of the trunk, the momentum of the whole body would have fluctuated through approximately the same range as that of the legs' contribution.

Although the variability was too large to provide any conclusive statements about the arms, the mean relative momentum patterns of the arms were somewhat similar to those of the trunk. Hence the arms may be capable of producing the same beneficial effect as the trunk.

Angular Momentum.   The results presented thus far have not dealt with the opposing action of the arms and legs in running (i.e., when the left leg swings forward, so does the right arm and vice versa). Although most coaches, physical educators, and runners no doubt appreciate this fact (as well as take it for granted), it is likely that they do not understand why it occurs.

Elftman (1939) was the first to point out that, in walking, this opposing action is a reflection of the interaction of angular momentum of the arms and of the rest of the body about a vertical axis passing through the body CM. Hinrichs and coworkers (1983) also showed the arm swing to be essential to the balance of angular momentum about the vertical axis. In the same study, which was based on a sample of 21 subjects running overground at a speed of 3.6 m • s^{-1} it was also found that the arms contributed very little to the angular momentum of the body about the other two coordinate axes. Similar results were shown for treadmill running at faster speeds (Hinrichs, 1987). For this reason the discussion will center on the vertical component only.

Figure 4.11 shows the algorithm used to calculate angular momentum. In the figure **H** is the total angular momentum vector of a given segment about the body CM (point G), [I] is the segment inertia tensor describing the mass distribution of the segment about its own CM (point C). ω is the segment angular velocity vector, **r** is the vector locating C relative to G, **ṙ** is the time derivative of **r**, and m is the mass of the segment. The body segment possesses two forms of angular momentum: local and remote. The local term is the angular momentum inherent in the segment's rotation about its own CM. The remote term is the angular momentum

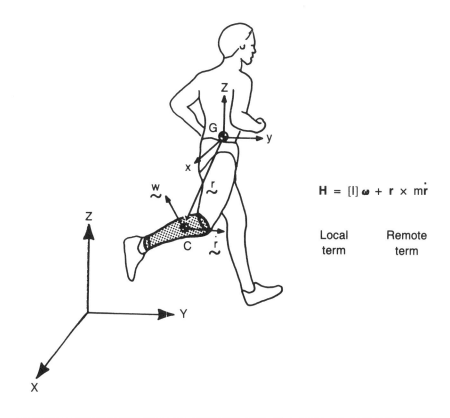

$$H = [I]\, \underset{\sim}{\omega} + r \times m\dot{r}$$

Local   Remote
term   term

Figure 4.11. Algorithm for calculating the angular momentum of a body segment relative to the body center of mass. *Note.* From "Upper Extremity Contributions to Angular Momentum in Running" by R.N. Hinrichs, P.R. Cavanagh, and K.R. Williams, 1983, in H. Matsui and K. Kobayashi (Eds.), *Biomechanics VIII-B*, p. 642, Champaign, IL: Human Kinetics. Copyright 1983 by Human Kinetics Publishers. Reprinted by permission.

inherent in the movement of the segment CM relative to the CM of the whole body.

The angular momentum of a system of segments, such as an entire upper extremity or the whole body, was calculated by adding up the contributions of the individual segments. To facilitate comparisons between subjects, the absolute angular momentum (expressed in kg • m² • s⁻¹) was normalized by dividing through by the mass of the subject (in kilograms) and the square of the subject's standing height (in meters). Because the resulting numbers are rather small, they have been expressed in units of 10^{-3} s⁻¹ (or 0.001/s).

The vertical component of angular momentum (H_z) turned out to be mostly a result of the remote terms of body segments. The farther the segment was from the Z axis, the greater potential it had to generate this remote form of H_z. The arms, in this respect, were able to compete favorably with the more massive legs in generating H_z because the shoulders are generally broader than the hips, placing the arms further from the Z axis than the legs. Moreover, the forward swing of one arm and the backward swing of the other, though tending to cancel out their combined linear momentum, both created angular momentum in the same direction. Their effects were additive. This was also true for the legs moving in the opposite direction. This effect can be seen in Figure 4.12, which

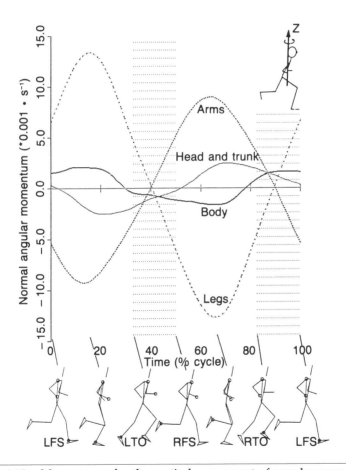

Figure 4.12. Mean curves for the vertical component of angular momentum (H_z) of the arms, head-plus-trunk, legs, and whole body for running at 4.5 m • s⁻¹. *Note*. From "Upper Extremity Function in Running. II: Angular Momentum Considerations" by R.N. Hinrichs, 1987, *International Journal of Sport Biomechanics*, **3**, p. 258. Copyright 1987 by Human Kinetics. Reprinted by permission.

shows the mean H_z curves of the arms, HT, legs, and the whole body at the medium speed. The arms were found to generate an alternating positive then negative H_z pattern during the running cycle. This tended to cancel out an opposite pattern of H_z in the legs. Along with a small contribution of the HT, the resulting whole-body H_z remained small throughout the running cycle.

Because the upper trunk was found to rotate in the same direction as the arms and the lower trunk with the legs, the upper body was defined as the head plus arms plus upper trunk. The lower body was defined as the lower trunk plus legs. When the shuffling around of H_z components was complete, the nature of the angular momentum balance became clearer. Figure 4.13 shows the mean H_z curves for the upper body and lower body at the medium speed. Although the addition of the lower trunk did not create a lower body H_z substantially different from the already large H_z of the legs, the result of adding the upper trunk to the arms produced an upper body H_z of a magnitude comparable to that of the lower body. The balance was not complete, however, as the lower body possessed roughly 10% more H_z than the upper body, depending on running speed. As the legs increased the magnitude of their H_z at faster running speeds, so did the arms and upper trunk such that there was no increase in the H_z of the whole body.

The nature of this balancing role that the arms and upper trunk play appears to be in the generation of internal torques between upper and lower body about the Z axis within the trunk segment. The lower body appeared to acquire most of the H_z needed to move the legs through their strides from the angular impulses received from the upper body instead of the ground. During the airborne phases, where the total-body H_z is constant in the absence of external torques, the lower body obviously received 100% of its angular impulse from the upper body. If the upper body were not present, the legs could not change direction while airborne, and a recognizable swing phase would not be possible. The runner would have to go through a series of leaps in the air, waiting until footstrike to begin swinging the recovery leg through while using the ground as the source of angular impulse.

General Discussion

It is interesting that what appear to be the two main functions of the arms in running both involve the vertical axis. Through their acceleration upward relative to the trunk, the arms help the legs in propelling the body upward (lift). Through their forward and backward swings, the arms (with the help of the upper trunk) provide the vast majority of the torque about the vertical axis needed to put the legs through their alternating patterns of stance and swing.

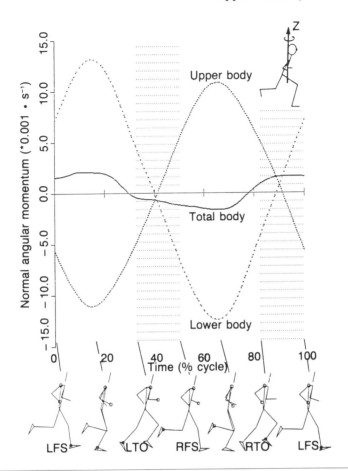

Figure 4.13. Mean curves for the vertical component of angular momentum (H_z) of the upper body, lower body, and whole body for running at 4.5 m • s⁻¹. *Note.* From "Upper Extremity Function in Running. II: Angular Momentum Considerations" by R.N. Hinrichs, 1987, *International Journal of Sport Biomechanics*, **3**, p. 259. Copyright 1987 by Human Kinetics Publishers. Reprinted by permission.

The role of the arms in the forward propulsion of the body has not been well defined. Based on the current definition of drive (i.e., the forward acceleration of both arms combined relative to the trunk), the arms generally do not contribute to the forward propulsion of the body. Hopper (1964) had some early insight into this aspect of running. He basically said that at "constant speed" running, the main aim of the runner is not to speed up but rather to get off the ground. The arms, by providing lift, help the runner achieve this goal. Whether the arms contribute to the forward propulsion of the body in situations where the runner is speeding up is unknown. This is a topic for further research.

Secondary functions of the arm swing include the reduction of excursions of the body CM from side to side and, at least in the case of constant speed running studied here, in the direction of progression. Thus the arms help the runner achieve a more constant horizontal velocity, which could lead to a reduction in energy cost.

The subject of energy cost is a topic for further research as well. Williams (1980) reported that runners who were more economical in terms of oxygen consumption tended to use less vigorous arm swings. Thus there may be a compromise between mechanical and physiological factors. A vigorous arm swing may provide more lift yet require more energy to execute. Further study is warranted in the area of uphill running where lift is of utmost importance.

Another topic for future research is in the area of sprinting. Some sprinters such as Carl Lewis demonstrate what is referred to as the classical style in which each arm swings nearly directly forward and back with little or no crossover. Perhaps this classical style has an advantage in sprinting that we don't understand yet. Most distance runners, however, do not adopt this form. The runners in this study all crossed over toward the body midline to some extent. On the average, the hands were brought 7 cm medial to the ipsilateral shoulder. This crossover did not increase with running speed.

It may require more energy to swing the arms directly forward and backward. The crossover of the arms in front of the chest, in fact, seems to be advantageous to the runner by reducing the side-to-side excursions of the body CM. From an angular momentum standpoint, just as much vertical angular momentum can be generated with a crossover as with the classical style. The classical style would seem to have an advantage when it comes to generating drive, but, at least in distance running, drive does not appear to be a factor.

There have been no comparable studies investigating upper extremity function in sprinting, although a current investigation by the author is under way. Previously Mann (1981) and Mann and Herman (1985) examined the kinematics of the arm swing in sprinting; however, they did not investigate lift, drive, angular momentum, or the effect of the arm swing on the path of the sprinter's CM. They concluded that the arms are simply used for balance and do not play a significant role in dictating the quality of sprinting performance. It is clear that more research is needed here. A two-dimensional analysis is not adequate for studying the mechanics of the arm swing.

Summary

The information presented in this chapter represents what little is known about upper extremity function in running. As in most research, more questions have been raised than answered. We do know a few things about the mechanics of the arm swing. First, the arm swing is a product of fairly intensive muscular action, especially in the extensor muscles, where the level of EMG activity reaches 60 %MAX at times. (Who says that running exercises only the legs?) The upper extremity muscles demonstrated stretch-shorten cycles similar to those reported by Elftman (1966) for lower extremity muscles during locomotion.

We also know that the arms provide lift and help the runner maintain a more constant horizontal velocity. We can also conclude that, at least in distance running, the arms do not provide any direct forward propulsion. There does not appear to be any advantage in the classical style of swinging the arms directly forward and backward compared to the style that most distance runners adopt of letting the forearms and hands cross over toward the body midline. Coaches might be advised not to *force* runners to use the classical style of arm swing until we know more about it.

Most important, however, is the role that the arms play in the balance of vertical angular momentum. Through internal torques about the trunk's long axis, the arms and upper trunk provide angular impulses to the lower body needed for the legs to alternate through stance and swing phases. Without these torques, one could not run at all.

Aristotle was right when he said that "runners run faster if they swing their arms" (1961, p. 489). He may not have understood exactly why, but to have written about it when he did made him a scientist truly ahead of his time.

References

Abdel-Aziz, Y.I., & Karara, H.M. (1971). Direct linear transformation from computer coordinates into object space coordinates in close-range photogrammetry. In *Proceedings of the American Society of Photogrammetry and University of Illinois Symposium on Close-Range Photogrammetry* (pp. 1-18). Falls Church, VA: American Society of Photogrammetry.

Ae, M., & Shibukawa, K. (1980). A biomechanical method for the analysis of the body segments in human movement. *Japanese Journal of Physical Education, 25*, 233-243.

Aristotle. (1961). *Progression of animals* (E.S. Forster, Trans.). Cambridge, MA: Harvard University Press.

Cavanagh, P.R., & Komi, P.V. (1979). Electromechanical delay in human skeletal muscle under concentric and eccentric contractions. *European Journal of Applied Physiology, 42*, 159-163.

Cavanagh, P.R., Pollock, M.L., & Landa, J. (1977). A biomechanical comparison of elite and good distance runners. *Annals of the New York Academy of Sciences, 301*, 328-345.

Elftman, H. (1939). The function of the arms in walking. *Human Biology*, 11, 529-535.

Elftman, H. (1966). Biomechanics of muscle with particular application to studies of gait. *Journal of Bone and Joint Surgery*, 48A, 363-377.

Gardiner, E.N. (1910). *Greek athletic sports and festivals*. London: Macmillan.

Hinrichs, R.N. (1982). Upper extremity function in running. Unpublished doctoral dissertation, The Pennsylvania State University, University Park.

Hinrichs, R.N. (1985). A three-dimensional analysis of the net moments at the shoulder and elbow joints in running and their relationship to upper extremity EMG activity. In D.A. Winter, R.W. Norman, R.P. Wells, K.C. Hayes, & A.E. Patla (Eds.), *Biomechanics IX-B* (pp. 337-342). Champaign, IL: Human Kinetics.

Hinrichs, R.N. (1987). Upper extremity function in running. II: Angular momentum considerations. *International Journal of Sport Biomechanics*, 3, 242-263.

Hinrichs, R.N., & Cavanagh, P.R. (1981). Upper extremity function in treadmill walking. Paper presented at the 1981 Annual Meeting of the American College of Sports Medicine, Miami, FL.

Hinrichs, R.N., Cavanagh, P.R., & Williams, K.R. (1983). Upper extremity contributions to angular momentum in running. In H. Matsui & K. Kobayashi (Eds.), *Biomechanics VIII-B* (pp. 641-647). Champaign, IL: Human Kinetics.

Hinrichs, R.N., Cavanagh, P.R., & Williams, K.R. (1987). Upper extremity function in running. I: Center of mass and propulsion considerations. *International Journal of Sport Biomechanics*, 3(3), 222-241.

Hopper, B.J. (1964). The mechanics of arm action in running. *Track Technique*, 17, 520-522.

Jackson, K.M. (1979). Fitting of mathematical functions to biomechanical data. *Institute of Electrical and Electronics Engineers Transactions of Biomedical Engineering*, BME-26, 122-124.

Mann, R.V. (1981). A kinetic analysis of sprinting. *Medicine and Science in Sports and Exercise*, 13, 325-328.

Mann, R., & Herman, J. (1985). Kinematic analysis of Olympic sprint performance: Men's 200 meters. *International Journal of Sport Biomechanics*, 1, 151-162.

Miller, D.I., & Nelson, R.C. (1973). *Biomechanics of sport*. Philadelphia: Lea & Febiger.

Murray, M.P., Sepic, S.P., & Bernard, E.J. (1967). Patterns of sagittal rotations of the upper limbs in walking, a study of normal men during free and fast walking. *Physical Therapy*, 47, 272-284.

Up in arms. (1980, April). *Runner's World*, pp. 53-57.

Wickstrom, R.L. (1970). *Fundamental motor patterns*. Philadelphia: Lea & Febiger.

Williams, K.R. (1980). A biomechanical and physiological evaluation of running efficiency. Unpublished doctoral dissertation. The Pennsylvania State University, University Park.

Winter, D.A., Sidwall, H.G., & Hobson, D.A. (1974). Measurement and reduction of noise in kinematics of locomotion. *Journal of Biomechanics, 7*, 157-159.

Chapter 5

Rearfoot Motion in Distance Running

Christopher J. Edington
E.C. Frederick
Peter R. Cavanagh

Control of movement in the joints of the foot and ankle, in particular the subtalar joint, is of interest to clinicians and biomechanists because it has been hypothesized that excessive pronation of the subtalar joint during the support phase of running is linked with various injuries of the hip, knee, Achilles tendon, and foot (Brody, 1980; Hlavac, 1977; James, Bates, & Osternig, 1978; Segesser & Nigg, 1980; Taunton, Clement, & McNicol, 1982). Research intended to influence the design of running shoes has focused on both medial and lateral stability in an attempt to control excessive rearfoot movement (Bauer, 1970; Krahenbuhl, 1974), but medial stability is of greatest interest because of its role in pronation (Figure 5.1a–c). Some pronation appears to be an entirely normal component of distance running style, but because of the myriad of problems that have been hypothesized to result from excessive pronation, researchers and sport shoe designers have sought ways to modify shoe design to limit this movement.

This chapter will concentrate on the methods of measuring rearfoot motion and factors influencing the degree to which rearfoot motion occurs. In addition, the effects that factors such as shoe design and orthoses have on controlling the extent of motion will be discussed.

Methods of Measurement

Subtalar Joint Function

The subtalar joint is located between the talus and the calcaneus (Figure 5.2). The talus is a unique structure in several ways. It is one of the few

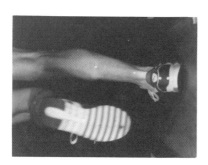

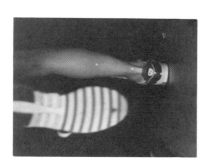

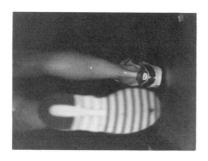

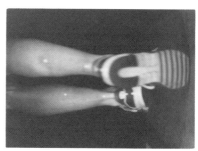

Figure 5.1a.   Rear-view photo sequence of a normal rearfoot pattern.

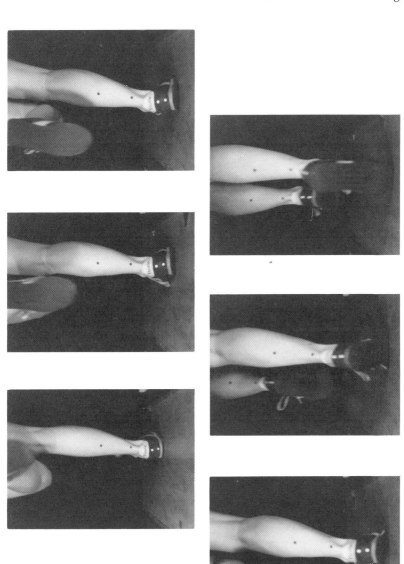

Figure 5.1b.   Rear-view photo sequence of an excessive pronator.

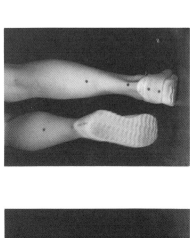

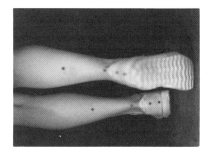

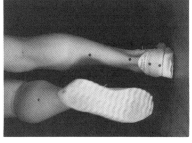

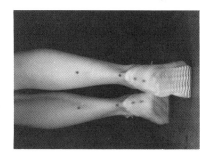

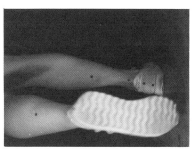

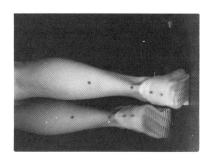

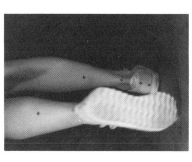

Figure 5.1c.   Rear-view photo sequence of a supinator.

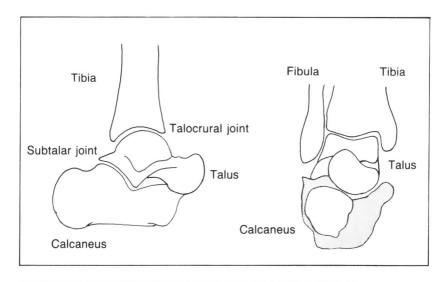

Figure 5.2. The bones of the foot and ankle showing the location of the talocrural and subtalar joints.

bones in the body that does not have any muscular attachments. Also, in the various movements of the foot and ankle, the talus appears functionally to be part of the foot during dorsiflexion and plantar flexion (which occur at the talocrural joint) but part of the leg during pronation and upination (which occur at the subtalar joint). The superior surface of the calcaneus, with which the talus articulates, usually has two facets, with the posterior facet being by far the largest (Figure 5.3).

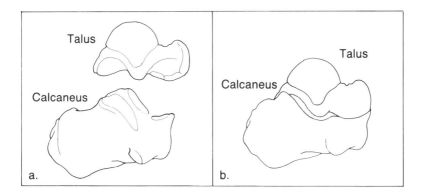

Figure 5.3. The talus and calcaneus in a disarticulated (a) and an articulated (b) state. (Note the two facets on the calcaneus with which the talus articulates.)

The function of the subtalar joint is frequently misunderstood, largely because of the unique orientation of its axis and its limited range of motion. In almost every textbook of foot mechanics the phrase "triplanar motion" is never more than one sentence away from a mention of the joint. This suggests that the joint has some unique ability to rotate about several axes at once, which is certainly not the case. The motion of the subtalar joint is triplanar only if the planes being considered are the cardinal planes of the body, and this characterization means simply that the axis of the subtalar joint is not aligned with any of the three cardinal planes.

The classic study of the subtalar joint is that of Vern Inman (1976). This remarkable work established that the subtalar joint is an extremely variable structure. The mean orientation of the axis with respect to the sagittal and transverse planes is shown in Figure 5.4 together with an indication of the range in 46 cadaver feet. This mean orientation indicates an axis that is directed approximately 23 degrees medially and tilted 42 degrees upward. Thus when pronation occurs it will, if viewed from a perspective of the cardinal planes, appear to have components of external rotation, dorsiflexion, and calcaneal eversion. Conversely, supination would appear to be a combination of internal rotation, plantar flexion, and calcaneal inversion. It is important to note that the frequently made statement that

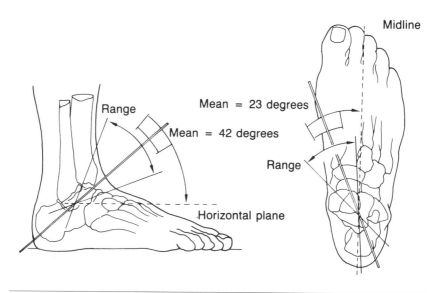

Figure 5.4.   The orientation of the axis of the subtalar joint with respect to the sagittal and coronal planes. Note the large variability in the 46 cadaver feet. *Note.* Adapted from *The Joints of the Ankle* (p. 37) by V.T. Inman, 1976, Baltimore: Williams and Wilkins. Copyright 1976 The Williams and Wilkins Co. Reprinted by permission.

pronation is a combination of external rotation, dorsiflexion, and eversion can be somewhat misleading if incorrectly interpreted: Pronation does not imply any motion of dorsiflexion at the talocrural joint. The position of the foot does become dorsiflexed in relation to the leg, but in pure pronation it is entirely as a result of motion at the subtalar joint.

One consequence of the oblique orientation of the subtalar joint axis when viewed in a sagittal plane is that pronation and supination tend to cause internal and external rotation of the leg and vice versa. This has been likened by Inman and Mann (1973) to the action of a mitered hinge (Figure 5.5), and the consequences of this relationship are potentially important in the etiology of knee injuries in runners (James et al., 1978).

It must be pointed out that, despite the widespread acceptance of the mitered hinge theory in the orthopedic literature, more recent work has questioned Inman's interpretation of the link between foot and leg movements (Lundberg, 1988). In particular, Lundberg found that external tibial rotation and supination were much more closely coupled than internal rotation and pronation.

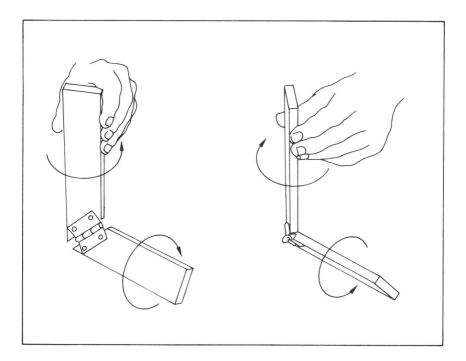

Figure 5.5. A representation of the left subtalar joint as a mitered hinge. Supination (left) and pronation (right) will cause external and internal tibial rotation respectively. *Note.* Reproduced by permission from Mann, Roger A: "Biomechanics of the foot and ankle." In Mann, Roger A., editor: *Surgery of the Foot*, ed. 5, St. Louis, 1986, The C.V. Mosby Co.

The orientation of the subtalar joint axis makes study of the motion of the joint by conventional planar analysis rather difficult. The component of subtalar joint motion that is the most independent of motion at other joints and the simplest to measure is calcaneal inversion and eversion. Therefore, in most of the experiments that study the effects of excessive pronation, subtalar joint supination and pronation have been approximated by calcaneal inversion (varus) and eversion (valgus), respectively, and the observed results are usually said to relate to rearfoot motion because of their incomplete representation of subtalar joint kinematics. The accuracy of this approximation has recently been examined (Soutas-Little, Beavis, Verstraete, & Markus, 1987).

Huson (1985, 1987) has suggested that inversion of the foot is a much more appropriate term than supination and also that the terms *valgus* and *varus* be reserved for abnormal positional deviations and not be used in the description of normal motions. Clearly it will take some years before clarity and unity is obtained on this confused issue. In keeping with most of the publications being reviewed, we will refer to calcaneal valgus and varus in relation to the leg as *pronation* and *supination* respectively, being fully conscious of the limitations of such descriptions.

Quantifying Rearfoot Motion

Nigg, Eberle, Frey, Segesser, & Weber (1977) first reported using this technique and subsequent work on the subject (Nigg, 1986; Nigg & Bahlsen, 1988; Nigg, Eberle, Frey, Luethi, Segesser, & Weber, 1978; Nigg & Luethi, 1980; Nigg & Morlock, 1987) along with that of Bates (Bates, 1985; Bates, Osternig, Mason, & James, 1978; Bates, Osternig, Mason, & James, 1979b), Cavanagh (Cavanagh, 1980; Cavanagh et al., 1985; Cavanagh, Clarke, Williams, & Kalenak, 1978), Clarke (Clarke, Frederick, & Hamill, 1983; Clarke, Frederick, & Hamill, 1984; Clarke, Frederick, & Hlavac, 1983; Clarke, Lafortune, Williams, & Cavanagh, 1980), and others (Frederick, Clarke, & Hamill, 1983; Frederick, Robinson, & Hamill, 1987; Smith, Clarke, Hamill, & Santopietro, 1986; Taunton, Clement, Smart, Wiley, & McNicol, 1982; Williams & Ziff, 1984) have created substantial literature on rearfoot motion. These reports involved a variety of subjects running at various speeds either on the treadmill or overground, wearing either no shoes, a variety of shoe designs, or shoes with orthoses. These studies are summarized in Tables 5.1 and 5.2. In general, data for at least three footstrikes were averaged for each subject and condition to account for variability.

The most traditional means of measuring rearfoot motion is through the use of high-speed film taken of the posterior aspect of the frontal plane

Table 5.1 Key Angular Rearfoot Kinematic Parameters From Various Studies

Studies	Speed (m · s⁻¹)	N	TD (°)	MP (°)	TRM (°)	AT (°)	ATM (°)	TO (°)	MVP (°/s)
Barefoot									
Bates, Osternig, Mason, & James (1978)	3.3-4.5	10	1.9	-8.6	10.5				
Nigg and Luethi (1980)	3.0	54	0.8				3.9	2.9	
Nigg (1986)		47				-8.3		0.1	
Smith et al. (1986)	3.8	11	1.8	-9.1	10.9				-406
Shoes									
Andrew (1986)	3.6	15	8.1	-7.6	15.7				-407
Andrew (1986)	4.5	15	9.5	-7.9	17.5				-504
Andrew (1986)	6.0	15	11.8	-8.0	19.9				-636
Bates, Osternig, Mason, & James (1978)	3.3-4.5	10	10.4	7.2	17.6				
Bates, Osternig, Mason, & James (1978)	3.3	10	8.8	-6.8	15.6				
Bates, Osternig, Mason, & James (1979b)	4.5	11		-9.1					
Cavanagh et al. (1978)	4.5	4			16.5				-789
Clarke et al. (1980)	4.5	15	3.7	-10.8	14.5				
Clarke, Frederick, & Hlavac (1983)	3.8	8	5.7	-11.4	17.1				
Clarke, Frederick, & Hamill (1983)	3.8	10	4.9	-11.7	16.6				-532
Nigg and Luethi (1980)	3.0	45	7.5			-3.8	6.3	12.3	
Nigg (1986)		47						7.8	
Smith et al. (1986)	3.8	7	5.6	-12.2	17.8				-498
Smith et al. (1986)	4.5	7	5.1	-13.5	18.6				-567

(Cont.)

Table 5.1 (Continued)

Studies	Speed (m · s⁻¹)	N	TD (°)	MP (°)	TRM (°)	AT (°)	ATM (°)	TO (°)	MVP (°/s)
Shoes with orthoses									
Cavanagh et al. (1978)	4.5	4			10.0				-240
Clarke, Frederick, & Hlavic (1983)	3.8	8	6.9	-8.9	15.8				
Nigg (1986)	3.8	44				-0.1		12.9	
Smith et al. (1986)	3.8	9	5.8	-10.1	15.9				-464

Note. TD = touchdown angle; MP = maximum pronation; TRM = total rearfoot movement; AT = Achilles tendon angle; ATM = Achilles tendon movement; TO = takeoff angle; MVP = maximum velocity of pronation.

Table 5.2   Key Temporal Rearfoot Kinematic Patterns From Various Studies

Studies	Speed (m · s⁻¹)	N	T-MP (ms)	T-MVP (ms)	PP (%)
Barefoot					
Bates, Osternig, Mason, & James (1978)	3.3-4.5	10	95		75.8
Shoes					
Andrew (1986)	3.6	15	94	32	
Andrew (1986)	4.5	15	80	29	
Andrew (1986)	6.0	15	66	25	
Bates, Osternig, Mason, & James (1978)	3.3-4.5	10	82		53.9
Bates, Osternig, Mason, & James (1978)	3.3	10	99		53.6
Bates, Osternig, Mason, & James (1979b)	4.5	11	72		69.0
Cavanagh et al. (1978)	4.5	4	45	15	
Clarke et al. (1980)	4.5	15	94		
Clarke, Frederick, & Hlavac (1983)	3.8	8	94		
Clarke, Frederick, & Hamill (1983)	3.8	10	94	27	
Shoes with orthoses					
Cavanagh et al. (1978)	4.5	4		12	
Clarke et al. (1983)	3.8	8	102		

Note. T-MP = time to maximum pronation; T-MVP = time to maximum velocity of pronation; PP = period of prona-tion, in % of contact time.

as subjects run overground or on a treadmill. High-speed video and opto-electric technology are two routinely used alternatives that have shown advantages in the automation of the digitizing process. This approach also brings the measurement and reporting of rearfoot motion into the realm of a clinical tool.

Absolute Versus Relative Measurement

There are two approaches to the placement of markers for the study of rearfoot motion. The first approach, a relative method, involves the arbitrary placement of four markers—two on the posterior aspect of the foot or shoe, and two on the posterior aspect of the leg. The subject first stands in the field of view of the camera or optoelectronic device, and the angle between the leg and foot markers is measured and subtracted from all subsequent dynamic measurements. Thus the position of the leg and rearfoot in some standardized standing posture is designated as a rearfoot angle of 0 degrees. The second approach, which we shall call an absolute method, involves placing similar markers, but this time in relation to anatomical landmarks, and measuring the segmental and joint angles as absolute values. The absolute approach is preferable, because if there is a deviation from normal values in standing—if, for example, the subject stands in calcaneal valgus—this will be apparent. Using the relative approach, such a deviation could not be distinguished in a subject who stands in calcaneal varus. As far as total excursion or the determination of angular velocity is concerned, both methods will yield the same results.

Typically, in the absolute approach, two reference markers are placed on the rear of the lower leg approximately 15 to 20 cm apart (Clarke et al., 1984; Smith et al., 1986). The distal landmark is placed in the midline of the Achilles tendon between the medial and lateral malleolus, and the proximal marker is placed below the belly of the gastrocnemius on a line joining the distal marker with the bisection of the leg at the level of the popliteal fossa. Two reference markers are then placed on a line that approximates the bisection of the posterior aspect of the calcaneus as well as can be estimated in the presence of the shoe. It is important to note that this line is drawn in relation to the calcaneus, not to the vertical. Clark et al. and Smith et al. both suggested the standardization of heel marking by using a wooden block that guides foot placement into a repeatable stance position, with the subject's heels 5 cm apart and with the feet externally rotated 7 degrees.

In the scheme of Cavanagh/Clarke (Cavanagh et al., 1978; Clarke et al., 1984) angles formed by each of the segments with the vertical are calculated, and the rearfoot angle is taken as the difference between these two component angles. In the Cavanagh/Clarke convention, pronation is a negative angle and supination positive. Bates, Osternig, Mason, and

James (1978) use a similar measurement scheme but reverse the sign of the angles. Nigg (1986) calculates a heel bone angle from the medial horizontal plane and an Achilles tendon angle between the heel and leg angles on the medial side. When Nigg reports an Achilles tendon angle of 192 degrees, this compares with − 12 degrees of pronation in the Cavanagh/Clarke convention and a + 12-degree angle in the Bates scheme. No particular scheme is inherently superior; however, for the sake of comparison all results discussed further in this paper have been converted to the Cavanagh/Clarke convention, which is shown in Figure 5.6

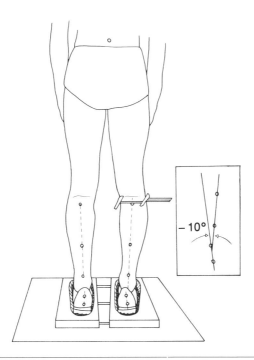

Figure 5.6.   A scheme for placement of markers for the measurement of rearfoot motion in absolute terms.

Calcaneal Movement Within the Shoe

Because there is bound to be some relative movement between foot and shoe, it is often questioned whether markers on the shoe can give an accurate representation of calcaneal position. Clarke et al. (1980) and Nigg (1986) have reported the results of studies in which windows were cut in the rear of shoes so that markers on both the heel and the rear of the shoe could be tracked simultaneously. Nigg found an offset of 2 to 3

degrees between the actual calcaneal movement and the movement of
the rear part of the shoe with the shoe being in the more inverted position.
The error appeared to be systematic, and, especially if measurement is
relative, it should not affect the validity of this method. It should be
pointed out that a small number of subjects were used in both these
studies, and the average amount of pronation observed was relatively
small. Further studies may reveal shifts in this systematic error in subjects
with greater or lesser degrees of pronation than those represented in these
first validation studies. Because it seems likely that too little or too much
pronation or significant departures in shoe design may result in less cor-
respondence of this method that depends on shoe measurement, care
should be taken in extrapolating this method beyond its currently proven
limits.

Typical Results

Despite the considerable attention that has been devoted to the measurement of rearfoot motion, no single study can be referenced to define the
mean pattern and the confidence limits for this mean pattern in a group
of distance runners. Thus the definition of an excessive pronator is by
no means established; yet each group of investigators and clinicians has
its own empirical criterion.

Parameters Measured

The rearfoot angle versus time curves in Figure 5.7 show examples of patterns similar to those shown pictorially in Figure 5.1. The typical normal
pattern (curve a) is characterized by approximately 10 degrees of supination at footstrike and then movement to a position of about 10 degrees
of pronation in 30 to 50 ms. The pattern shown in curve b would be classified as that of an excessive pronator by almost all definitions. Even at
footstrike, the calcaneus is everted, and it continues to a value of approximately 26 degrees of pronation. The pattern shown in curve c is from
a functionally rigid foot that becomes increasingly inverted beyond the
initial angle after footstrike and then remains inverted throughout the
entire contact.

From these basic curves, the most frequently reported values are for
touchdown angle (TD), the rearfoot angle at footstrike; maximum pronation (MP), the greatest negative rearfoot angle during contact; total
rearfoot movement (TRM), the sum of the absolute values for TD and
MP; and time to maximum pronation (TMP), or the time after initial contact when MP occurs. Interesting insights are provided by other parameters such as the Achilles tendon angle and the relative change of the

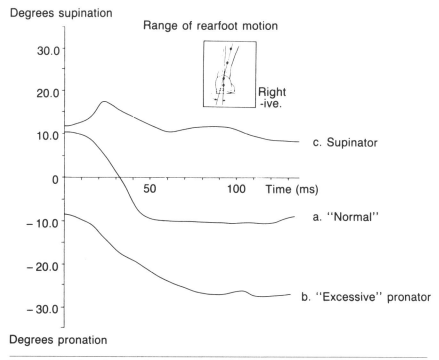

Figure 5.7. Rearfoot angle–time graphs for three runners with rearfoot patterns similar to those shown in Figure 5.1.

Achilles tendon angle within the first 1/10th of foot contact (Nigg, 1986); period of pronation, the time the foot spends in a pronated position (Bates, Osternig, Mason, & James, 1979b); and the takeoff angle, or rearfoot angle as the foot leaves the ground (Bates, Osternig, Mason, & James, 1979b; Nigg, 1986). Maximum velocity of pronation (MVP), the peak angular velocity of the rearfoot angle, is calculated by differentiating the displacement data. Extreme care must be taken in the calculation of this parameter because of the problems associated with the need to filter data before differentiation (Pezzack, Norman, & Winter, 1977; Winter, Sidwell, & Hobson, 1974).

Measurement Error

Because of the obliquity of the subtalar joint axis mentioned earlier and because extreme angles of flexion, external rotation, or both can result in parallax distortions in the measured rearfoot angles, some attention must be paid to the development of three-dimensional filming techniques. Both Engsberg and Andrews (1987) and Soutas-Little et al. (1987) have developed a protocol for the detailed three-dimensional analysis of foot motion during running. Comparisons of traditionally measured rearfoot motion with three-dimensional joint coordinate system eversion/inversion by Soutas-Little et al. showed that it was possible to have a constant phase difference and up to 20% difference in maximum eversion, depending upon the amount of rotation about an axis other than that designated for eversion/inversion. Similar results followed for the calculation of angular velocity and acceleration, with angular acceleration calculated by the joint coordinate method being consistently lower than that calculated by two-dimensional rearfoot measurement.

Clarke et al. (1984) present an extensive discussion of the technical pitfalls encountered with the existing methodology. Neither their cautions nor their proposed solutions need be repeated here, although it is clear from their analysis of the problem that the most accurate and reproducible measurements are those that are relative, such as total rearfoot movement, rather than absolute values, such as maximum pronation. Yet it may be that the absolute measures have more clinical relevance, and therefore methods to improve the accuracy of absolute methods should be explored.

A simple analysis of possible errors in absolute measurement is shown in Figure 5.8. In this model, it was first assumed that the base condition was one in which the markers on the leg were placed exactly on the anatomical points of interest and were separated by 20 cm, and those on the posterior aspect of the foot were separated by 4 cm. For the purpose of illustration, the markers have been placed mediolaterally so that the rearfoot angle is 0 degrees. Next, the mediolateral position of the distal marker of each pair was systematically varied to simulate an error in marker placement or measurement during digitization. Figure 5.8b shows the resulting errors in rearfoot angle that would result from the errors in marker location. Calculations have been made separately for each segment, assuming that only one error at a time was made.

The diagram shows that the major source of error in the calculation of absolute rearfoot angle is due to the location of the markers on the shoe. For example, in this base condition an error of only 3 mm in marker location on the shoe would cause an error of 4.3 degrees in rearfoot angle, whereas the same error in location of the markers on the leg would result in a rearfoot angle error of 0.9 degrees.

In Figure 5.8c, the spacing of the markers in a proximal-distal direction has been varied at each mediolateral error value. As the spacing between the markers on the shoe is reduced, the error increases drastically. For example, at a marker spacing of 2 cm, the rearfoot angle error for a 3-mm error in location of one marker is 8.5 degrees. It should be pointed out that these errors are close to best-case conditions because each calculation involves the mislocation of only a single marker. In reality, all four markers

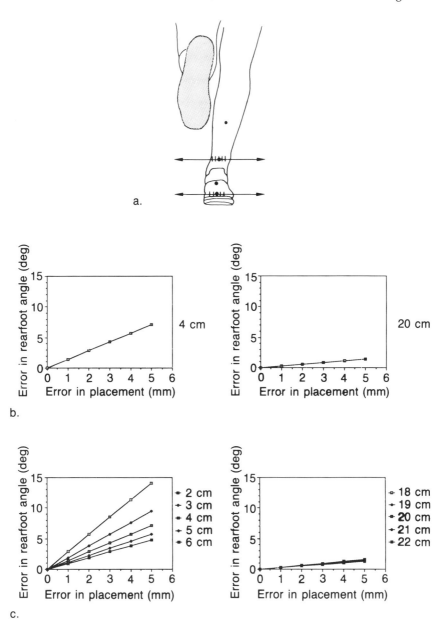

Figure 5.8. Diagram showing the model used for simulation of marker error (a), the resulting error in rearfoot angle that results from errors in medio-lateral location at the base condition (b), and the effect of the vertical distance between markers on rearfoot angle error (c).

may be placed some distance from the ideal placement based on anatomical reference. It is therefore not suprising that the measurement of absolute rearfoot angle has been found to be less reliable than relative measurements. If a standing calibration is taken and relative calculations are made, the placement of the markers is entirely arbitrary and does not result in error, but the same analysis shown above always applies on successive measurements for digitization error.

Factors Affecting Rearfoot Kinematics

Speed of Running

A number of factors have been found to affect rearfoot motion, and reports of such experiments should routinely be examined to determine the conditions under which data were collected. Although Bates, Osternig, Mason, and James (1978) reported no significant changes in four of the key parameters of pronation as running speed was increased from 3.3 to 4.5 meters per second, Nigg (1986), Smith et al. (1986), and Andrew (1986) all reported significant increases in pronation with increased running speed. Andrew investigated several rearfoot variables at speeds of 3.6, 4.5, and 6.0 m • s^{-1} (Figure 5.9a,b). As the velocity of running increased, touchdown angle, the total rearfoot movement, and the maximum velocity of pronation increased whereas the time to maximum pronation and the time to maximum pronation velocity decreased. It would clearly be advantageous in all studies of running mechanics if researchers at least included a common speed in their experimental series. A survey of the recent literature reveals that this may be happening, as a speed of 3.8 m • s^{-1} is apparently becoming a de facto standard for rearfoot kinematic studies.

Overground vs. Treadmill

Dillman's review of the biomechanics literature on overground versus treadmill running concluded that no consistent kinematic differences have been shown for the two modes of running (Dillman, 1974). Clarke, Frederick, and Hamill (1983) extended this observation to rearfoot kinematics with a comparison of 10 subjects running overground and on the treadmill at 3.8 m • s^{-1}. No consistent differences in any of the parameters of pronation were found between the two modes. This leads to the cautious conclusion that the results from easier to control treadmill tests can be extended to overground running.

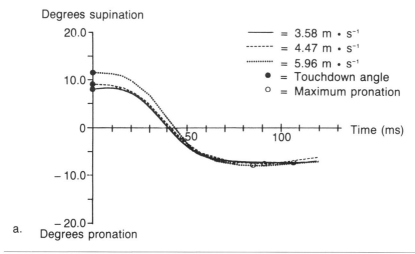

Figure 5.9a.   Mean rearfoot angle–time curves for 15 subjects running at speeds of 3.6, 4.5, and 6.0 m • s⁻¹. Note that although maximum pronation is similar at the three speeds, the difference in the angles at touchdown results in increased total range of movement as speed increases. *Note:* From *The Effect of Running Velocity on Rearfoot Motion and Mediolateral Placement of the Feet* (p. 65) by G.C. Andrew, 1986, an unpublished master's thesis, University Park: The Pennsylvania State University. Reprinted with permission.

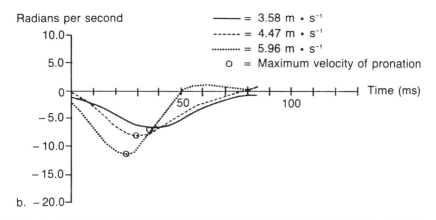

Figure 5.9b.   Rearfoot velocity–time curves for the data shown in Figure 5.9a. The clear trend of increasing maximum pronation velocity as speed increases was significant at $p < 0.05$. *Note.* From *The Effect of Running Velocity on Rearfoot Motion and Mediolateral Placement of the Feet* (p. 66) by G.C. Andrew, 1986, an unpublished master's thesis, University Park: The Pennsylvania State University. Reprinted with permission.

Leg Angle at Foot Contact

Clarke et al. (1984) observed that the mean leg angle (with the vertical) has a value of between 6 and 10 degrees of varus and within an individual varies by only 1 to 2 degrees during foot contact. This underlines the major contribution made by rearfoot movement to pronation. Despite these relatively small changes in leg angle during contact, there appears to be a significant positive correlation between leg angle at footstrike and the maximum amount of pronation that follows (Clarke et al. 1980). This makes sense because a foot attached to a leg with greater varus would have to pronate more simply to reach a flat foot position. Williams and Ziff (1984) have examined this question experimentally and have shown that subjects trained to run with a wider stance and therefore a decreased leg angle at contact show a measurable decrease in pronation.

Crossover

A number of authors (Dyson, 1977; Slocum & Bowerman, 1962; Slocum & James, 1968) have expressed the opinion that the optimum position of the medial edge of the shoe to maintain equilibrium is along the line of progression (as established, for example, by the navel or by a marker placed on the midline of the trunk). Williams (1980) studied the foot position of 31 subjects during the midsupport phase of unfatigued running at 3.6 m • s^{-1} and showed that the mean placement of the centerline of the rearfoot was +1 cm lateral to the midline of the trunk with a range of −1.4 cm to 6 cm (crossing over the midline is designated by a negative number). Andrew (1986) established that there was a greater tendency toward crossover of the feet with increasing running speed (3.6, 4.5, and 6.0 m • s^{-1}). As the speed of running increased, the mean mediolateral foot placements from the midline were 1.27 cm, 1.05 cm, and −0.04 cm, respectively. Andrew suggested that the tendency toward crossover with increased running velocity could be caused by increased external rotation of the leg and foot during the swing phase, which would cause the heel to rotate inward and increase the tendency toward crossover. He offered an additional possibility that increased frontal plane trunk sway could cause a need for decreased stride width to counteract the shift in the center of body mass. A pattern of exaggerated crossover at any speed could result from either structural factors (such as genu or tibial varum) and/or functional factors (such as preferred running style).

Barefoot Running

Some general observations can be made from the data presented in Tables 5.1 and 5.2. Several kinematic parameters are markedly changed during barefoot running. Bates, Osternig, Mason, and James (1978), Nigg and

Luethi (1980), and Smith et al. (1986) all found a tendency for barefoot runners to exhibit less of a touchdown angle. In addition, Bates was able to show that removing the shoe resulted in a significant increase in the period of pronation and less total rearfoot movement. Nigg and his co-workers (1986) found that during barefoot running the takeoff angle was less, and there was less calcaneal movement in the first 10% of foot contact. Smith et al. (1986) were also able to show a significant decrease in total rearfoot movement and maximum pronation velocity without shoes. Wearing shoes for running, therefore, appears to cause greater pronation and increases in many of the associated angular and temoral aspects of pronation.

Temporal Events

With the use of simultaneous frontal and sagittal plane analysis, we can compare rearfoot movement to other temporal kinematic events. Bates (Bates, James, & Osternig, 1979; Bates, Osternig, Mason, & James, 1979a) found that maximum knee flexion and maximum pronation occurred at approximately the same time (35% to 40% of foot contact). This topic deserves much more attention than it has received to date because the effect of mistiming certain temporal events is not well understood.

Fatigue

The effects of fatigue on rearfoot motion have also not been examined in detail. Our own pilot studies of four elite runners have shown mixed results. One subject showed a marked increase in total rearfoot motion, one showed a marked decrease, and the remaining two showed little trend. For our study, these runners were not willing to run much more than 50% of the time that they could have sustained during a race at a similar pace, and therefore the effects may be different under actual race conditions.

Shoe Design Effects

Sole Hardness, Heel Height and Lift, and Heel Flare

Clarke, Frederick, and Hamill (1983) looked at the relationship between three features of shoe design and rearfoot control: sole hardness, heel height and lift, and the angle of flare of the sole. It was found that soles softer than 35 Shore A durometer allow significantly more maximum pronation (MP) and total rearfoot movement (TRM). Also, increased flare results in a significant reduction in MP and TRM, although increases in flare beyond 15 degrees are only marginally helpful. Heel height was found to have little influence on the amount of pronation in this study, although other studies of this feature contradict such a conclusion.

Both Bates, Osternig, Mason, and James (1978) and Stacoff and Kaelin (1983) have reported that changes in heel dimensions affect rearfoot kinematics. Bates showed that slight increases in heel lift (the height to which the heel is raised relative to the forefoot) reduced MP and the period of pronation. Stacoff and Kaelin analyzed a range of heel heights (the total distance between the plantar surface of the rearfoot and the ground) from 1.8 to 4.3 cm and reported that pronation was reduced in the range from 2.3 to 3.3 cm but increased above as well as below that range. Stacoff and Kaelin evidently did not base their conclusions on statistical tests of significance, so we have no way of knowing the statistical validity of their findings. In contrast to these findings, Clarke, Frederick, and Hamill (1983) compared heel lifts of 0, +1.0 cm, and +2.0 cm, as well as heel heights of 1.0, 2.0, and 3.0 cm and found no statistically significant trends in the data with the exception of an increase in the average angular acceleration of the rearfoot with lower heel height.

Nigg (1986) performed an analysis similar to that of Clarke, Frederick, and Hamill (1983) on the effects of sole hardness but with slightly different results. Nigg found that total pronation at a range of running speeds was consistently lower in shoes made with soles of 25 Shore A hardness when compared with shoes of 35 durometer. Clarke et al. (1983), on the other hand, found an increase in MP when a sole hardness of 25 durometer was used versus 35 durometer.

It is likely that both results are correct because of the similarlity of test procedures, but the apparent discrepancies in the data are likely due to physical differences in the shoes tested. For example, as Nigg has noted (1986), the overall hardness of the shoe is not determined by a durometer test of the sole material. Other factors such as the construction technique (e.g., board vs. slip lasting) and geometry of the sole can also exert significant influences on the net hardness. Nigg has suggested that the 25 Shore A shoe used in his study is similar in overall hardness to the 35 Shore A shoe of Clark, Frederick, and Hamill (1983). These issues should be borne in mind when one compares the results of experiments from different laboratories where such variables are not controlled uniformly.

Lateral Heel Flare

Because initial contact between the foot and the ground occurs on the outside border of the shoe in almost all runners, it would appear reasonable to suppose that the elimination of the knife edge that lateral flare represents would have a positive influence on rearfoot motion. A running shoe with a rounded lateral border (Figure 5.10) was proposed by

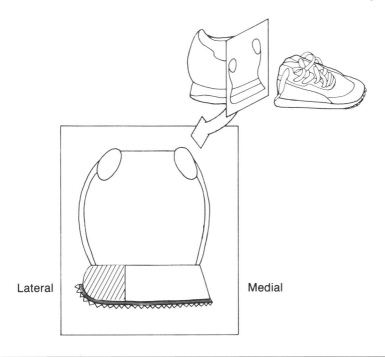

Figure 5.10. The "Roll-Sole" shoe proposed by Cavanagh (1984).

Cavanagh (1984), and the influence of this design and other modifications to the lateral flare on rearfoot motion was subsequently examined by Nigg and Morlock (1987) and Nigg and Bahlsen (1988). Nigg and Morlock used three identical shoes with varying degrees of lateral heel flare that included a conventional flare (16%), a neutral flare (0%), and a negative flare (rounded off). It was found that decreasing the heel flare caused a significant decrease in initial pronation and initial pronation velocity (reducing both within the first 10% of foot contact) but did not influence total pronation or ground reaction forces. A lateral flare acts as a lever arm that actually increases the touchdown angle and initial rearfoot angular velocity. Nigg and Morlock concluded that the use of a rounded heel would have an influence on injury risk because excessive initial pronation and pronation velocity are assumed to be associated with a number of injuries that include overloading of the tibia and the anterior and medial part of the tibiofemoral joint.

Nigg and Bahlsen's study was similar but these investigators also varied the midsole construction of the shoe. They found results similar to those of the earlier study with respect to pronation and the degree of lateral heel flare. Significant differences were also found for both lateral heel flares and shoe sole constructions with respect to ground reaction forces and the time of peak force. The highest vertical impact force peaks occurred in shoes with the softest midsoles, and the lowest vertical peaks were with the hardest midsoles. Impact forces with respect to heel flare did not show a clear trend, but the time of occurrence of the peak force decreased more than 15% with a rounded heel as compared to a conventional flare. The maximum vertical loading rate was also highest for shoes with the softest midsoles and lowest for those with the hardest midsoles. From their findings the authors suggested that if it is assumed that a shoe should be built so that initial pronation, impact forces, and loading rates are low, then an optimally designed shoe would have a neutral heel flare, a relatively hard midsole of 45 shore A, and a heel stabilizer.

Kinematically Mediated Effects

Frederick (1986), in his review of sport shoe design, summarized work in four main areas: cushioning, stability, friction, and energetic consequences. The emphasis of this work was to stress the idea of kinematically mediated effects that are the indirect result of sports shoe–induced adaptations in movement. A particular shoe characteristic can cause a kinematic adjustment that will in turn cause a further change in kinetics. As an example, Clarke, Frederick, and Cooper (1983) suggest that the results of studies in rearfoot motion are influenced by individual responses to shoes. In their explanation, they cite studies in which shoes with hardnesses differing by as much as 50% in their impact characteristics have had no effect on impact on the force platform. They also found a significant positive correlation between shoe hardness and knee flexion velocity immediately following heelstrike. This kind of kinematic adaptation, then, would most likely work to modify ground reaction forces.

Sports shoe manufacturers have proposed many shoe modifications in an effort to control rearfoot movement. Cavanagh (1980) has discussed the evolution of running shoe design and the relationship between shoe design and rearfoot control. Most modern running shoes are equipped with motion control features such as a thermoplastic heel counter, dual density midsole or wedge, and/or other motion control devices (Figure 5.11). The extent to which these devices are effective in controlling rearfoot motion is frequently difficult to ascertain.

Orthoses

Several authors have examined the effects of orthoses (orthotic devices) on rearfoot kinematics in running. Cavanagh et al. (1978), Clarke, Frederick, and Hlavac (1983), Nigg (1986), Smith et al. (1986), and

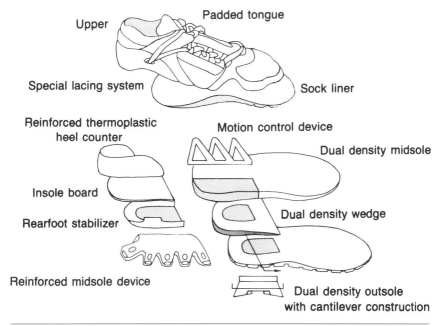

Figure 5.11. Various devices built into running shoes in an attempt to control rearfoot motion.

Taunton, Clement, Smart, Wiley, and McNicol (1982) all found slight but significant reductions in maximum pronation (MP) in subjects using orthoses of various designs and placements. Nigg (1986) has also reported changes in the rearfoot angle at toe-off when an orthosis is worn, although the biomechanical and clinical significance of this change is not well understood. Bates, Osternig, Mason, and James (1979a) and Rodgers and LeVeau (1982), however, did not find a significant change in maximum pronation. Reviewing the mixed results from earlier studies of orthoses, Smith et al. (1986) stated that the studies that have not shown a significant change in maximum pronation did not contain information about necessary foot and leg measurements, the type of orthoses tested, or the degree of posting. The importance of accurate clinical measurements, correct prescription of orthoses, and periodic modifications that must be done over an extended trial period cannot be underestimated.

Orthosis-induced adjustments in initial pronation and in the rate of pronation are similarly provocative but not well understood as of this writing. Even though reductions in MP of only 8% to 11% (Clarke, Frederick, & Hlavac, 1983; Nigg, 1986; Smith et al., 1986) have been found, proportionately greater changes in the maximum velocity of pronation (MVP) (20% to 70%) have been reported by Cavanagh et al. (1978) and Smith et al. (1986). In addition, Nigg has also shown that the most pronounced effect of orthosis use is on initial pronation, a parameter directly related to MVP. It should be noted that even though MVP and initial pronation are strongly linked, Clarke et al. (1984) did not find a significant correlation between MVP and MP. Andrew (1986) did, though, find a moderate correlation between MVP and touchdown angle (TD) and a high correlation between MVP and total rearfoot motion (TRM).

Bieber, Coates, Lohmann, and Danoff (1988) looked at the effect of orthoses on ground reaction forces under the foot during walking in 11 subjects. The center of pressure was established to be significantly more medial at heel contact and more lateral at 25% of contact and beyond with orthoses than without. A significant decrease in the vertical component of ground reaction force was also found with the use of orthoses in the first 25% of contact while the center of pressure was moving laterally. The authors suggested that a weight-bearing surface was created by the medial posting that comes in contact with the floor earlier and positions the calcaneus in more inversion. This causes the lateral shift in the center of pressure. In previous studies neither Cavanagh et al. (1978) nor Clarke et al. (1980) was able to establish a significant trend for center of pressure migration with respect to rearfoot motion. Further research in this area is needed to determine if the force plate could be utilized to quantify rearfoot motion or give further insight into rearfoot mechanics.

If orthoses do prove to have positive effects on the prevention and treatment of sports injuries, their modes of action may likely involve the poorly understood parameters of rate of pronation and initial pronation. An elucidation of the importance of adjustments in the rate of initial pronation must also await the results of clinical and biomechanical studies that uncover the influence of different variables on the etiology of sports injury.

Concluding Remarks

There is little doubt among clinicians that excessive rearfoot motion may be an important factor in the etiology of running injuries. In this chapter we have resisted speculation on this relationship but have examined some of the many factors that may cause changes in rearfoot motion. Further work is needed to extend this rather narrowly focused approach to a view that includes the remainder of the lower extremity. If abnormal rearfoot motion does indeed result in lower extremity pain, then we should strive for methods that can quantify the musculoskeletal changes thought to be injurious. If information on tibial rotation is the goal, then effort may be best expended on experiments to directly measure this motion rather than simply on making inferences from rearfoot motion.

References

Andrew, G.C. (1986). *The effect of running velocity on rearfoot motion and mediolateral placement of the feet.* Unpublished master's thesis, The Pennsylvania State University, University Park.

Bates, B.T. (1985). Testing and evaluation of running shoes. In D.A. Winter, R.W. Norman, R.P. Wells, K.C. Hayes, & A.E. Patla (Eds.), *Biomechanics IX-B* (pp. 128-132). Champaign, IL: Human Kinetics.

Bates, B.T., James, S.L., & Osternig, L.R. (1978). Foot function during the support phase of running. *Running,* 3(4), 24-31.

Bates, B.T., James, S.L., & Osternig, L.R. (1979). Foot function during the support phase of running. *American Journal of Sports Medicine, 7,* 328.

Bates, B.T., James, S.L., Osternig, L.R., Sawhill, J.A., & Hamill, J. (1981). Effects of running shoes on ground reaction forces. In A. Morecki, K. Fidelus, K. Kedzior, & A. Wit (Eds.), *Biomechanics VII-B* (pp. 226-233). Baltimore: University Park.

Bates, B.T., Osternig, L.R., Mason, B., & James, S.L. (1978). Lower extremity function during the support phase of running. In E. Asmussen & K. Jorgensen (Eds.), *Biomechanics VI-B* (pp. 30-39). Baltimore: University Park.

Bates, B.T., Osternig, L.R., Mason, B., & James, S.L. (1979a). Foot orthotic devices to modify selected aspects of lower extremity mechanics. *The American Journal of Sports Medicine, 7*(6), 338-342.

Bates, B.T., Osternig, L.R., Mason, B., & James, S.L. (1979b). Functional variability of the lower extremity during the support phase of running. *Medicine and Science in Sports and Exercise, 11,* 328-331.

Bauer, H. (1970). The effect of high-top and low-cut football shoes on speed and agility. *Athletic Journal, 50,* 74.

Bieber, J.M., Coates, J.C., Lohmann, K., & Danoff, J. (1988). The effects of pronation-controlling orthotic devices on pressure and force under the foot during dynamic stance. *Physical Therapy Journal, 68,* 805.

Brody, D.M. (1980). Running injuries. *CIBA Clinical Symposia, 32*(4), 1-36.

Cavanagh, P.R. (1980). *The running shoe book.* Mountain View, CA: Anderson World.

Cavanagh, P.R. (1984). Running shoe sole construction. U.S. patent number 4,449,306.

Cavanagh, P.R., Andrew, G.C., Kram, R., Rodgers, M.M., Sanderson, D.J., & Hennig, E.M. (1985). An approach to biomechanical profiling of elite distance runners. *International Journal of Sports Biomechanics, 1,* 36-62.

Cavanagh, P.R., Clarke, T.E., Williams, K.R., & Kalenak, A. (1978, June). *An evaluation of the effect of orthotics on pressure distribution and rearfoot movement during running.* Paper presented at the meeting of the American Orthopedic Society for Sports Medicine, Lake Placid, NY.

Clarke, T.E., Frederick, E.C., & Cooper, L.B. (1983). Biomechanical measurement of running shoe cushioning properties. In B.M. Nigg & B.A. Kerr (Eds.), *Biomechanical aspects of sports shoes and playing surfaces* (pp. 25-34). Calgary, AB: University of Calgary.

Clarke, T.E., Frederick, E.C., & Hamill, C.L. (1983). The effect of shoe design upon rearfoot control in running. *Medicine and Science in Sports and Exercise,* **15**(5), 376-381.

Clarke, T.E., Frederick, E.C., & Hamill, C.L. (1984). The study of rearfoot movement in running. In E.C. Frederick (Ed.), *Sport shoes and playing surfaces* (pp. 166-189). Champaign, IL: Human Kinetics.

Clarke, T.E., Frederick, E.C., & Hlavac, H.F. (1983). The effects of a soft orthotic upon rearfoot movement in running. *Podiatric Sports Medicine,* **1**(1), 20-23.

Clarke, T.E., Lafortune, M.A., Williams, K.R., & Cavanagh, P.R. (1980). The relationship between center of pressure and rearfoot movement in distance running. *Medicine and Science in Sports and Exercise,* **12**(2), 192.

Dillman, C.J. (1974). Kinematic analysis of running. *Exercise and Sports Science Reviews,* **2**, 193-218.

Dyson, G.H.G. (1977). *The mechanics of style.* New York: Holmes and Meier.

Engsberg, J.R., & Andrews, J.G. (1987). Kinematic analysis of the talocalcaneal/talocrural joint during running support. *Medicine and Science in Sports and Exercise,* **19**(3), 275-284.

Frederick, E.C. (1986). Kinematically mediated effects of sport shoe design: A review. *Journal of Sports Science,* **4**, 169-184.

Frederick, E.C., Clarke, T.E., & Hamill, C.L. (1983). Shoe design and rearfoot control in running. *Medicine and Science in Sports and Exercise,* **15**(2), 176.

Frederick, E.C., Robinson, J.R., & Hamill, C.L. (1987). Rearfoot kinematics and ground reaction forces in elite caliber twin runners. In B. Jonsson (Ed.), *Biomechanics X-B* (pp., 809-812). Champaign, IL: Human Kinetics.

Hlavac, H. *The foot book.* (1977). Mountain View, CA: World.

Huson, A. (1985). Biomechanics of the Subtalar joint: A re-appraisal. *Chirurgia del piede,* **9**, 389-398.

Huson, A. (1987). Joints and movement of the foot: Terminology and concepts. *Acta Morphologica Neerlando-Scandinavica,* **25**, 117-130.

Inman, V.T. (1976). *Joints of the ankle.* Baltimore: Williams and Wilkins.

Inman, V.T., & Mann, R.A. (1973). Biomechanics of the foot and ankle. In V.T. Inman (Ed.), *Duvries' Surgery of the Foot* (pp. 3-22). St. Louis: C.V. Mosby.

James, S.L., Bates, B.T., & Osternig, L.R. (1978). Injuries to runners. *American Journal of Sports Medicine*, **6**, 40-50.

Krahenbuhl, G.S. (1974). Speed of movement with varying footwear conditions on synthetic turf and natural grass. *Research Quarterly*, **45**, 28-33.

Lundberg, A. (1988). *Patterns of motion of the ankle/foot complex*. Unpublished M.D. dissertation. Department of Orthopedics, Karolinska Hospital, Stockholm, Sweden.

Mann, R.A. (1986). Biomechanics of the foot and ankle. In R.A. Mann (Ed.), *Surgery of the foot* (5th ed.). C.V. Mosby: St. Louis.

Nigg, B.M. (Ed.) (1986). *Biomechanics of running shoes*. Champaign, IL: Human Kinetics.

Nigg, B.M., & Bahlsen, H.A. (1988). Influence of heel flare and midsole construction on pronation, supination, and impact forces for heel-toe running. *International Journal of Sport Biomechanics*, **4**, 205-219.

Nigg, B.M., Eberle, G., Frey, D., Luethi, S., Segesser, B., & Weber, B. (1978). Gait analysis and sport-shoe construction. In E. Asmussen & K. Jorgensen (Eds.), *Biomechanics VI-A* (pp. 303-309). Baltimore: University Park.

Nigg, B.M., Eberle, G., Frey, D., Segesser, B., & Weber, B. (1977). Bewegungsanalyse für Schuhkorrekturen [Movement analysis for corrective shoe prescription]. *Medita*, **9a**, 160-163.

Nigg, B.M. & Luethi, S. (1980). Bewegunsanalysen beim Laufschuh [Movement analysis of sport shoes]. *Sportwissenschaft*, **3**, 309-320.

Nigg, M., & Morlock, M. (1987). The influence of lateral heel flare of running shoes on pronation and impact forces. *Medicine and Science in Sports and Exercise*, **19**, 294-302.

Pezzack, J.C., Norman, R.W., & Winter, D.A. (1977). Technical note: An assessment of derivative determining techniques used for motion analysis. *Journal of Biomechanics*, **10**, 177-182.

Rodgers, N.M., & LeVeau, B.F. (1982). Effectiveness of foot orthotic devices used to modify pronation in runners. *The Journal of Orthopaedic and Sports Physical Therapy*, **4**, 86-90.

Segesser, B., & Nigg, B.M. (1980). Insertionstendinosen am Schienbein, Achillodynie und Überlastungsfolgen am Fuss: Ätiologie, Biomechanik, therapeutische Möglichkeiten [Tibial insertion tendinoses, achillodynia, and overuse damage of the foot—etiology, biomechanics, and therapy]. *Orthopäde*, **9**, 207-214.

Slocum, D.B., & Bowerman, B. (1962). The biomechanics of running. *Clinical Orthopaedics*, **23**, 39-45.

Slocum, D.B., & James, S.L. (1968). Biomechanics of running. *Journal of the American Medical Association*, **205**, 721-728.

Smith, L., Clarke, T., Hamill, C., & Santopietro, F. (1986). The effect of soft and semi-rigid orthoses upon rearfoot movement in running. *Podiatric Sports Medicine*, **76**(4), 227-233.

Soutas-Little, R.W., Beavis, G.C., Verstraete, M.C., & Markus, T.L. (1987). Analysis of foot motion during running using a joint coordinate system. *Medicine and Science in Sports and Exercise*, **19**(3), 285-293.

Stacoff, A., & Kaelin, X. (1983). Pronation and sportshoe design. In B.M. Nigg & B.A. Kerr (Eds.), *Biomechanical aspect of sport shoes and playing surfaces* (pp. 143-151). Calgary, AB: University of Calgary.

Taunton, J.E., Clement, D.B., & McNichol, K. (1982). Plantar fasciitis in runners. *Canadian Journal of Applied Sports Science*, **7**, 41-44.

Taunton, J.E., Clement, D.B., Smart, G.W., Wiley, J.P., & McNicol, K.L. (1982). A triplanar electrogoniometer investigation of running mechanics in runners with compensatory overpronation. *International Journal of Sports Medicine*, **2**(2), 31.

Williams, K.R. (1980). *A biomechanical and physiological evaluation of running efficiency*. Unpublished doctoral dissertation, The Pennsylvania State University, University Park.

Williams, K.R., & Ziff, J.L. (1984). *Changes in rearfoot movement associated with systematic variations in running style*. Paper presented at the American Society of Biomechanics meetings, Tucson, AZ.

Winter, D.A., Sidwell, H.G., & Hobson, D.A. (1974). Measurement and reduction of noise in kinematics of locomotion. *Journal of Biomechanics*, **7**, 157-159.

Chapter 6

Muscle Activity in Running

Irene S. McClay
Mark J. Lake
Peter R. Cavanagh

A knowledge of muscle action in the lower extremity is of fundamental importance to the understanding of human gait. It is one of the wonders of the human machine that brief, carefully sequenced bursts of electrical activity hardly longer than one or two 10ths of a second are sufficient to cause the smooth, coordinated action of the runner.

It has been known since the work of Galvani in the 18th century that "animal electricity" is responsible for the activation of muscles. However, techniques of instrumentation lagged behind this knowledge, and devices to accurately record, amplify, and measure the electrical signals causing muscle activity were not available until the early 1920s. The lack of such methods did not stop the creative investigators of the 19th century from recording muscle activity. Carlet (1872) and Marey (1895) used a pneumatic method to record the phases of muscle activity in running, walking, and other activities. Tambours with tightly stretched diaphragms were placed over the muscle bellies. A tube of air leading back to a recording device was pressurized by the subject from a hand-held rubber bulb, and, as the muscle changed shape during activity, a signal was recorded that indicated the presence of muscle activity. Surprisingly, this method was still used by Hubbard in 1939 in a study of muscle activity in running. But Hubbard (1939) also used the "kymograph" that had been described earlier by Hudgkins and Stetson (1932), and this appears to be the first time that electromyographic (EMG) recordings were made of lower extremity muscles during running. Hubbard studied the rectus femoris, the vastus medialis, the gastrocnemius, and the gluteus maximus. He found that these muscles acted in brief bursts to accelerate and retard the body segments and that there were long periods of silence as the movements of the limb segments continued in a ballistic manner. A further refinement of the experiment was the study of rapid unloaded limb oscillation that was increased to a rate at which breakdown of the movement occurred. Through these experiments, Hubbard was among the first to implicitly postulate the "stretch-shortening" cycle that has received much subsequent attention (Cavagna, 1977).

The purpose of this chapter is to review the literature that has appeared since the time of Hubbard's work on the timing of muscle activity during running. We make no attempt to enter the controversy over the relationship of EMG and muscle force, but we do discuss the use of kinematic data to enhance the value of EMG recording. In addition, we present some new information concerning the exact timing of quadriceps activity in relation to stance phase knee flexion and extension during distance running. It is hoped that these observations will stimulate further work in what is a somewhat underrepresented area in the study of biomechanics of running.

Review of the Literature

Methodology

A variety of configurations of instrumentation are apparent in the literature with the major differences being in the electrode system and placement used. Surface electrodes have been most widely used (Elliot & Blanksby, 1979; Mann & Hagy, 1980a; Mann & Hagy, 1980b; MacIntyre & Robertson, 1987; Nilsson, Thorstensson, & Halbertsma, 1985) probably because they are the easiest to use and the most comfortable for the subject. As Basmajian and Deluca (1985) have repeatedly pointed out, surface electrodes are limited to the recording of activity from superficial muscle groups resulting in a gross representation of muscle action. They are nevertheless appropriate for the large lower extremity muscles that are frequently the object of study during running.

During dynamic activities such as running, movement artifact is a continuing problem with all types of electrodes. In addition, with surface electrodes the relative motion between the skin and muscle during movement results in recordings from different parts of the muscle at various points in the range of motion. Signal attenuation can be high with surface electrodes as a result of subcutaneous fat. However, subjects with well-defined muscles and low body fat—a description that fits most distance runners—lend themselves well to surface EMG recording.

A standardized method for electrode placement has yet to be agreed upon. The "motor point" of the muscle is sometimes used because it is easy to locate and provides for repeatability of placement (Elliot & Blanksby, 1979; MacIntyre & Robertson, 1987). The motor point is identified in one of two ways in the literature. Warfel (1974) identifies the anatomical regions where there is a high density of neuromuscular junctions, and these regions have been termed "motor points." The second definition of motor point is that point at which the largest twitch is produced by the least current via electrical stimulation (Cohen & Brumlik, 1968). It should be noted that these regions are not always coincident, with the latter sometimes depending on the relative proximity of the motor nerve to surface stimulation, that is, it may reflect indirect rather than direct stimulation.

Basmajian and Deluca (1985) suggest electrode placement between the center of the innervation zone and the distal tendon, which would locate the electrode in the distal muscle belly. Others recommend placement over the center of the muscle belly (Nilsson et al., 1985). There is still some debate as to the optimal orientation of the electrodes. Zuniga, Truong, and Simons (1969) experimented with both longitudinal and transverse orientation of the electrodes with respect to muscle fiber orientation and reported that the greatest signals (from both the longitudinal and transverse orientations) were recorded from electrodes overlying the center of the muscle belly. Kramer, Kuchler, and Brauer (1972) recommended that the electrodes be placed longitudinally along the center of the muscle belly for the best signal. Other variables such as skin resistance, electrode size, and interelectrode distance also contribute to variability in surface EMG signal.

Bipolar, fine-wire electrodes have also been used in studies of running (Brandell, 1973; Paré, Stern, & Schwartz, 1981; Schwab, Moynes, Jobe, & Perry, 1983). Fine wires were threaded through a hollow hypodermic needle that was then inserted and withdrawn, leaving the wires (whose tips had been bent back) embedded in the muscle. This method alleviated some of the problems encountered with surface electrodes but introduced other limitations. Because fine-wire electrodes are inserted into the muscle, they are recording activity from a very discrete area that may not be representative of the overall muscle activity. In addition, there may be movement of the electrodes within the muscle during activity, creating movement artifact. Placement of the indwelling electrodes also creates variability in the signal, in relation not only to position, but also to depth of insertion. This is more difficult to standardize than with the surface electrodes.

Various options for processing and display of the raw signals have been used. Many investigators have simply made subjective judgments as to magnitude and phase characteristics of raw EMG signal (Brandell, 1973; Mann & Hagy, 1980a; Mann & Hagy, 1980b; Paré et al., 1981; Schwab et al., 1983). Rectification of the signal to create only positive components is done in one of two ways. The negative portion of the signal can simply be eliminated (half-wave rectification), or an absolute value can be taken (full-wave rectification). The signal can then be processed with a low-pass filter that eliminates the higher frequency components to produce a linear envelope. Winter (1979) reports that the linear envelope generally follows the trend of the EMG signal and represents a clear and easily understood description of the muscle activity. This method was used by Nilsson et al. (1985) and MacIntyre and Robertson (1987).

The next level of complexity in EMG analysis is integration of the area under the linear envelope. There are three common forms of integration to be considered. The simplest method is to integrate over single or multiple muscular actions. A second method, which allows amplitude trends to be followed, involves integrating for a preset time interval and has been examined by Norman, Nelson, and Cavanagh (1978). The sum of all the ''bins'' represents the total integrated signal. The third method involves integrating to a preset voltage level. Once that level is reached, the integrator is reset to zero and begins again. The frequency of the resets lends insight into the strength of the contraction. Figure 6.1 demonstrates the results of these various processing methods on a typical phasic EMG from Winter (1979).

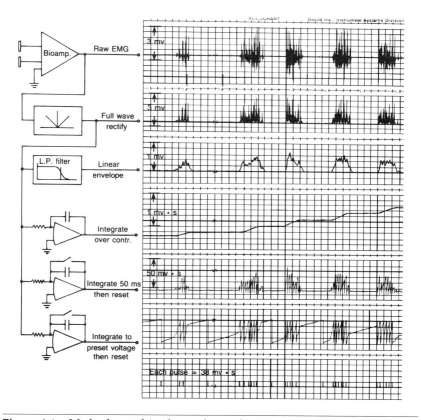

Figure 6.1. Methods used in the analysis of EMG signals. From *Biomechanics of Human Movement* (p. 140) by D.A. Winter. Copyright © 1979 by John Wiley & Sons. Reprinted by permission of John Wiley & Sons, Inc.

Review of Methods Used by Various Authors

Compared to the volume of studies on topics such as economy or ground reaction forces, there are relatively few investigations of muscle activity during running. This may be due to the exacting nature of the experiments, which are frequently complicated by extraneous noise, movement artifact, and instrumentation problems. The following account is a chronological description of the methods used by previous investigators in the area of EMG and running. The value of EMG data is greatly enhanced when it is associated with simultaneous kinematics, and many of the studies have incorporated some form of kinematic analysis. A summary of results of these investigations will be presented in the next section.

Brandell (1973) examined the EMG patterns of the lower extremity musculature during running. He studied seven males and three females during walking and running at a free cadence for 30 laps of a 22-yard course. Kinematic and EMG data were collected simultaneously during laps 2, 4, 15, 17, 28, and 30. An electrically driven camera running at 80 frames per second was utilized to track the movements of the subject. Twelve muscles were recorded using fine-wire indwelling electrodes; however, only the raw EMG signal from the semimembranosus, semitendinosus, biceps femoris, vastus medialis, vastus lateralis, medial and lateral head of the gastrocnemius, and soleus muscles were reported.

In 1979, Elliot and Blanksby studied 10 females between 20 and 28 years of age who were active joggers and experienced treadmill runners. The subjects ran at speeds of 2.5 m $\cdot$ s^{-1} (V1) and 3.5 m $\cdot$ s^{-1} (V2) while the EMG activity of the rectus femoris, vastus lateralis, vastus medialis, semitendinosus, semimembranosus, long head of the biceps femoris, tibialis anterior, and triceps surae muscles was collected via surface electrodes. Both an integrated and a raw signal were collected. The integrated EMG (IEMG) was averaged over 40 cycles producing an averaged IEMG (AIEMG). Kinematics were collected utilizing a motor-driven, 16-mm high-speed camera. Speed effects were analyzed and discussed.

Mann and Hagy (1980a) studied walking, running, and sprinting. They utilized 13 runners: two male sprinters, five experienced joggers (three males, two females) and six elite long-distance runners (three males, three females). The subjects traversed a 150-foot runway at speeds of 1.4 m $\cdot$ s^{-1} (walk), 5.4 m $\cdot$ s^{-1} (run), and 7.7 m $\cdot$ s^{-1} (sprint). Using surface electrodes, the experimenters studied the raw EMG signals of the gluteus maximus, quadriceps, hamstrings, anterior tibialis, and gastrocsoleus complex. They also collected sagittal plane kinematics utilizing high-speed cinematography and made comparisons between the three activities. These same authors also performed a similar study (Mann & Hagy, 1980b) in which five subjects were utilized, three females and two males. The speeds investigated were 1.4 m $\cdot$ s^{-1} (walk), 3.5 m $\cdot$ s^{-1} (jog), and 5.4 m $\cdot$ s^{-1} (run). High-speed photography together with force plate analysis were utilized. Surface electrodes were again used, and the same muscles were studied with the addition of the adductor, abductor, and sacrospinalis groups.

Paré et al. (1981) were interested in the specific function of the tensor fasciae latae during locomotor activities to determine the potential of this muscle for compensation in anterior cruciate ligament insufficiency. They utilized four fine-wire, bipolar electrodes placed transversely across the muscle belly. Ten subjects (five males and five females) between the ages of 20 and 36 were studied during walking, jogging, and running at freely chosen cadences. The raw EMG data were superimposed on a video image of the subject running so that it could be matched to the events of heel-strike, midstance, and toe-off.

In 1983, Schwab et al. studied seven males (ages 23-32) with no previous history of knee pathology. The experimenters collected EMG from the distal muscle bellies of the vastus medialis oblique, long head of the biceps femoris, and medial gastrocnemius. This study was unique in that simultaneous bipolar surface and indwelling electrodes were used and both treadmill and overground running were studied. The raw EMG signals were transmitted via a battery-powered telemetry pack worn by the subject. Telemetered footswitches also monitored gait events, and kinematic data were collected with a high-speed 16-mm camera. The treadmill was operated at a slow speed of 4 mph while overground running was performed at a free cadence and speed. There was a marked difference in the resultant mean velocities tested. The mean overground speed was 4.5 m • s^{-1} whereas the mean treadmill speed was 1.78 m • s^{-1}, but comparisons between the two were still made by the authors.

Nilsson et al. (1985), in the most comprehensive investigation to date, studied 10 males (ages 19 to 29) during locomotion over a wide range of speeds. Using bipolar surface electrodes, these authors collected myoelectric data from the major muscle groups of the lower extremity including gluteus maximus, rectus femoris, vastus lateralis, vastus medialis, semitendinosus, semimembranosus, lateral gastrocnemius, and tibialis anterior. The raw EMG signal was rectified and then passed through a low pass filter resulting in a linear envelope. These authors studied both walking (0.4 to 3.0 m • s^{-1}) and running (1.4 to 9.0 m • s^{-1}) and collected kinematic data using an optoelectronic system. This study provided a broad perspective on the adaptation of both EMG and kinematics to changes in both speed and mode of progression.

Most recently, MacIntyre and Robertson (1987) studied the function of knee muscles during running. They tested 11 normal subjects (ages 21 to 36) with no known lower extremity pathology and collected 10 foot-strikes during treadmill running at a speed of 3.3 m • s^{-1}. Linear envelope EMG signals were recorded from vastus medialis, vastus lateralis, rectus femoris, medial hamstrings, and lateral gastrocnemius via surface electrodes placed over the "motor points," which were not precisely defined.

A footswitch was used to determine gait cycle endpoints, and each stride was normalized to 100% and then ensemble averaged over the 10 trials for each subject. Ensemble averages of the five muscles from the 11 subjects were computed to obtain a grand ensemble average for each muscle. This was the only report located in which cycle-to-cycle variability in the EMG data was reported.

The Phasic Action of Individual Muscles

This section describes the phasic activity of major muscle groups of the lower extremity during running as reported by the authors mentioned previously. If the authors reported a range of speeds, a medium speed—closest to 3.8 m • s⁻¹—has been chosen for the graphic presentation. A summary of all available results is presented in Figures 6.2, 6.3, and 6.4 except for those of Elliot and Blanksby (1979), whose method of analysis did not lend itself to this type of presentation. It must be stated that the results shown in these figures represent different subjects running at different speeds and should therefore be interpreted with caution.

Gluteus Maximus

Only one phase of activity has been reported for the gluteus maximus (GM) during the gait cycle. It becomes active in late swing and continues through the first third of stance (Mann & Hagy, 1980a; Mann & Hagy, 1980b). Nilsson et al. (1985) noted that at slower speeds of running, onset did not occur until footstrike; however, with increasing speeds, the pattern was similar to that reported by Mann and Hagy, with activation beginning in late swing. During late swing, GM is apparently acting eccentrically to decelerate the thigh (especially at the higher speeds). At footstrike, it may also be assisting with the stabilization of the thigh and pelvis while initiating hip extension throughout early stance.

Tensor Faciae Lata

Paré et al. (1981) were the only investigators to study tensor fasciae latae (TFL) during running. They found two distinctly different phasic patterns from different regions of the muscle and concluded that there are two functionally different parts of the TFL—the anteromedial (AM) fibers and the posterolateral (PL) fibers. The PL fibers appeared to be hip internal rotators and abductors, whereas the AM fibers were apparently primarily hip flexors. During running, the PL fibers were found to be active just prior to and shortly after footstrike. The authors noted that GM was also active during this time as speed increased and suggested a synergism between GM and the PL fibers of TFL, as the former externally rotates the thigh while the latter internally rotates it. As speed increased, there was a second phase of activity at toe-off and well into swing. The AM fibers were active during toe-off to midswing in both jogging and running, indicative of their hip flexor function. Unlike the PL fibers, they were inactive during heelstrike. This was necessary in light of their hip flexor function and the hip extension phase that follows footstrike.

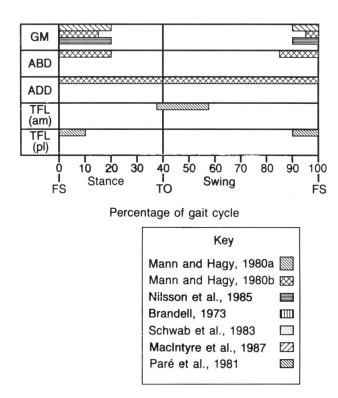

Figure 6.2. Summary of the literature on the phasic activity of the hip musculature during running. GM—gluteus maximus, ABD—abductors, ADD—adductors, TFL (am)—Tensor fasciae latae (anteromedial fibers), TFL (pl)—tensor fasciae latae (posterolateral fibers).

Abductors

Mann and Hagy (1980b) reported that the abductors (ABD) were active in late swing and early stance. It is likely that this muscle group is acting to prepare for contact and provide hip stabilization in early stance. In addition, it may have a synergistic role with the GM and PL fibers of TFL as these muscles were also active at this time as noted previously. The anterior fibers of the ABD assist in internal rotation whereas the posterior fibers assist in external rotation. However, the authors did not differentiate between these fibers.

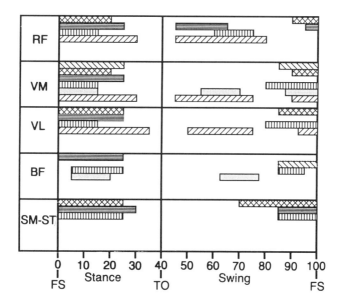

Percentage of gait cycle

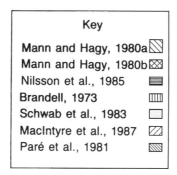

Figure 6.3. Summary of the literature on the phasic activity of the knee musculature during running. RF—rectus femoris, VM—vastus medialis, VL—vastus lateralis, BF—biceps femoris, SM-ST—semimembranosus-semitendinosus.

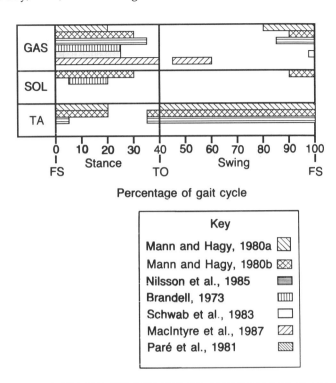

Figure 6.4. Summary of the literature on the phasic activity of the ankle musculature during running. GAS—gastrocnemius, SOL—soleus, TA—tibialis anterior.

Adductors

Mann and Hagy (1980a) found that the adductors (ADD) demonstrated continuous activity throughout the entire gait cycle at all speeds of running tested. They suggested that these muscles stabilized the pelvis with respect to the thigh during support and the thigh to the pelvis during swing.

Hamstrings

The hamstrings consist of the semitendinosus (ST) and semimembranosus (SM) medially and the biceps femoris (BF) laterally. The hamstrings are two joint muscles (with the exception of the short head of the biceps femoris) and act to extend the hip and flex the knee. Although some authors have measured the activity of the separate muscles, results were most often reported for the entire muscle group as one functional unit.

Most investigators were in agreement as to the activity of the hamstrings, which were generally noted to be active through the last 25% to 40% of swing. During early swing to midswing, because there is no hamstring activity, the knee appears to be passively flexed as a result of the rapid forward acceleration of the thigh (Mann & Hagy, 1980a). During late swing, the hamstrings are apparently working to decelerate hip flexion and shortly thereafter to control knee extension, which occurs as the thigh is decelerated and momentum is transferred to the shank. Once the thigh has completed its forward swing and begins moving backward prior to footstrike, the hamstrings act concentrically to extend the hip and flex the knee. There was consensus among investigators that the hamstrings remain on through the first half of stance (Brandell, 1973; Mann & Hagy, 1980b; Nilsson et al., 1985). At footstrike, the hamstrings and quadriceps co-act, apparently to provide stability during impact. Mann and Hagy (1980a) suggested that during support, the hamstrings were acting isometrically to maintain a constant length as the hip and knee were simultaneously extending. Elliot and Blanksby (1979) reported increasing hamstring activity from footstrike throughout the entire stance phase until toe-off. They noted that BF achieved maximal levels at heel-off and toe-off. Both SM and ST also acted strongly, although their activity was not at a maximum at this time. MacIntyre and Robertson (1987) reported that the SM and ST exhibited the greatest variability in phasic activity of the knee muscles that they examined.

Quadriceps

Almost all authors found quadriceps activity during the last half of swing phase and into the first half of stance. Nilsson et al. (1985) found that the rectus femoris (RF) exhibited a burst of activity during early swing that was associated with its hip flexor function. MacIntyre and Robertson (1987) found a similar burst of activity during early swing in RF, vastus medialis (VM), and vastus lateralis (VL), but this may have been a function of their selection of running speed (3.3 m • s^{-1}). Unlike RF, VM and VL do not cross the hip joint and therefore, as they cannot be assisting in hip flexion, they are probably acting to control knee flexion. Brandell (1973) believed that swing phase knee extension was brought about via a slowing of the forward motion of the thigh as momentum carried the leg rapidly forward. He found that quadriceps activity began almost coincident with hamstring action when the knee flexed in preparation for footstrike.

Elliot and Blanksby (1979) observed that VM, VL, and RF all recorded their highest levels during the period between heelstrike and heel-off. Nilsson et al. (1985) stated that the main activity occurred during eccentric muscle action during early stance while the knee was undergoing flexion. Brandell (1973) reported that the quadriceps often reached maximal activity at the transition between knee flexion and extension but was quiet during the extension phase. He suggested that, as a result of electromechanical delay, the contractile force was delayed long enough to cause the initiation of the extension that followed. This apparently paradoxical situation, in which quadriceps activity has usually ceased prior to the beginning of knee extension, will be examined later in this chapter.

Anterior Tibialis

Anterior tibialis (AT) is reported to be active throughout most of the gait cycle. Elliot and Blanksby (1979) reported that maximal activity occurred at heelstrike where AT was acting eccentrically as the foot lowered to the ground, an observation supported by angular velocity measurements. During support when the foot was flat on the ground, AT apparently acted concentrically to bring the shank forward. Electrical silence has been observed shortly after footstrike (Nilsson et al., 1985), and during the last half of stance (Mann & Hagy, 1980a, 1980b). All authors noted constant activity throughout swing to control plantar flexion and initiate dorsiflexion.

Gastrocsoleus

Although gastrocnemius (GAS) and soleus (SOL) are two separate muscles, they are often considered as one functional unit and referred to as gastrocsoleus (GS). In general, gastrocsoleus (GS) has been reported to be active from late swing through to 50% to 80% of the stance phase. In late swing it is co-acting with AT to stabilize the foot into position for footstrike and then begins to act eccentrically to decelerate the forward-moving shank. The function again becomes concentric as active plantar flexion begins at the ankle. There is some controversy concerning the action of GS during support. Mann and Hagy (1980b) reported that, at faster speeds, GS activity ceased prior to the beginning of plantar flexion, which would indicate purely an eccentric function at this time. However, Elliot and Blanksby (1979) reported maximum GS activity at toe-off for faster speeds of running, which they believed was a result of concentric action of the muscle. At slow running speeds, GS activity persisted only through 30% to 50% of plantar flexion, according to Brandell (1973) and Mann and Hagy (1980a).

Changes in EMG With Running Speed

Nilsson et al. (1985) have looked the closest at the effect of running velocity on EMG. In absolute terms, the duration of the signal was inversely proportional to velocity. However, in relative terms, muscles were active for a greater percentage of the gait cycle at higher velocities. In addition, they tended to become active earlier in the cycle. An example of this was the action of the hamstrings in late swing: As the shank velocity increased with increasing running speeds, activity in the hamstrings was required earlier to effectively decelerate the leg. At times these increased demands resulted in an extra burst of activity that was not present at the slower speeds. VM and VL were normally quiescent in late swing; however, at the faster running speeds they exhibited a burst of activity during this time. Finally, it appears that velocity can also affect the relative magnitudes of activity. This point is demonstrated in rectus femoris as the burst during early swing becomes larger than that during support.

The Stretch-Shortening Phenomenon

In subjectively reviewing the phases of muscle activity in relation to joint motion, one sees many apparent examples of concentric action being preceded by eccentric actions. In the ankle, the gastrocsoleus muscle group is active during initial lengthening through early stance prior to shortening during active plantar flexion. The gluteus maximus acts eccentrically to assist in deceleration of the thigh prior to its action of extending the hip. This natural sequence of lengthening-shortening, often referred to as the stretch-shortening cycle, allows for preloading of the muscle, which has been shown to enhance force output (Cavagna, 1977) for a given neural innervation.

Quantitative confirmation of such observations requires estimates of muscle lengths to be made simultaneously with EMG acquisition. Equations for the prediction of gastrocnemius length have been presented by Grieve, Pheasant, and Cavanagh (1978) and used to calculate muscle length changes during running by Milliron and Cavanagh (this volume). We have used these results here to estimate muscle length–EMG relationship in a single subject running at $3.4 \text{ m} \cdot \text{s}^{-1}$.

Surface EMG was quantified using time reset integration in 5% units of the running cycle. The mean length and velocity for gastrocnemius were calculated in the same time interval. The results of plotting all of the variables as functions of time are shown in Figure 6.5a, and IEMG is shown as a function of rate of change of length for the part of the cycle surrounding footstrike in Figure 6.5b. The first point on the curve, labeled 19, is 5% of cycle prior to footstrike; the point labeled 20 is footstrike; and the point labeled 6, a time shortly before toe-off.

It is apparent from Figure 6.5b that during most of the time when EMG activity is increasing, the muscle is lengthening—the curve is in the positive X quadrant. Shortly after the IEMG reaches a maximum between

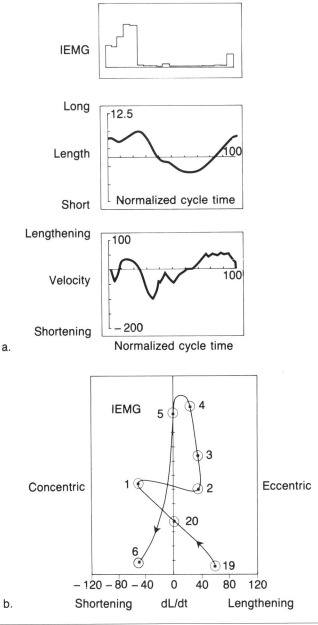

Figures 6.5a and 6.5b. The relationship between IEMG and gastrocnemius muscle length and rate of change of length during level running at 3.4 m • s⁻¹. In (a), IEMG, dL, and dL/dt have been plotted separately over an entire normalized cycle. In (b), dL/dt is plotted against the IEMG and represents only the period of late swing through early stance (see text for details).

points 4 and 5—approximately 25% of the cycle—the muscle action has halted the lengthening and concentric action occurs. However, the muscle is turned off so rapidly that only 5% of cycle after peak activity the IEMG is indistinguishable from residual noise. This is a clear graphic demonstration of the stretch-shortening cycle and illustrates how muscle length information lends further insight into the interpretation of the EMG signal.

The Extensor Paradox Experiment

It is well known that knee flexion occurs just before and immediately after footstrike during running to cushion the impact of landing (Milliron & Cavanagh, this volume). Once the downward movement of the center of gravity associated with this cushioning phase has finished, knee extension begins and the propulsive phase of the cycle continues.

There is evidence from Brandell (1973) and Mann and Hagy (1980b) that the quadriceps are generally silent during the phase of knee extension following the cushioning. Few experiments have focused on this puzzling aspect of knee joint action during running. The purpose of the experiment described in this section was to examine the activity of the three heads of the quadriceps that are amenable to surface recording during distance running and to simultaneously measure the angle of the knee joint.

Subjects and Speed

Six male recreational runners, ages 19 to 26, experienced in treadmill running with no history of recent injury, volunteered for the study. Each subject ran at a constant speed of 4.0 m · s^{-1} on a motorized treadmill. This speed was chosen as it was in the middle of the range used by previous workers.

Equipment and Method of Analysis

To investigate knee extensor muscle activity during the stance phase of running, EMG of the vastus medialis, vastus lateralis, and rectus femoris muscles of one leg were recorded using a battery-powered GCS 67 Electromyographic Processor. Silver–silver chloride electrodes with on-site preamplifiers were placed in the middle of the muscle belly after thorough preparation of the skin. An inertia switch attached to the heel was used to define the cycle endpoints and knee angle was recorded simultaneously with a self-aligning ULGN-67 Electrogoniometer. This design compensates for errors in placement and does not assume a fixed center of rotation for the joint. The electrogoniometer was calibrated for knee angle by comparing voltage output against knee angle measured by a protractor.

The EMG processor, together with the goniometer and footswitch signals, were interfaced with an SMS 1000 computer, which sampled at a rate of 500 Hz per channel. The raw EMG signal was prefiltered using a high pass filter of 75 Hz cut-off frequency. Custom software allowed for storage, processing, and display of the data. An example of the raw data for the complete 5-second sampling period is shown in Figure 6.6a, and the region surrounding footstrike is shown with greater resolution in Figure 6.6b.

Five-second samples were collected after each subject had undergone a warm-up period at the test speed. This allowed at least six full cycles of running to be recorded for each individual. For each period of stance, the phasic activity of all three muscles was subjectively determined by comparison with a noise-free baseline. Data from six footstrikes were examined, and mean values were obtained for the time at which rectus femoris, vastus lateralis, and vastus medialis muscle activity ceased. The beginning and end times of the knee extension phase following initial flexion were also determined.

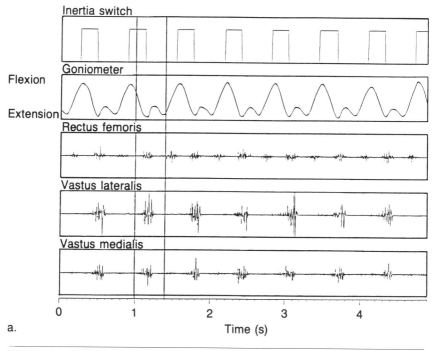

Figure 6.6a.   A 5-s raw experimental record.

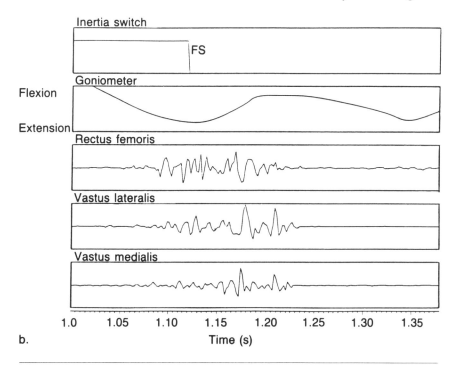

Figure 6.6b. A portion of the same experimental record surrounding foot-strike shown with greater time resolution.

Results

Figure 6.7a illustrates the mean results of six footstrikes for a typical subject. It can be seen that approximately 85 milliseconds before footstrike, muscle activity begins while knee extension is under way. Vastus lateralis is the first to show activity, some 25 milliseconds before vastus medialis and 60 milliseconds before rectus femoris. This period of muscle activity appears to help in stabilizing the leg in preparation for footstrike. All three muscles are active through footstrike while knee flexion occurs, but they cease activity simultaneously approximately 20 milliseconds after peak knee flexion has been achieved. In this subject knee extension continues for a further 150 milliseconds.

The mean results for the group as a whole are presented in Table 6.1 and shown schematically in Figure 6.7b. The mean time of knee extension that was not accompanied by quadriceps EMG was 133.7 milliseconds

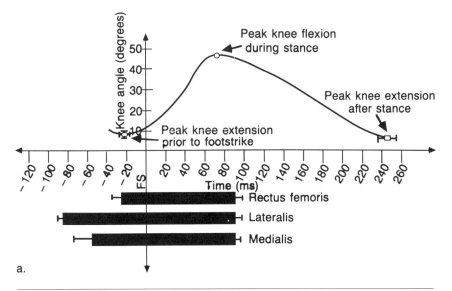

Figure 6.7a. Results of phasic quadriceps EMG and knee angle for a typical subject averaged over six footstrikes.

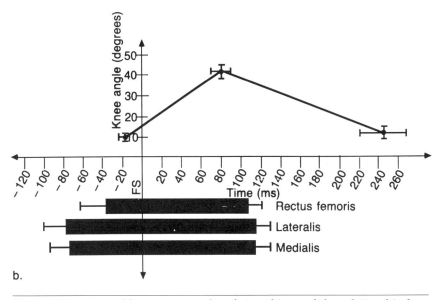

Figure 6.7b. Ensemble average results of six subjects of the relationship between phasic quadriceps EMG and knee angle. The values of peak knee extension prior to footstrike, peak knee flexion during stance, and peak knee extension after stance have been joined by straight lines as the mean curve was not determined.

Table 6.1   Mean Values (in Milliseconds) for Critical Phases During Knee Extension During Running at 4 m • s⁻¹

	Flexion-extension duration	Mean all muscle off after peak flexion	Mean duration of silence during extension
Mean for group	162.8	29.2	133.7
±SD	19.5	10.4	16.5

($SD = 16.5$). These results are further illustrated in Figure 6.8, where electrical activity is indicated by the presence of shading over the muscle. The amplitude of the activity is also schematically indicated by the intensity of the shading. The large amount of knee extension that occurs in the absence of muscle activity is readily apparent from this figure.

Discussion

For the group of runners examined in this study, it is clear that the quadriceps cease their activity shortly after peak stance phase knee flexion has occurred. A phase of knee extension of approximately 130 milliseconds continues without the assistance of the quadriceps. The function of the

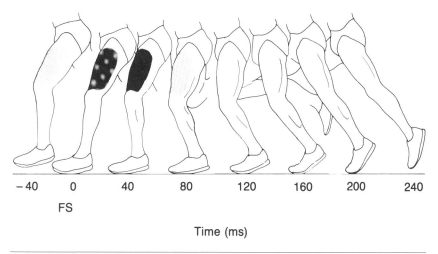

```
-40      0      40      80      120      160      200      240
 FS
```

Time (ms)

Figure 6.8.   The amplitude of EMG activity throughout the stance phase of running. (The intensity of shading indicates relative amount of activity.)

quadriceps must therefore be described as principally controlling the descent of the body center of gravity after landing. Certainly they help to initiate knee extension, but they rapidly become quiescent when knee extension has been under way for only about 30 milliseconds, a time during which less than 5 degrees of extension has been achieved. The duration of electrical silence in extension is large enough to exclude the possibility that electromechanical delay (EMD) between EMG activity and force production may explain the paradox. EMD time in concentric muscle action has been determined to be 40 to 55 milliseconds (Cavanagh & Komi, 1979; Norman & Komi, 1979), and in rapid movements it may be possible for EMG activity to have terminated before force can be detected.

A reasonable hypothesis may be that hip extensor action during the second half of the stance phase is causing the knee joint to extend. However, if one examines the co-activation of the quadriceps and hamstrings in Figure 6.3, it is apparent that many investigators have found these muscle groups to cease activity at about the same time in the cycle. Neither does there appear to be a prolonged period of gluteus maximus activity that would provide an explanation. Figure 6.4 indicates that the last extensor muscle to cease activity during stance appears to be the gastrocnemius, which is of course also a knee flexor. Because only the quadriceps were measured in the present study, it is not possible to say with certainty what patterns of activity were exhibited in other muscles in these particular subjects. These experiments have, however, shown that the notion of an extensor thrust—with plantar flexors, knee extensors, and hip extensors all being active in late support to generate forward and upward thrust—is in need of modification. They also indicate that the problem is worthy of further investigation using a kinetic approach in addition to multichannel EMG so that the joint moments can be determined.

References

Basmajian, J.V., & Deluca, C.J. (1985). *Muscles alive* (5th ed.). Baltimore: Williams and Wilkins.

Brandell, B.R. (1973). An analysis of muscle coordination in walking and running gaits. In S. Cerquiglini, A. Venerando, & J. Wartenweiler (Eds.), *Medicine and Sport: Biomechanics III* (pp. 278-287). Basel, Switzerland: Karger.

Carlet, M. (1872). Essai experimental sur la locomotion humaine: Étude de la marche [Experimental test on human locomotion: Study of walking]. *Annales des Sciences Naturelles*, Sect. Zool., XV.

Cavagna, G.A. (1977). Storage and utilization of elastic energy in skeletal muscle. *Exercise and Sport Sciences Reviews*, 5, 89-129.

Cavanagh, P.R., & Komi, P.V. (1979). Electromechanical delay in human skeletal muscle under concentric and eccentric contractions. *European Journal of Applied Physiology, 42*, 159-163.

Cohen, H.L., & Brumlik, J. (1968). *A manual of electroneuromyography.* New York: Harper and Row.

Elliot, B.C., & Blanksby, B.A. (1979). The synchronization of muscle activity and body segment movements during a running cycle. *Medicine and Science in Sports, 11*(4), 322-327.

Grieve, D.W., Pheasant, S., & Cavanagh, P.R. (1978). Prediction of gastrocnemius length from knee and ankle joint posture. In E. Asmussen & K. Jorgensen (Eds.), *Biomechanics VI-A* (pp. 405-412). Baltimore: University Park.

Hubbard, A.W. (1939). An experimental analysis of running and of certain differences between trained and untrained runners. *Research Quarterly of the American Association of Health and Physical Education, 10*(3), 28-38.

Hudgkins, C.V., & Stetson, R.H. (1932, July 15). A unit for kymographic recording. *Science,* p. 60.

Kramer, H., Kuchler, G., & Brauer, D. (1972). Investigations of the potential distribution of activated skeletal muscles in man by means of surface electrodes. *Electromyography and Clinical Neurophysiology, 12*, 19-26.

MacIntyre, D.L., & Robertson, D.G.E. (1987). EMG profiles of the knee muscles during treadmill running. In Bengt Jonsson (Ed.), *Biomechanics X-A* (pp. 289-294). Champaign, IL: Human Kinetics.

Mann, R.A., & Hagy, J.L. (1980a). Biomechanics of walking, running, and sprinting. *American Journal of Sports Medicine, 8*(5), 345-350.

Mann, R.A., & Hagy, J.L. (1980b). Running, jogging and walking: A comparative electromyographic and biomechanical study. In J.E. Bateman & A. Trott (Eds.), *The foot and ankle* (pp. 167-175). New York: Thieme-Stratton.

Marey, E.J. (1972). *Movement.* New York: Arno. (Original work published 1895)

Nilsson, J., Thorstensson, A., & Halbertsma, J. (1985). Changes in leg movements and muscle activity with speed of locomotion and mode of progression in humans. *Acta Physiologica Scandinavica, 123*, 457-475.

Norman, R.W., & Komi, P.V. (1979). Electromechanical delay in skeletal muscle under normal movement conditions. *Acta Physiologica Scandinavica, 106*, 241-248.

Norman, R.W., Nelson, R.C., & Cavanagh, P.R. (1978). Minimum sampling time required to extract stable information from digitized EMGs. In E. Asmussen & K. Jorgensen (Eds.), *Biomechanics VI-A* (pp. 237-243). Baltimore: University Park.

Paré, E.B., Stern, J.T., & Schwartz, J.M. (1981). Functional differentiation within the tensor fasciae latae. *Journal of Bone and Joint Surgery*, **63-A**(9), 1457-1471.

Schwab, G.H., Moynes, D.R. Jobe, F.W., & Perry, J. (1983). Lower extremity electromyographic analysis of running gait. *Clinical Orthopedics and Related Research*, **176**, 166-170.

Warfel, J.H. (1974). *The extremities* (4th ed.). Philadelphia: Lea & Febiger.

Winter, D.A. (1979). *Biomechanics of human movement*. New York: John Wiley & Sons.

Zuniga, E.M., Truong, X.T., & Simons, D.G. (1969). Effects of skin electrode position on averaged electromyographic potentials. *Archives of Physical Medicine and Rehabilitation*, **50**, 264-271.

Chapter 7

The Lessons From Animal Studies
R. McNeill Alexander

Human legs and gaits are in some ways very different from those of other mammals, but they are also in some ways very similar. The same basic principles apply to human running as apply to the running of other animals. Much of our understanding of human running depends on research on other species. Also, comparisons with other species are sometimes illuminating. This chapter aims to show how.

Size, Speed, and Gait

We walk at low speeds and run to go faster. Within each gait we increase speed partly by increasing our stride frequency and partly by taking longer strides. Is there a basic rule that specifies the speed at which we should break into a run, and the stride frequency and stride length that we should use at any particular speed? How does any rule depend on body size? Should tall runners and short ones take strides proportional to their statures?

Studies of other species help to answer these questions, for two reasons: First, they offer a much wider range of sizes than the human species does. Second we can be more confident that we have found a general rule reflecting basic principles if we can show that it applies to diverse species.

Theory

The hypothesis that will be presented (Alexander & Jayes, 1983) depends on the concept of dynamic similarity, which is an extension of the more familiar concept of geometric similarity. Two shapes are described as geometrically similar if one can be made identical to the other by multiplying all linear dimensions by some constant factor. For example, all cubes are geometrically similar to each other. Two systems of moving bodies are dynamically similar if the motion of one can be made identical to that of the other by multiplying

1. all linear dimensions by some constant factor,
2. all time intervals by another factor, and
3. all forces by a third factor.

For example, pendulums of different lengths swinging through equal angles have dynamically similar motions. Notice that Condition (1) implies geometric similarity. Mammals of different size (for example, cats and horses) are not geometrically similar, but they are often sufficiently near to it for predictions based on dynamic similarity to be useful.

Systems of different sizes can move in dynamically similar fashions only if their speeds are in appropriate ratio. If gravitational forces are important (as for swinging pendulums or running mammals) the condition is that the systems must have equal values of the Froude number u^2/gl. Here u and l are a characteristic speed and a characteristic length, measured in the same way for the systems being compared, and g is the acceleration of free-fall. In the discussion that follows, u will be the speed of running and l will be leg length (defined as the height of the hip joint from the ground in normal standing).

The dynamic similarity hypothesis (Alexander & Jayes, 1983) predicts that animals of different sizes tend to move in dynamically similar fashions when they walk or run at speeds that give them equal Froude numbers. There is no necessity for them to do so, but there is a good reason why they should. Suppose that an animal has adjusted its gait to minimize the mechanical power needed for running at a particular speed. Another, geometrically similar animal wishing to minimize its power requirement, when running with the same Froude number, must run in dynamically similar fashion.

On earth, g is almost constant, and the speeds required for equal Froude numbers are proportional to the square roots of leg length. For example, a cat of leg length 0.2 m running at 1 m $\cdot$ s^{-1} has the same Froude number as a camel of leg length 1.8 m running at 3 m $\cdot$ s^{-1}.

The dynamic similarity hypothesis predicts that animals traveling with equal Froude numbers will use the same gait. They will change from walking to running at the same Froude number. Another of its predictions concerns stride length. Animals moving in dynamically similar fashions have stride lengths proportional to the leg length l, so the hypothesis predicts

$$\lambda / l = F(u^2/gl) \qquad (7.1)$$

where F is the same function for animals of different sizes and species.

Results

The predictions of the hypothesis are reasonably accurate, even for very different animals. Gait changes occur at approximately equal Froude numbers. Humans, kangaroos, and various quadrupedal mammals all break into a run (or hop) at Froude numbers close to 0.6 (Alexander & Jayes, 1983). For a woman of leg length 0.8 m, this means a speed of 2.2 m • s^{-1}.

Figure 7.1 is a graph of relative stride length (λ/l) against Froude number for running gaits of humans, kangaroos, dogs, ferrets, and horses. Equation 7.1 predicts that all the points should lie along a single line. In fact, they cluster reasonably closely around the line

$$\lambda/l = 2(u^2/gl)^{0.4}. \tag{7.2}$$

More extensive surveys show differences between major groups of animals. At equal Froude numbers, small mammals that run on bent legs (e.g., rats and ferrets) use longer relative stride lengths than large mammals that run on straight legs (e.g., dogs and horses) (Alexander & Jayes, 1983). Monkeys use even longer relative stride lengths (Alexander & Maloiy, 1984).

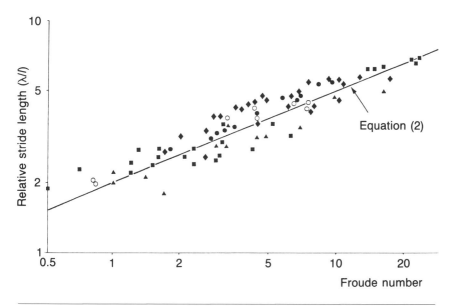

Figure 7.1. A graph on logarithmic coordinates of relative stride length (λ/l) against Froude number (u^2/gl) for running. Data from Alexander and Vernon (1975a), Alexander and Jayes (1983), and Alexander and Maloiy (1984).

Implications

How should runners adjust their stride lengths to suit their different statures? It seems likely that the rule expressed by Equation 7.2 minimizes energy costs. Zarrugh and Radcliffe (1978) showed that the rates of oxygen consumption of walking humans were less when they used their preferred stride lengths than when they took longer or shorter strides. The equation can be rewritten

$$\lambda = 2\ u^{0.8} l^{0.6} g^{-0.4}. \tag{7.3}$$

Thus if people of different statures run at the same speed, they should use stride lengths proportional to (leg length)$^{0.6}$. Leg length is an almost constant fraction of stature. Thus if a short (1.5 m) woman runs with a fairly tall (1.8 m) man, she should take strides $(1.5/1.8)^{0.6} = 0.9$ times as long, at 1.1 times the frequency. If they run in step, one or both of them must use a suboptimal gait.

Stresses in Muscles

The skeletal muscles of vertebrates are very uniform in some aspects of their structure and physiology. In particular, all seem capable of exerting maximum stresses of about 0.3 MPa, in isometric contraction (Close, 1972). The physiological experiments that supply these data are feasible only for small muscles and have been performed on small species, but it seems highly likely that human muscles and muscles of other large mammals have the same capability.

Which muscles exert their maximum stresses in running and which are less fully used? To answer the question we need data about the forces exerted on leg segments and about the dimensions of muscles. The dimensions should be from healthy animals, not ones wasted by senility or disease. Such data are difficult to obtain for humans.

Methods

The data presented in this section depend on analysis of cinefilms supplemented, in some cases, by force platform records. It is in principle possible to calculate the force and torque acting on every body segment, if the linear and angular accelerations of the segments have been measured from films (see Alexander, 1981). In practice, potentially serious sources of error can be avoided if force platform records are also available. The forces exerted by individual muscles cannot be calculated without further assumptions, because the number of active muscles generally exceeds the number of degrees of freedom of the leg. For lack of information it will be assumed (where necessary) that cooperating muscles exert equal stresses. Stresses are calculated from the forces and the dimensions of the muscles, taking account, where necessary, of pennate fiber arrangement.

Further details of methods can be found in the papers referred to in this section.

Results

In an investigation of human running, Alexander & Vernon (1975b) calculated peak stresses of about 0.4 MPa, both for the extensors of the ankle (gastrocnemius and soleus) and for those of the knee (rectus and vastus). These stresses occurred at the midpoint of the stance phase as the center of mass passed over the supporting foot. They refer to quite slow running, at 3.5 to 4.0 m • s^{-1}. Faster running would involve larger stresses. However, the calculated stresses are probably too high. The runners were normal, healthy men, though not athletes. The anatomical data, however, came from the cadaver of a man who had been prevented by illness from working normally for several years before his death. His muscles were probably smaller than those of the runners, though he matched them well in stature and body mass.

Some data for other species are shown in Table 7.1. They are probably more reliable, because the dissected animals had been killed while still healthy or (in the case of the wallaby) after a few days' illness. From the films we had available, we chose those that showed the animals running fastest, but most of the animals were running at well below maximum speeds. Only the greyhound (which was filmed while training for racing)

Table 7.1 Peak Stresses in Leg Muscles of Running Animals

Type of animal	Speed (m • s^{-1})	Range of stresses (MPa)
Wallaby (*Macropus rufogriseus*)	3	0.08 - 0.15
Kangaroo (*Macropus rufus*)	6	0.11 - 0.17
Greyhound (*Canis familiaris*)	15	0.26 - 0.38
Buffalo (*Syncerus caffer*)	5	0.15 - 0.30
Elephant (*Loxodonta africana*)	4	0.14[a]

Note. Data are from Alexander and Vernon (1975); Alexander, Maloiy, Hunter, Jayes, and Hturibi (1979); and Jayes and Alexander (1982).
[a]No range available.

seems to have been running at maximum speed. The general impression given by the table is that peak muscle stresses tend to be around 0.15 MPa during running at moderate speeds but about 0.3 MPa at the highest speeds of greyhounds.

Most of the data in Table 7.1 refer to the extensor muscles of the hip, knee, and ankle. These muscles exert maximum forces during the stance phase, when they balance the moments about the joints of the force exerted by the ground on the foot. Some other muscles, studied only in the greyhound, may exert their maximum forces when large torques are needed to give angular accelerations to the legs, at the forward and backward extremes of their swing. One of the aims of the study of the greyhound was to discover whether muscles with different functions exert equal peak stresses in fast locomotion. Does a racing greyhound use all its leg muscles to the maximum?

Successive strides of the greyhound were very uniform, and only one was analyzed in detail (Figure 7.2). Frame 366 was the midpoint of the stance phase of the right hind leg, and it was calculated that the stress in the gastrocnemius and plantaris muscles was then 0.34 MPa. (In dogs and many other mammals, unlike in humans, the plantaris is a major extensor of the ankle.) Frame 377 was midstance for the right forelimb, and it was calculated that the stress in the serratus muscles (which attach the shoulder blade to the rib cage) was then 0.30 MPa. The right forelimb

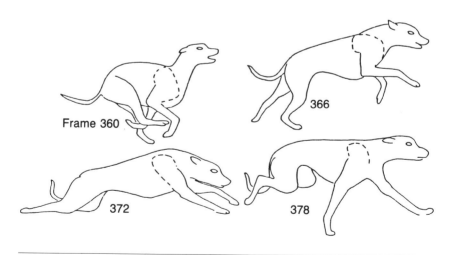

Figure 7.2. Outlines traced from selected frames of a film of a greyhound galloping at 15 m • s⁻¹. (The framing rate was 100 Hz.) *Note.* From ''Estimates of Mechanical Stresses in Leg Muscles of Galloping Greyhounds'' by A.S. Jayes and R.McN. Alexander, 1982, *Journal of Zoology, London,* **198,** p. 320. Copyright 1982 by The Zoological Society of London. Reprinted with permission of The Zoological Society of London.

had its greatest clockwise and counterclockwise accelerations at frames 371 and 380, and it was calculated that the shoulder muscles principally responsible exerted 0.26 to 0.38 MPa and 0.31 to 0.37 MPa, respectively. The right hind limb had its greatest counterclockwise acceleration at frame 373, and the sartorius, rectus, and tensor fasciae latae were calculated to exert 0.31 MPa. Clockwise acceleration of this limb does not involve high stresses because the hamstring muscles, which are presumably responsible, have to exert less force then than during the stance phase. Notice that all the stresses mentioned in this paragraph lie in the range of 0.26 to 0.38 MPa.

Implications

The legs of greyhounds seem to be constructed so that about equal peak stresses act in all the principal muscles in fast running. The same may be true of human legs, but it should be remembered that people (unlike greyhounds) have not been bred selectively for running speed. It is more likely to be true of trained runners than of other people. The peak stresses calculated for human leg muscles are probably too high because the cadaver's muscles were wasted. Those for greyhounds are probably more reliable: The animals were killed for veterinary reasons but had been in training for racing until shortly before death. By analogy with the greyhounds and the other mammals in Table 7.1, it seems likely that peak stresses in human leg muscles during sprinting are about 0.3 MPa. They must be lower in distance running, when the foot is on the ground for a larger fraction of the stride and exerts correspondingly low forces.

Tendon Elasticity

The center of mass of a runner is highest at the midpoint of the floating phase of the stride and lowest at the midpoint of the stance phase. The velocity of the center of mass falls during the first half of the stance phase (when the foot exerts a braking action) and rises again during the second half. The total of kinetic and gravitational potential energy is therefore highest during the floating phase and falls to a minimum during the stance phase. Work must be done when this total rises, and energy must be dissipated (i.e., "negative work" must be done) when it falls. The positive and negative work can be done entirely by muscle action, but it can be done at less metabolic cost if elastic mechanisms are involved. The principle is that of a bouncing ball: A perfect ball in a perfect (i.e., frictionless) world would continue bouncing forever, without the need for any fresh input of energy. When a ball hits the ground its kinetic energy is converted to elastic strain energy, which is reconverted to kinetic energy in the rebound.

The first indication that the elastic mechanisms might be important in running seems to have come from experiments on humans. The metabolic energy consumption of runners, calculated from rates of oxygen consumption, is lower than would be expected if no such mechanisms were involved (Cavagna, Saibene, & Margaria, 1964). Much of our further knowledge of elastic mechanisms in running comes from studies on other species. The most complete series of studies is on wallabies (small kangaroos, Macropodidae).

Methods

Alexander and Vernon (1975a) made force platform records and films simultaneously of a wallaby hopping in a large room. These data have already been referred to in Table 7.1. Marks on the skin showed approximate joint positions. Instantaneous centers of joints and moment arms of muscles were obtained from X-ray photographs. The forces exerted by major muscle groups and the length changes of individual muscles were calculated.

Morgan, Proske, and Warren (1978) exposed the gastrocnemius muscle of an anesthetized wallaby. They freed its tendon from the heel and subjected it to rapid, controlled stretches while stimulating the muscle electrically. They paralyzed the muscle to varying degrees during their experiments by infusing a drug, and they made further experiments after shortening the tendon. From the results of these experiments they were able to distinguish between the elastic properties of the tendon and those of the active muscle fibers.

Ker, Dimery, and Alexander (1986) used a servohydraulic tensile testing machine to investigate the mechanical properties of major leg tendons removed from wallaby carcasses. With this machine they were able to exert forces like those that would act in hopping at realistic frequencies. They measured the strain energy stored in the tendons under the forces (calculated from the force platform experiments) that act in hopping.

Results

Alexander and Vernon (1975a) showed that, during the stance phase of the wallaby's hop, the hip extensor muscles shorten while exerting tension: They do work. The knee extensor muscles lengthen while exerting tension, doing negative work. Neither of these muscle groups behaves like an elastic body. The gastrocnemius and plantaris muscles lengthen as their tension rises and shorten as it falls, like elastic bodies stretching and recoiling. This does not prove that their behavior is principally elastic, but merely admits the possibility. The length changes could be mainly due to inelastic lengthening and shortening of the contractile apparatus of the muscles. However, calculations suggested that quite a lot of the length change was due to stretching and recoil of the tendons.

Muscle has elastic properties as well as contractile properties. Alexander and Bennet-Clark (1977) and Alexander (1984) discussed the relative importance of muscle and tendon as strain energy stores in running and hopping. They showed that if the muscle fibers were much longer than the tendons, the muscle would be the more important strain energy store. If the tendon were much longer than the muscle fibers, it would be more important. The gastrocnemius and plantaris muscles of wallabies have been identified as potentially useful strain energy stores. They have tendons 270 and 440 mm long (in an 11-kg wallaby; Alexander & Vernon, 1975a) and muscle fibers only 20 to 25 mm long. Thus the tendons are clearly the more important strain energy stores.

These arguments depended on the results of physiological experiments on frog muscle, but Morgan, Proske, and Warren (1978) confirmed the conclusion by their experiments with wallaby gastrocnemius muscles. They showed that when the muscle exerts its maximum isometric tension the tendon stretches eight times as much (and stores eight times as much strain energy) as the muscle fibers.

Ker, Dimery, and Alexander (1986) showed that wallaby tendon, like tendon from other mammals, has excellent elastic properties. Figure 7.3 shows a record from one of our tests. The narrow hysteresis loop shows that very little of the work done stretching the tendon was dissipated: Nearly all of it was returned in the elastic recoil. We calculated from the results of the tests that 33% of the negative and positive work required during the stance phase in slow hopping is performed by tendons that are stretched and then recoil. The remaining 67% has to be performed (almost entirely inelastically) by muscles. At higher speeds, however, the tendons would do a bigger percentage of the work.

Similar studies of donkeys and deer indicate that tendon elasticity gives very large energy savings in fast running. Ungulates like these seem much more specialized than wallabies to exploit tendon elasticity. Some of the distal leg muscles have muscle fibers so short that their length changes cannot cause significant joint movement. In an extreme case, the plantaris tendon of a small horse is about 900 mm long, but its muscle fibers are vestiges only 1 to 2 mm long (Dimery, Alexander, & Ker, 1986). The plantaris as a whole stretches by about 50 mm in galloping, but this must be almost entirely due to stretching of the tendon with very little contribution from the muscle fibers. Such a plantaris presumably functions with hardly any demand for metabolic energy.

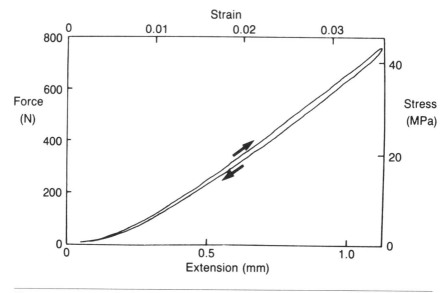

Figure 7.3. Record of a dynamic tensile test on a short section of a wallaby gastrocnemius tendon, at a frequency of 2.2 Hz. *Note.* From *Elastic Mechanisms of Animal Movement* (p. 103) by R.McN. Alexander, 1988, Cambridge: Cambridge University Press. Copyright 1988 by Cambridge University Press. Reprinted by permission.

Implications

Tendon elasticity must be important in human running, and recent experiments show that the elasticity of ligaments in the arch of the foot is also important (Ker, Bennett, Bibby, Kester, & Alexander, 1987). Measurements of oxygen consumption suggest that elastic mechanisms save about 50% of the metabolic energy that would otherwise be needed (Cavagna, Saibene, & Margaria, 1964). This is better than the savings calculated from mechanical tests for wallabies hopping slowly, but even larger savings can be calculated from the oxygen consumptions of kangaroos hopping fast (Cavagna, Heglund, & Taylor, 1977).

People have relatively long muscle fibers in the leg muscles that are likely to be important as strain energy stores. For example, an adult man had muscle fibers 36 to 65 mm long in his gastrocnemius and soleus muscles (Alexander & Vernon, 1975b). In this respect we resemble wallabies rather than ungulates. We are relatively unspecialized for saving energy by elastic mechanisms.

Padded Feet

Runners use shoes with soles that are padded to various degrees. Dogs and many other mammals have soft pads on their paws. Studies of paw pads may cast light on the function of shoes.

Theory

Figure 7.4a shows a simple model used by Alexander, Bennett, and Ker (1986) in a discussion of paw pads. It represents the body of a biped and either one leg (if it is running) or the pair of legs (if it is hopping, like a kangaroo). The body has mass M and the foot (or feet) has mass m. The spring of stiffness K represents the compliance of the leg, which flexes and extends (either elastically or by muscle action) while the foot is on the ground. The spring of stiffness k and the dashpot of damping constant c represent the viscoelastic properties of the paw pad.

The model is supposed to fall until the foot hits rigid ground. If there were no paw pad (no spring k or dashpot c), mass m would be brought to rest instantaneously by an infinitely large transient force. It seems likely that one of the functions of paw pads is to prevent excessive impact forces that might cause damage. With spring k and dashpot c present, however, the model behaves more or less as in Figure 7.4b. This is a schematic graph of the force on the ground against time. The initial fluctuations of force are due to mass m being set oscillating by the impact. The oscillations are damped by dashpot c. The oscillating foot may leave the ground briefly (indicated by f_{min} reaching zero). Indeed, it may leave the ground several times before settling for the duration of the step. Alexander, Bennett, and Ker (1986) described such behavior as "chattering."

Chattering would presumably be disadvantageous, because the foot would be apt to shift its position on the ground. Alexander, Bennett, and Ker (1986) used their model to discover the conditions needed to prevent chattering. A crude argument gave the condition

$$k/K < 5 \; \dot{Y}_o/e\dot{y}_o \qquad (7.4)$$

where $-\dot{Y}_o$ and $-\dot{y}_o$ are the downward velocities of M and m at the instant of impact, and e is the coefficient of restitution for the impact of the paw with the ground (a function of c). More rigorous investigation by numerical simulation showed that Equation 7.4 is reasonably accurate, provided e is not too low (i.e., the pad is not too heavily damped). The condition says that for given velocities at impact, the paw stiffness must not be too much larger than the leg stiffness unless the pad is severely damped.

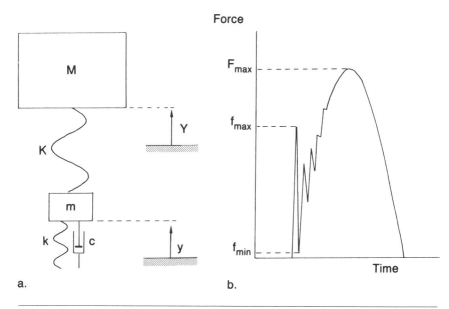

Figure 7.4. (a) The model used in the discussion of paw pads. (b) A schematic graph of force against time, showing force that the model might exert on the ground. *Note*. From "Mechanical Properties and Function of the Paw Pads of Some Mammals" by R.McN. Alexander, M.B. Bennet, and R.F. Ker, 1986, *Journal of Zoology, London, A,* **209**, p. 407. Copyright 1986 by The Zoological Society of London. Reprinted by permission of The Zoological Society of London.

Experiments

Paw pads dissected from dogs, a wallaby, and other mammals were mounted between flat steel surfaces in a servohydraulic dynamic testing machine. They were subjected to dynamic compressive tests at frequencies up to 11 Hz to discover their viscoelastic properties. Peak loads were approximately equal to those that would act in running. A typical record is shown in Figure 7.5. The stiffness (indicated by the gradient of the loop) increases with increasing load. This is as expected because no load, however great, can reduce pad thickness to zero. The area of the loop indicates moderate damping. Stiffness and energy dissipation changed little with frequency over the range tested. Force platform records of dogs running and of the wallaby hopping show components of frequency up to about 100 Hz.

The velocities of paws were measured from films of running dogs and a hopping kangaroo. The vertical component of foot velocity of the kangaroo was -1.4 to -3.1 m $\cdot$ s^{-1} just before hitting the ground. At the

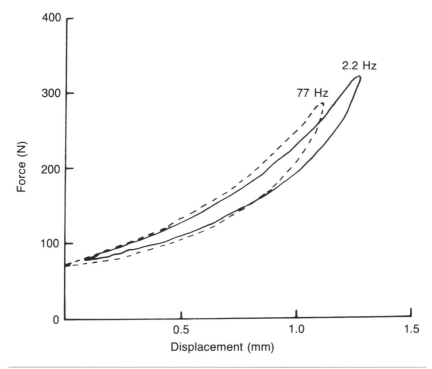

Figure 7.5. Record of dynamic compressive tests on a dog metatarsal pad, at two frequencies. *Note.* From *Elastic Mechanisms of Animal Movement* (p. 103) by R.McN. Alexander, 1988, Cambridge: Cambridge University Press. Copyright 1988 by Cambridge University Press. Reprinted by permission.

same time, the vertical component of velocity of a point on the trunk close to the center of mass was -1.1 to -1.4 m • s^{-1}. Thus the ratio $\dot{Y}_o/\dot{y}_o$ that occurs in the condition for no chattering (Condition [7.4]) was sometimes less than 0.5.

Force platform records of dogs running and a wallaby hopping were examined. The oscillations in the vertical component of force were quite small in every case: f_{max} (Figure 7.4b) was always smaller than F_{max}. Also, f_{min} was always greater than zero, indicating that chatter was not occurring.

Simulations of the behavior of the model (Figure 7.4a) were run, using values of M, m, K, Y_0, and y_0 that seemed to be realistic for dogs and kangaroos. (The value of K was chosen to give a realistic duration of contact with the ground.) Various values of the paw properties k and c were tried. To prevent chatter, it was necessary either to make k much smaller than the measured stiffness of pads or to make c much larger than the values suggested by the tests. This makes it difficult to explain why real paws do not chatter.

Discussion

There seem to be three possible explanations. First, real ground is not rigid, as assumed in the model. However, modern force platforms are exceedingly stiff, and animals do not chatter on them. Second, dogs and kangaroos have digital pads on their toes as well as metacarpal or metatarsal pads further back on the foot. The digital pads are set down first. This increases the distance over which the foot is decelerated, reducing its effective stiffness and making chatter less likely. Similarly, human runners who land on their toes generate smaller impact forces (f_{max}, Figure 7.4b) than those who land on their heels (Cavanagh & Lafortune, 1980). Third, the model assumes that spring K is initially unstrained, but electromyograph records suggest that dogs may start developing tension in the antigravity muscles before the foot hits the ground.

This work suggests a new way of thinking about running shoes, but its relevance and significance for human running remain uncertain.

Conclusion

I have discussed stride length, muscle stresses, tendon elasticity, and paw pads. In each case, data from nonhuman mammals illuminate human running. In the study of stride lengths, the diversity and size range of mammals made it possible to establish a rule of wide applicability. Knowledge of human muscle stresses is poor because anatomical data are not available for cadavers that were healthy immediately before death; the data for greyhounds are probably more reliable. Humans seem much less well adapted than many other mammals to save energy by tendon elasticity, and experimental material is easier to obtain from other mammals. Finally, the study of paw pads suggests thoughts about running shoes.

References

Alexander, R.McN. (1981). Mechanics of tendon and skeleton. In V.B. Brooks (Ed.). *Handbook of physiology—the nervous system* (2nd ed., Vol. 2) (pp. 17-42). Bethesda, MD: American Physiological Society.

Alexander, R.McN. (1984). Elastic energy stores in running vertebrates. *American Zoologist*, **24**, 85-94.

Alexander, R.McN. (1988). *Elastic mechanisms in animal movement.* Cambridge: Cambridge University.

Alexander, R.McN., & Bennet-Clark, H.C. (1977). Storage of elastic strain energy in muscle and other tissues. *Nature*, **265**, 114-117.

Alexander, R.McN., Bennett, M.B., & Ker, R.F. (1986). Mechanical properties and function of the paw pads of some mammals. *Journal of Zoology, London A*, **209**, 405-419.

Alexander, R.McN., & Jayes, A.S. (1983). A dynamic similarity hypothesis for the gaits of quadrupedal mammals. *Journal of Zoology, London*, **201**, 135-152.

Alexander, R.McN., & Maloiy, G.M.O. (1984). Stride lengths and stride frequencies of primates. *Journal of Zoology, London*, **202**, 577-582.

Alexander, R.McN., Maloiy, G.M.O., Hunter, B., Jayes, A.S., & Nturibi, J. (1979). Mechanical stresses in fast locomotion of buffalo (*Syncerus caffer*) and elephant (*Loxodonta africana*). *Journal of Zoology, London*, **189**, 135-144.

Alexander, R.McN., & Vernon, A. (1975a). Mechanics of hopping by kangaroos (Macropodidae). *Journal of Zoology, London*, **177**, 265-303.

Alexander, R.McN., & Vernon, A. (1975b). The dimensions of knee and ankle muscles and the forces they exert. *Journal of Human Movement Studies*, **1**, 115-123.

Cavagna, G.A., Heglund, N.C., & Taylor, C.R. (1977). Mechanical work in terrestrial locomotion: Two basic mechanisms for minimizing energy expenditure. *American Journal of Physiology*, **233**(5), R243-R261.

Cavagna, G.A., Saibene, F.P., & Margaria, R. (1964). Mechanical work in running. *Journal of Applied Physiology*, **19**, 249-256.

Cavanagh, P.R., & Lafortune, M.A. (1980). Ground reaction forces in distance running. *Journal of Biomechanics*, **13**, 397-406.

Close, R.I. (1972). Dynamic properties of mammalian skeletal muscles. *Physiological Reviews*, **52**, 129-197.

Dimery, N.J., Alexander, R.McN., & Ker, R.F. (1986). Elastic extension of leg tendons in the locomotion of horses (*Equus caballus*). *Journal of Zoology, London A*, **210**, 415-425.

Jayes, A.S., & Alexander, R.McN. (1982). Estimates of mechanical stresses in leg muscles of galloping greyhounds (*Canis familiaris*). *Journal of Zoology, London*, **198**, 315-328.

Ker, R.F., Bennett, M.B., Bibby, S.R., Kester, R.C., & Alexander, R.McN. (1987). The spring in the arch of the human foot. *Nature*, **325**, 147-149.

Ker, R.F., Dimery, N.J., & Alexander, R.McN. (1986). The role of tendon elasticity in hopping in a wallaby (*Macropus rufogriseus*). *Journal of Zoology, London A*, **208**, 417-428.

Morgan, D.L., Proske, U., & Warren, D. (1978). Measurements of muscle stiffness and the mechanism of elastic storage of energy in hopping kangaroos. *Journal of Physiology*, **282**, 253-261.

Zarrugh, M.Y., & Radcliffe, C.W. (1978). Predicting metabolic cost of level walking. *European Journal of Applied Physiology*, **38**, 215-223.

Chapter 8

Ground Reaction Forces in Distance Running

Doris I. Miller

Over the last 15 to 20 years, ground reaction force (GRF) patterns associated with running have been reported in the literature. Early work was limited with regard to the number of subjects and speed ranges studied because investigators did not have access to the on-line data collection systems now available. In addition, the force platforms utilized were somewhat crude and tended to have low natural frequencies that were excited by the runner's footfall (e.g., Payne, 1968; Payne, Slater, & Telford, 1968). Although the initial outlay currently required for a commercial force platform and computerized on-line data acquisition system is substantial, once in place and supported by appropriate software, GRF data collection and analysis can be reasonably straightforward—at least when compared to the labor-intensive methodology associated with film analysis.

Perhaps as a result of the relative ease with which GRF-time histories can be collected and analyzed, some attempts have been made to answer research questions based on GRF data when these data are incapable of providing the type of answers sought. For example, efforts to deduce the nature and extent of pronation following footstrike, to identify potentially injurious running patterns, or to provide a definitive assessment of the adequacy of sport shoe construction on the basis of GRF records alone will most likely be frustrated. GRF has also been subject to other misconceptions that continually need to be dispelled. Examples are that ground reaction is solely the result of the action of the lower extremity musculature of the support limb, particularly ankle plantar flexors and knee extensors; that the center of pressure completely defines pressure distribution along the sole of the foot; and that center of pressure and the point of intersection of the line of gravity with the support surface are synonymous.

Consequently, it is important to recognize what GRF data can and cannot tell us about the biomechanical basis of running. Information in the current chapter is presented toward this end.

Theoretical Basis

Ground reaction, as the name suggests, is the force that reacts to the push *transmitted* to the ground by the foot of the runner. In accordance with Newton's third law, it is equal in magnitude and opposite in direction to the "action" force or push. Like weight and most other contact forces, ground reaction is actually a distributed force that acts over the entire contact surface (i.e., that part of the foot or shoe in contact with the running surface). As a force, ground reaction is a vector quantity that can most conveniently be defined in terms of the magnitude, direction, and point of application of its resultant.

For the purpose of analysis, it is simplest to decompose the resultant GRF into three orthogonal components that have functional significance in running. The directions of these components are vertical, backward-forward (termed braking-propulsion), and side-to-side (termed medial-lateral). By definition, these directions are at right angles to one another. This convention assumes that the individual is running on a horizontal surface. If this is true, the two horizontal components comprise the friction or shear force that opposes the potential (or impending) motion between the runner's foot and the running surface. It is important not to lose sight of the fact that the vertical, backward-forward, and medial-lateral GRF are three components of a single force that changes in magnitude, direction, and point of application during the course of support.

When one examines the free body diagram of the runner during support and writes the corresponding equations of motion (Figure 8.1), it is evident that GRF reflects the acceleration of the total body center of gravity (CG). Further, all segments of the body contribute to the total body acceleration in proportion to the acceleration of their own centers of gravity (cg) and to their relative masses. Thus, the linear acceleration of the cg of the head and trunk accounts for slightly more than half of the runner's acceleration whereas each upper extremity contributes about 5% and each lower extremity approximately 16% to 18%. This relationship indicates why it is incorrect to attribute the entire GRF-time history to the action of the foot-ankle-leg of the support limb. Rather, GRF reflects the acceleration of the total body and thereby the individual accelerations of the segments each contributing in direct proportion to its mass. The support limb simply transmits the force to the ground and, theoretically, need not make any contribution to the push. This can be demonstrated by standing on a force platform and simply swinging the arms. Even though the action is confined to the upper body, a change in the GRF is elicited. In running, however, the support limb does make a significant contribution to GRF (Ae, Miyashita, Shibukawa, Yokoi, & Hashihara, 1985), but it does not account for all of it. Herein lies the basis for some of the difficulties

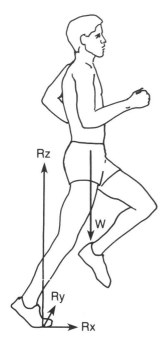

Equations of motion

$$\sum_{i=1}^{n} F = m \, a$$
$$R - W = m \, a$$
$$R = m \, a + W$$
$$R = \Sigma \, m_i \, a_i + W$$
$$\int F \, dt = \Delta \, m \, v$$
$$\int (R - W) \, dt = \Delta \, m \, v$$
$$\int R \, dt = m \, (v_f - v_i) + \int W \, dt$$

Where:

R is the ground reaction force
W is the body weight
m is the body mass
i represents each of n segments comprising the body
a is the acceleration of the center of gravity of the total body
v is the velocity of the center of gravity of the total body
t is the stance time

Figure 8.1. Free body diagram and associated equations of motion of a runner during stance.

encountered by those who attempt to directly relate ground reaction force patterns to differences in running shoe construction and/or lower extremity overuse injuries due to running.

Methodology and Instrumentation

Ground reaction force is most accurately measured with one of the proven force platforms available commercially (e.g., Kistler, which utilizes piezoelectric transducers; and AMTI, which uses strain gauge technology).

Early attempts to custom design force platforms resulted in inordinate amounts of time being devoted to design, construction, and calibration. Even then, the finished products generally did not produce satisfactory outputs in terms of linearity and natural frequency. Consequently, most would now agree that, if the data are going to be used for research, the relatively large financial outlay associated with the commercial models is justified.

In assessing the instrumentation, it is important to verify the natural frequency of the platform. It should be considerably higher (i.e., stiffer) than the highest frequency to be measured in the biological signal. Antonsson and Mann (1985) have described a simple method for doing this by dropping a steel ball onto the surface of the platform and measuring the "ringing." Calibration of the force platform, especially in the two horizontal directions, is not an easy task. Two indirect methods, however, can be readily used to check the adequacy of mounting and factory calibration. In the first case, outputs from the horizontal amplifiers are set at a very sensitive level. A static load is then placed on the platform. If the platform is mounted level, as it should be, the force outputs in the two horizontal directions will remain zero. Second, a point load can be applied at known locations on the platform surface and the calculated center of pressure coordinates compared with those where the load was actually applied.

Today most force platforms used in research are connected on-line to a mini- or a microcomputer. The signals from the force transducers, which are mounted in the corners of the platform, are each sampled in turn (i.e., multiplexed) at a given rate (usually 500 Hz) utilizing a custom-designed assembly language routine. Care must be taken to verify that the sampling rate is as high as assumed. This is particularly important when implementing a negative time feature (i.e., storing 20 values and then checking the magnitude of the vertical output. If it exceeds a given "trigger" level signifying that the runner is on the platform, then a designated number of force values are saved and the original 20 are appended to the beginning of the file. If not, another 20 values are stored, refreshing the previous 20, and the check is again run). It is also important to have a minimum of a 12-bit nonintegrating A/D converter. The latter provides 4,096 (2^{12}) divisions within the measurement range (i.e., 409.6 divisions per volt if a 10-volt range is being measured). Anything less will not provide sufficient sensitivity (Ford, 1985; Wyss, 1984).

Because almost all ground reaction force characteristics are running speed–dependent, in the process of data collection it is essential to record the running speed. This is usually accomplished by having the subject break light beams on either side of the platform. It is also important to have someone carefully observe the performance to determine whether the trial is "good" (i.e., representative of the normal running pattern of the individual). In addition, the trials should be recorded on videotape if film or high-speed video data are not being collected simultaneously. Qualitative analysis of the video records can be used subsequently to verify which trials should be retained and which discarded.

There are a number of obvious limitations in using the force platform to collect data on running. The relatively small size of the platform itself (0.6 to 0.9 m in the running direction) often makes it difficult for the runner to contact its surface without altering the running stride. This becomes even more of a problem as running speed increases. Camouflaging the location of the platform by covering the entire running surface may be one solution to obtaining more representative GRF records, but it will also increase the number of "missed" trials to the extent that perhaps only one in five would be "good." In addition to the small size of the platform, many laboratory areas in which the platform is installed place unnatural constraints on the runner. The result is that most speeds tested are necessarily limited to those achieved by distance runners. Further, it is possible to collect GRF data for only a single support phase for each running trial. Having multiple force platforms would not necessarily solve this problem, as accurate placement for consecutive footfalls would be extremely difficult to achieve.

Because of these limitations associated with force platform instrumentation, efforts have been made to install force transducers within the shoe itself or to attach them to the sole. To date, these attempts have met with limited success when applied to running (Harrington, Lippert, & Maritz, 1978; Spolek, Day, Lippert, & Kirkpatrick, 1975). Drawbacks have been related to the difficulty in transferring the instrumentation from one shoe to another, the weight or encumbrance associated with the equipment, and/or the nature of the output in terms of linearity and completeness. Research and development efforts continue along these lines (e.g., Nakhla & King, 1985) but most applications have focused upon walking rather than running.

Attempts to predict ground reaction force on the basis of position-time data from film have largely been futile in the case of running. In particular, it has not been possible to capture the initial impact peak characterizing most vertical components, nor have the two horizontal ground reaction force-time histories been predicted with acceptable accuracy. Reasons for this have been the difficulty of obtaining sufficiently accurate center of gravity locations from all the body segments when only a single camera is used and limitations in predicting the segment inertial characteristics on the basis of existing cadaver data. The most serious problem, however, is that of noise associated with the double differentiation of experimental data even when filtering routines are employed. Coupled with this is the fact that the rapid changes in force associated with footstrike and heelstrike are not followed well by smoothing routines.

Ground Reaction Force Characteristics

Center of Pressure

The center of pressure is the point of application of the resultant GRF and its location changes during the course of stance. Output from the force platform gives X and Y coordinates of the center of pressure with respect to the center of the platform surface, which represents the origin (0, 0, 0) of the measurement reference system.

To relate center of pressure coordinates to the running stance, the position of the runner's foot on the platform must be known. Cavanagh (1978; Cavanagh & Lafortune, 1980) has suggested putting colored chalk into two small holes drilled in the sole of the shoe 70% and 90% of the heel-toe distance measured from the heel. When the subject contacts the platform, two colored dots are left on athletic tape, which is used to cover the surface. Immediately following each trial, the coordinates of these two points are measured and recorded. A somewhat different method (but accomplishing the same purpose) is described by Munro, Miller, and Fuglevand (1987) in which lightly textured, fabric-backed vinyl wallpaper is taped to the surface of the platform. Prior to a trial, the subject steps into blue powdered tempera paint and thus leaves an imprint on the wallpaper when the platform is contacted. A cardboard template of the sole of each runner's shoe is then used to mark the center heel and center toe on the wallpaper along with the corresponding trial identification number. Heel and toe coordinates can be measured following data collection. With this method, 20 to 25 trials can be recorded on a single sheet that then provides a permanent hard copy of the raw data.

To quantify the location of the center of pressure at initial foot contact, Cavanagh and Lafortune (1980) developed what they termed a footstrike index. To determine this index, a straight line is drawn from midheel to midtoe along the longitudinal axis of the footprint (Figure 8.2). A perpendicular is then dropped from the initial center of pressure point to the longitudinal axis, and the distance from the heel to this intersection is measured and expressed as a percentage of total foot length. If the intersection lies in the rear third of the foot, the runner is classified as a rearfoot striker; if in the middle third, a midfoot striker, and if in the forward third (a rather rare occurrence), the runner is termed a forefoot striker. This system of footstrike quantification has been widely adopted.

The actual location of the initial center of pressure point is not without error, however. Because the center of pressure calculation is sensitive to very small vertical ground reaction force magnitudes, the center of pressure cannot be located accurately before a certain level of vertical force is reached. Therefore there is an element of subjectivity in identifying the first "real" center of pressure coordinate. Cavanagh and Lafortune (1980)

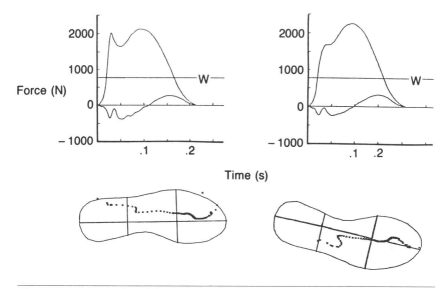

Figure 8.2. Vertical and braking-propulsion GRF components and center of pressure patterns during stance for one subject running at 4.2 m • s⁻¹.
Though a rearfoot strike with one foot and a midfoot strike with the other is unusual, somewhat lesser asymmetries than shown here are not uncommon.

have reported initial values outside the footprint record but in line with the direction the heel would be moving relative to the ground. They related this pattern to a scuffing or initial positioning of the foot. In our research (e.g., Munro et al., 1987), we have not consistently seen such a scuffing pattern.

It is important to recognize that center of pressure data provide only very restricted information on the overall pressure distribution on the sole of the foot. It is even possible to have the case in which there is no pressure acting on the foot at the center of pressure, as, for example, when equal pressure is applied to the forefoot and rearfoot but none under the midfoot region. With this fact in mind, it can be appreciated that in a so-called rearfoot or forefoot strike, the center of pressure location would in actuality reflect most of the pressure's being applied under the designated portion of the foot. By contrast, the functional significance of a midfoot strike is open to interpretation. It could indicate that the midfoot actually does contact the ground first, that the initial contact is along the entire border of the foot, or that the contact is with the rearfoot and forefoot simultaneously with no contact at all in the midfoot region.

Researchers are almost unanimous in their agreement that the initial contact with the ground for most skilled and recreational runners occurs on the lateral border of the foot. The precise location of the initial contact seems to be a function of the hardness of the midsole (Nigg, Bahlsen,

Luethi, & Stokes, 1987; Stacoff, Denoth, Kaelin, & Stuessi, 1988). The latter study concluded that for harder midsoles, the GRF was more laterally placed and that, at touchdown, the lever about the subtalar axis could be 100% greater (i.e., an increase from 5 mm to 10 mm) than for a softer midsole. It was maintained that this larger lever arm would increase initial pronation and stress the muscles on the medial side.

Because, by convention, center of pressure points are plotted at equal time increments throughout stance, points that are spread further apart signify a rapid change in center of pressure location. This is the case immediately following footstrike when the center of pressure path moves medially and initially also posteriorly or immediately anteriorly. Center of pressure points that are closer together imply a slower change. Thus, the density of center of pressure points in the forefoot region indicates that the resultant of the GRF acts in this region of the foot for the majority of the stance period. As stance progresses, the center of pressure in general migrates medially and anteriorly. It does not, however, continue right to the end of the foot prior to toe-off.

What is the influence of running speed on the center of pressure patterns and particularly on the initial contact point? To answer this question, it is necessary to examine a large number of right foot contacts and a large number of left foot contacts of individual sprint, middle distance, and distance runners over a large speed range. Sufficient information is not available as yet in published form to provide a definite answer to this question.

Aside from the footstrike index, it is extremely difficult to quantify center of pressure patterns for intersubject comparisons.

Stance Time

Theoretically, stance time should be a very straightforward value to determine from ground reaction force records. It begins when the GRF time curve deviates from the zero line and terminates when the GRF curve returns to and remains at zero. The vertical component of GRF is usually selected as the basis upon which to make these determinations. Because there is a certain amount of drift or random fluctuation above and below the baseline voltage emitted by the force platform amplifiers, it is customary to designate some arbitrary value slightly above zero volts to designate when the footstrike and toe-off events occur. Munro et al. (1987) have shown that there will be a 15- to 20-ms increase in stance time if this so-called trigger value is set at 15 N rather than at 50 N.

It is well established that stance time and running speed are negatively related. In general, the faster the speed of the run, the shorter the stance time. Munro et al. (1987) have provided reference standards for stance time based on the data of 20 adult men running at paces from 2.5 to 5.5

m • s⁻¹ across a Kistler Z4582 platform that was 0.9 m in length in the running direction. These and other standards developed in the aforementioned study (Table 8.1) are intended to provide a frame of reference for GRF data, taking the running speed factor into account.

Vertical Ground Reaction Component

Of all the GRF characteristics, the vertical component has received the greatest attention from a research standpoint. Because of its magnitude, it dominates the resultant GRF and can barely be distinguished from that of the resultant if one is superimposed on the other. Therefore, if one were to be chosen to characterize GRF, it would be the vertical. In addition, its force-time history appears to be more straightforward than that of the other two components and hence is easier to quantify for comparative purposes. Characteristics such as loading rate, impact peak, relative minimum, thrust maximum, decay rate, impulse, average vertical GRF, and change in vertical velocity of the CG can be readily quantified and have a functional relationship to the performance.

Loading Rate. Because of the common absence of an impact peak in midfoot strikers (Cavanagh & Lafortune, 1980) and in special populations such as below-knee amputees (Miller, Enoka, McCulloch, Burgess, & Frankel, 1981), quantification of the initial part of the vertical GRF curve may be effectively characterized by the loading rate. The absolute units for this variable, N • s⁻¹, can be normalized to BW • s⁻¹ (body weights per second) to facilitate comparison across runners. Because the foot may still be undergoing some positioning at the very beginning of stance, it is wise to delay the start of this computation until the vertical GRF reaches a magnitude of approximately 50 N and then to make the computation over the time it takes to increase a further body weight. The selection of a 1 BW change over which to calculate the loading rate is, of course, arbitrary. Other force ranges in this initial, sharply rising portion of the curve could be used equally well.

The literature (Clarke, Frederick, & Cooper, 1983a, 1983b; Dickinson, Cook, & Leinhardt, 1985; Snel, Delleman, Heerkens, & van Ingen Schenau, 1985) has shown that for running speeds between 4.0 and 4.5 m • s⁻¹ the relative degree of shoe hardness is negatively related to the rise time of the initial portion of the vertical GRF curve and thus positively to the loading rate (i.e., the harder the sole, the higher the loading rate). A study that is in disagreement with these findings is that of Nigg and Bahlsen (1988). Their somewhat counterintuitive findings were that shoes with the hardest midsoles showed the lowest maximal vertical loading rate whereas softer midsoles were associated with the highest rates. They also showed a relationship between maximal loading rate and the

Table 8.1 Ground Reaction Force Reference Standards

Speed (m · s⁻¹)	Stance time (ms)	Loading rate (BW/s)	Impact maximum (BW)	Relative minimum (BW)	Thrust maximum (BW)	Average vertical GRF (BW)	Decay rate (BW/s)	Change in vertical velocity (m · s⁻¹)	Braking impulse (BWI)	Propulsive impulse (BWI)
3.00	270	77.2	1.57	1.28	2.51	1.40	14.6	.99	−.15	.14
	(20)	(26.7)	(.35)	(.24)	(.21)	(.11)	(2.1)	(.21)	(.05)	(.01)
3.25	258	77.4	1.69	1.34	2.56	1.44	15.8	1.01	−.17	.16
	(18)	(19.0)	(.21)	(.22)	(.17)	(.08)	(1.8)	(.46)	(.03)	(.01)
3.50	247	80.0	1.76	1.40	2.62	1.49	16.9	1.07	−.18	.17
	(17)	(16.9)	(.19)	(.22)	(.16)	(.08)	(1.8)	(.51)	(.02)	(.01)
3.75	238	84.6	1.86	1.46	2.67	1.53	18.0	1.15	−.20	.19
	(15)	(17.1)	(.20)	(.23)	(.16)	(.09)	(1.8)	(.42)	(.02)	(.01)
4.00	229	90.5	1.95	1.52	2.72	1.57	19.2	1.23	−.21	.20
	(14)	(18.3)	(.21)	(.24)	(.17)	(.09)	(1.7)	(.28)	(.02)	(.01)
4.25	221	97.1	2.05	1.57	2.76	1.61	20.3	1.31	−.23	.21
	(13)	(20.3)	(.23)	(.25)	(.17)	(.09)	(1.6)	(.16)	(.02)	(.02)
4.50	214	103.6	2.15	1.63	2.79	1.65	21.5	1.38	−.24	.23
	(13)	(23.0)	(.25)	(.26)	(.18)	(.09)	(1.7)	(.17)	(.02)	(.02)
4.75	206	109.2	2.25	1.69	2.81	1.68	22.7	1.43	−.25	.24
	(13)	(26.7)	(.27)	(.27)	(.18)	(.08)	(1.7)	(.26)	(.03)	(.02)
5.00	199	113.0	2.32	1.75	2.83	1.70	23.9	1.47	−.25	.25
	(13)	(29.4)	(.28)	(.27)	(.17)	(.08)	(1.9)	(.40)	(.03)	(.02)

Note. SD indicated in parentheses. Adapted with permission from *Journal of Biomechanics*, 1987, **20**, by C.F. Munro, D.I. Miller, and A.J. Fuglevand, "Ground Reaction Forces in Running: A Reexamination," pp. 149 and 153, Copyright 1987, Pergamon Press.

type of heel flare. Nigg et al. (1987) had previously shown that impact load rate was dependent on running velocity—a finding in agreement with Munro et al. (1987), who also demonstrated that loading rate is positively related to running speed, increasing from an average of 77 BW • s⁻¹ at 3.0 m • s⁻¹ to 113 BW • s⁻¹ at 5.0 m • s⁻¹. No attempt was made to control for running shoe type in the latter study.

Intraindividual differences in loading rate would be particularly important to monitor in specific cases such as during a runner's recovery from a lower extremity injury. It would seem logical to expect it to increase in magnitude as healing progressed.

Impact Peak.   Particular interest in the impact force has been motivated by concern about the transmission of shock waves upward through the musculoskeletal system. It has been hypothesized that there may be a positive relationship between the magnitude of the impact force and overuse injuries in running, degenerative changes in joints, and low back pain.

Although GRF-time histories are associated with a certain amount of individual variability, one reasonably characteristic configuration for the vertical component has been reported for most rearfoot strikers and another for most midfoot strikers (Figure 8.2). Those who initially contact the ground with the rear portion of the foot tend to elicit a high force of short duration that has been termed an *impact peak*. For running speeds in the range of 3 to 6 m • s⁻¹, it is usually between 2 and 3 BW in magnitude and has been attributed to the passive characteristics of the contacting limb (Nigg, 1983; Nigg, Denoth, & Neukomm, 1983). For example, if a bone were simply dropped onto the platform, the reaction elicited would be a high force peak of short duration. In the vertical GRF configuration associated with midfoot strikers, only a vestige if anything remains of the impact peak. Thus, initial contact with the heel does not appear to incorporate soft tissue and linked bony segment shock absorption mechanisms to as great an extent as landing with initial contact in the midfoot or forefoot region.

It is difficult to collect comparable data on barefoot running while maintaining a heel-contact pattern, given the fact that with the exceptions of notable competitors such as Zola Budd, few individuals are accustomed to running barefoot. However, the limited data that are available do show that impact peaks are either higher when running barefoot than when wearing shoes (Dickinson et al., 1985) or are among the highest recorded (Snel et al., 1985). It would therefore seem logical to assume that the magnitude of the impact peak for heelstrikers would be in some way related to heel cushioning and shock-absorption properties of the shoe.

Attempts to detect differences among running shoes in this regard, however, have met with little success. In comparing the impact forces elicited by 10 males running at 4.5 m • s^{-1} wearing running shoes representing the ends of the spectrum in terms of hard and soft midsoles, Clarke et al. (1983a, 1983b) did not find any statistically significant difference in the magnitude of the impact peak between the two conditions. Similarly, Snel et al. (1985) were unable to identify significant differences among the impact peak magnitudes for nine different shoes varying in sole hardness. In 1987, Nigg and his colleagues recorded vertical impact forces at four running speeds with shoes of varying midsole hardnesses. They concluded that impact peaks increased linearly with running velocity but did not correlate with midsole hardness. When compared to a later study conducted in the same laboratory, this seems surprising—Nigg and Bahlsen (1988) reported that vertical impact force peaks showed significant differences for various midsole constructions (as well as for three different lateral heel flares). In the latter study the shoes with the softest midsoles showed the highest peak force for all flare conditions whereas the hardest shoes had the lowest peak force. With regard to the effects of insoles, however, Nigg, Hertzog, and Read (1988) reported no significant difference in maximum vertical impact force for viscoelastic and regular insoles. It must be borne in mind that it is not only the nature of the shoe that determines the characteristics of the impact force but also the technique of running. Nigg (1986) has suggested that the impact peak will be reduced as the knee becomes more flexed at touchdown.

An interesting observation with below-knee amputees, all of whom were rearfoot strikers, would also seem relevant in this context. When they were just beginning to relearn this skill, the vertical GRF elicited during ground contact with the amputated limb fitted with a prosthesis lacked the typical heelstrike impact force. The latter, however, was evident in the GRF records of the intact limb. As the amputees became more comfortable with running and after they had accumulated considerable mileage, an impact peak began to appear in the GRF patterns associated with stance on the prosthesis. This was taken to be a positive sign of progress.

Thrust Maximum.   For most midfoot strikers, the vertical GRF curve rises more or less directly to the thrust peak. For rearfoot strikers running at slow to moderate speeds, the thrust maximum is usually the second peak in their vertical GRF record and is less than that recorded at impact. Munro et al. (1987) found that it generally occurred between 35% and 50% of the total stance time, and its magnitude increased from 2.5 to 2.8 BW as running speed increased from 3.0 to 5.0 m • s^{-1}. These data were in agreement with Roy (1982), who presented mean thrust peaks ranging from 2.6 to 3.0 BW for 20 subjects running at speeds between 3.5 and 5.4 m • s^{-1}. Similar values have been reported by Hamill, Bates, Knutzen, and Sawhill (1983) (2.8 to 2.9 BW for running speeds of 4.0 and 5.0 m • s^{-1}) and by

Cavanagh and Lafortune (1980) (2.8 and 2.7 BW at 44% and 43% stance for rear and midfoot strikers respectively running at approximately 4.5 m • s⁻¹).

Decay Rate. Following the thrust peak, the vertical GRF drops to zero. The decay rate is higher for the faster running speeds than for the slower ones (Table 8.1) and is smaller than the loading rate by a factor of approximately 5. In cases in which the decay rate is protracted, resulting in an extended period prior to toe-off when the vertical GRF is abnormally low (e.g., for below-knee amputees), there may be a problem with slipping. This situation occurs because the low normal reaction (vertical GRF when running on a level surface) reduces the maximum static friction force that can be supported, and this condition coincides temporally with the propulsion phase during which the foot is pushing backward against the running surface. In such instances, an individual would be well advised to avoid running on smooth surfaces such as gymnasium floors.

Average Vertical GRF. If the vertical GRF had to be described by a single variable, the most meaningful selection would be the average vertical GRF exerted throughout the entire stance phase. Because it is an extremely stable indicant (unlike impact peak magnitude, for example), the average vertical GRF can be used to monitor treatment or training programs that result in changes in the vertical acceleration of the total body center of gravity. In the study of Munro et al. (1987), the average vertical GRF increased significantly from 1.40 BW at a running speed of 3.0 m • s⁻¹ to 1.70 BW at 5.0 m • s⁻¹ (Figure 8.3).

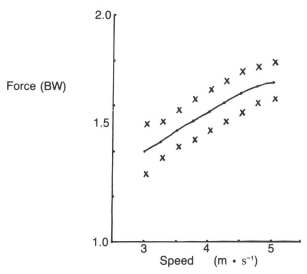

Figure 8.3. Mean and standard deviation (X) of average vertical ground reaction force expressed as a function of running speed.

From the linear impulse–linear momentum relationship (Figure 8.1), the role played by the average vertical GRF in changing the velocity of the runner's CG can be appreciated. It must be emphasized, however, that from GRF records alone, it is not possible to determine the runner's velocity at initial or final contact but only the change in velocity that occurs during the stance period. Change in vertical velocity provides information on a runner's ability to reverse the downward velocity of the CG at initial ground contact to an upward velocity at final contact with the ground just preceding the flight. Holden (1984) reported a change of 1.55 m • s^{-1} for 10 males running at 4.5 m • s^{-1}. This variable has also been shown to increase from 1.0 to 1.5 m • s^{-1} over the running speed range of 3.0 to 5.0 m • s^{-1} (Munro et al., 1987), which is not unexpected given the positive relationship between the average vertical GRF and running speed.

Braking-Propulsion Ground Reaction Component

The posterior-anterior GRF elicited during running is predominantly biphasic (Figure 8.2). During the initial phase (termed braking), the GRF direction opposes forward motion. During the latter phase (termed propulsion), its direction is consistent with forward motion. Although the force-time configuration is reasonably consistent for a given runner, the pattern of the braking force varies among individuals. It exhibits single, double, or multiple peaks that do not appear linked in any simple way to footstrike classification as was previously believed (Cavanagh & Lafortune, 1980; Hamill et al., 1983; Munro et al., 1987; Payne, 1983). In a number of instances, the braking component remains near zero for a brief period before displaying a substantial negative deflection (e.g., Figure 8.4). This pattern, which is particularly discernible in GRF records of below-knee amputee runners, may reflect the time during which there is a slight forward movement of the foot within the shoe associated with initial contact.

Maximum Braking and Propulsion Forces.   The magnitudes of the maximum braking and propulsion force, like most other GRF characteristics, are running-speed–dependent. The published literature is equivocal on whether the two maxima are nearly equal or whether one exceeds the other. This may be a reflection of runners' accelerating or decelerating while in contact with the platform. If the runner were accelerating, one would anticipate the propulsive impulse (and likely but not necessarily the peak) to be the larger of the two. Mason (1980), who tested 24 skilled runners, reported average braking and propulsive maxima of 38% and 50% BW at 4.7 m • s^{-1} and 70% and 75% BW at 7.6 m • s^{-1} respectively. For speeds of 4.5 m • s^{-1}, Cavanagh and Lafortune (1980) found braking maxima averaging 43% and 45% BW for rearfoot and midfoot strikers and propulsion maxima of 50% BW at 4.9 m • s^{-1}. For the lower running speed

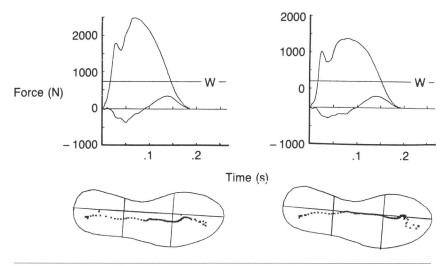

Force (N)

Time (s)

Figure 8.4. Ground reaction force–time histories of two different subjects running at 5.0 m • s⁻¹.

range, Munro (1984) recorded values of 25% to 30% BW at 3.0 m • s⁻¹ to approximately 45% to 55% BW at 5.0 m • s⁻¹ for both braking and propulsion.

The point at which maximum braking occurs varies with the force-time configuration. For individuals who tend to elicit a single braking peak, maximum braking occurs at approximately 22% stance (Cavanagh & Lafortune, 1980; Hamill, Bates, & Knutzen, 1984; Hamill, Bates, Knutzen, & Sawhill, 1983). The time of occurrence of the maximum propulsive force seems to fall consistently within 70% to 76% of stance over the speed ranges reported in the literature.

Zero Fore-Aft Shear. Although theoretically there are three points of zero fore-aft shear in the backward-forward GRF-time history, this term is usually reserved for the transition between braking and propulsion that customarily occurs slightly before midstance. Both Cavanagh and Lafortune (1980) and Munro et al. (1987) have reported a value of 48% across a range of roughly 3.5 to 5.0 m • s⁻¹. Mason (1980) indicated transition occurred at 46% and 44% of stance for speeds of 4.7 and 7.6 m • s⁻¹ respectively, and Hamill et al. (1983) reported values from 50% to 43% over the range of 4.0 to 7.0 m • s⁻¹. Both Roy (1982) and Hamill et al. (1984) cited values slightly in excess of 50%. Roy's subjects were running at between 3.4 and 5.4 m • s⁻¹ whereas 51% of Hamill's subjects had a pace of 4.9 m • s⁻¹. Although overall there appears to be a slight negative relationship between running speed and the time of transition, the trend may be somewhat accentuated by the possibility that subjects running in excess of 6.0 m • s⁻¹ may not have attained their maximum speed by the time they reached the platform and would therefore be spending a greater proportion of their time in propulsion (acceleration) than would be the case had they been running at a constant speed.

Impulses.   The relative magnitudes of the braking and propulsive impulses for a given trial can serve as an objective measure for verifying whether a runner has met a *constant velocity* criterion. For so-called constant velocity running averaged over a single step, it is necessary for the forward and backward impulses exerted by external forces upon the runner to be equal in magnitude; otherwise momentum (and thus velocity) in the running direction will change from step to step. In most instances, the negative impulse of the air resistance operative throughout both stance and flight phases of the step is disregarded upon the assumption that, within the controlled laboratory conditions under which GRF data are usually collected, it is negligible with respect to the other external impulses acting upon the runner. If this assumption holds, then the so-called braking and propulsive impulses of the horizontal GRF should be nearly equal in magnitude. If, however, air resistance is a significant influence (cf. Ward-Smith, 1985a, 1985b) then one would expect the magnitude of the propulsive impulse to exceed that of the braking impulse to provide necessary compensation.

Horizontal GRF impulses in N • s can be normalized by dividing each by the impulse of the individual's body weight over the entire stance time, yielding units of body weight impulse (BWI). In the Munro et al. study (1987), both braking and propulsive impulses increased as a function of running speed from approximately 0.15 BWI at 3.0 m • s^{-1} to 0.25 BWI at 5.0 m • s^{-1}.

Medial-Lateral Ground Reaction Component

Although it has been possible to identify readily quantifiable characteristics of the braking-propulsion and vertical GRF components, this has proven much more difficult in the case of the medial-lateral or side-to-side GRF component elicited during running stance. Variables most commonly reported are the overall maximum, maximum medial, maximum lateral, maximum peak-to-peak, and average medial-lateral GRF; medial, lateral, and resultant GRF impulse; and change in the velocity of the CG in a medial-lateral direction.

Magnitude.   The maximum medial-lateral GRF is the smallest of the three components, and virtually all researchers agree that it falls in the neighborhood of 0.10 to 0.20 BW (Bates, Osternig, Sawhill, & James, 1983; Cavanagh & Lafortune, 1980; Hamill et al., 1983, 1984). As evident from the force-time histories published in the literature and also from Figure 8.5, the average medial-lateral force is considerably smaller than the peak value recorded.

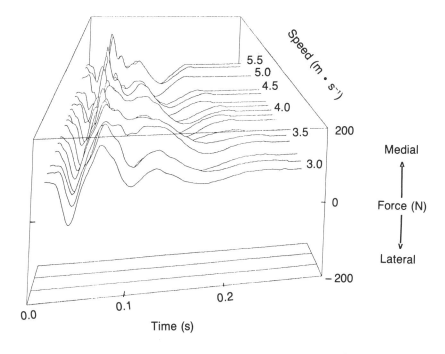

Figure 8.5. Medial-lateral component of ground reaction force elicited during left foot stance of a single subject illustrating similarity in the force-time configurations across a range of running speeds. *Note.* Adapted with permission from *Journal of Biomechanics,* 1987, **20**, by C.F. Munro, D.I. Miller, and A.J. Fuglevand, "Ground Reaction Forces in Running: A Reexamination," p. 150, Copyright 1987, Pergamon Press.

General Configuration. This component is generally characterized by high variability. Magnitudes, directions, and number of zero line crossings differ across subjects and often between feet of a given subject. In this connection, it is interesting to note that the functional significance of positive and negative on the force axis is not indicated on the figures of many published articles. The reader therefore can only guess which direction represents a medial and which a lateral reaction. Unlike walking, in which the initial reaction elicited is usually lateral and the dominant ground reaction during the latter part of stance is medial (similar to a skating push), when the records of a number of runners are combined, it is difficult to discern any consistent pattern. It is only when multiple force records from the right (or left) foot of a single runner are examined that consistency is apparent (Figure 8.5). Some runners initially elicit a medial reaction and others a lateral reaction. It is, in fact, during the initial 35 to 40 ms of support where the greatest variability is evident (Holden, 1984).

In recording this component, it is essential to take into consideration the direction of run over the platform as well as the foot contacting the platform. If the subject runs in a consistent direction, the configuration of the medial-lateral GRF elicited by the right foot should be *roughly* the mirror image of that elicited by the left foot (assuming no major gait asymmetries).

Impulse. If the runner is to maintain a straight-line running path, the resultant side-to-side GRF impulse elicited during right- and left-foot contacts will have to equal zero when averaged over several footfalls. Equivalent impulses can, however, be produced from markedly different force-time configurations, and thus an individual with seemingly different medial-lateral GRF-time histories may still be running in a straight path.

There is a dearth of data in the published literature on the magnitude of the change in velocity in the side-to-side direction during single foot contact. Unpublished sources (Holden, 1984; Munro, 1984), however, have provided some basic values. Holden reported the change in medial-lateral velocity to average $0.02 \text{ m} \cdot \text{s}^{-1}$ (lateral) for individuals running at $4.5 \text{ m} \cdot \text{s}^{-1}$. Of the 440 trials run by Munro's (1984) 20 subjects over a 2.5 to $5.5 \text{ m} \cdot \text{s}^{-1}$ speed range, 87% were characterized by medial-lateral velocity changes between $\pm 0.05 \text{ m} \cdot \text{s}^{-1}$ range. Half of these velocity changes were in the medial direction and half in the lateral direction for the group as a whole.

Relation to Pronation. Before this section is complete, it must be emphasized that it is *not* appropriate to attempt to discern rearfoot movement from medial-lateral ground reaction force records. First, a medial reaction force could be indicative of a decreasing lateral velocity. Second, what the ground reaction reflects is the acceleration of the total body center of gravity and not of the center of gravity of the foot or even of the support lower extremity.

Free Moment

In addition to force magnitudes in the three orthogonal directions and the coordinates of the point of application of their resultant (i.e., location of the center of pressure), it is also possible to determine from the force platform output the magnitude and direction of the free moment about a vertical axis. This moment acts in the plane parallel to the running surface and is thus termed a "free" moment. It results from the action of the shear forces that produce a transverse moment or force couple.

To date, this GRF characteristic has not yet been addressed in any comprehensive manner in the literature. Nigg (1986) has reported free moment curves for a single subject running at 3.5 m • s⁻¹, and in an unpublished study by Holden (1984), average curves for 10 subjects running at 4.5 m • s⁻¹ have been calculated. Like the medial-lateral GRF component, Holden found the free moment to exhibit within-subject repeatability for a given foot/shoe condition but diversity across foot/shoe conditions and particularly across subjects. Whether the free moment is sufficiently sensitive to provide clinically relevant data on pronation changes during running stance remains open to question.

Quo Vadis GRF???

Over the past decade, systems for effective collection, analysis, and display of GRF-time histories and characteristics have been developed. Descriptive data associated with the stance phase of distance running, at least in the lower speed range of 3 to 5 m • s⁻¹, have been published. In attempts to normalize and average, however, it becomes easy to overlook individual responses and to think only in terms of neat stereotypical GRF configurations that so often grace the pages of the published literature. For example, it is unlikely that right-left asymmetries will be found when GRF data across subjects are pooled. Within subjects, however, the picture is often a different one in which distinct bilateral asymmetries become apparent.

GRF analysis thus has the potential to be a valuable tool for individual diagnosis (Cavanagh, Andrew, Kram, Rodgers, Sanderson, & Hennig, 1985) and for examining the effect of various intervention strategies. Because GRF reflects the total body CG acceleration, it is not a particularly sensitive measure. Consequently, when statistically significant differences are noted, they are most likely to have a sound functional basis. Although the GRF records do not identify the reason(s) for asymmetries, they bear witness to their existence. Concurrent or subsequent photo or video analysis must then be undertaken to pinpoint their cause(s) and to gain insight into possible reasons why such compensatory mechanisms are being employed.

As it stands, GRF analysis provides three-dimensional data without the need to resort to the labor-intensive procedures associated with film or video analysis. This is good, provided its limitations in terms of interpretation are recognized. To achieve its full potential in the future, however, GRF analysis in clinical and field applications will have to incorporate miniaturized in-shoe GRF transducers to obtain data from a number of consecutive footfalls and to utilize three-dimensional high-speed video analysis to link the kinematics of the running strides with the kinetics as reflected in the GRF record.

References

Ae, M., Miyashita, K., Shibukawa, K., Yokoi, T., & Hashihara, Y. (1985). Body segment contributions during the support phase while running at different velocities. In D.A. Winter, R.W. Norman, R.P. Wells, K.C. Hayes, & A.E. Patla (Eds.), *Biomechanics IX-B* (pp. 343-349). Champaign, IL: Human Kinetics.

Antonsson, E.K., & Mann, R.W. (1985). The frequency content of gait. *Journal of Biomechanics*, **18**, 39-47.

Bates, B.T., Osternig, L.R., Sawhill, J.A., & James, S.L. (1983). An assessment of subject variability, subject-shoe interaction, and the evaluation of running shoes using ground reaction force data. *Journal of Biomechanics*, **16**, 181-191.

Cavanagh, P.R. (1978). A technique for averaging center of pressure paths from a force platform. *Journal of Biomechanics*, **11**, 487-491.

Cavanagh, P.R., Andrew, G.C., Kram, R., Rodgers, M.M., Sanderson, D.J., & Hennig, E.M. (1985). An approach to biomechanical profiling of elite distance runners. *International Journal of Sport Biomechanics*, **1**, 36-62.

Cavanagh, P.R., & Lafortune, M.A. (1980). Ground reaction forces in distance running. *Journal of Biomechanics*, **13**, 397-406.

Clarke, T.E., Frederick, E.C., & Cooper, L.B. (1983a). Biomechanical measurement of running shoe cushioning properties. In B.M. Nigg & B.A. Kerr (Eds.), *Biomechanical aspects of sport shoes and playing surfaces* (pp. 25-33). Calgary, AB: University of Calgary.

Clarke, T.E., Frederick, E.C., & Cooper, L.B. (1983b). Effects of shoe cushioning upon ground reaction forces in running. *International Journal of Sports Medicine*, **4**, 247-251.

Dickinson, J.A., Cook, S.D., & Leinhardt, T.M. (1985). The measurement of shock waves following heel strike while running. *Journal of Biomechanics*, **18**, 415-422.

Ford, L.E. (1985). Laboratory interfacing. *Byte*, **10**, 263-266.

Hamill, J., Bates, B.T., & Knutzen, K.M. (1984). Ground reaction force symmetry during walking and running. *Research Quarterly for Exercise and Sport*, **55**, 289-293.

Hamill, J., Bates, B.T., Knutzen, K.M., & Sawhill, J.A. (1983). Variations in ground reaction force parameters at different running speeds. *Human Movement Science*, **2**, 47-56.

Harrington, R.M., Lippert, F.G., & Maritz, W.W. (1978). A shoe for measuring foot-to-ground forces while running. In D. Jaron (Ed.), *Proceedings of the Sixth Annual New England Bioengineering Conference* (pp. 158-161).

Holden, J.P. (1984). *The free moment of ground reaction in distance running and its change with pronation*. Unpublished master of science thesis, The Pennsylvania State University, University Park.

Mason, B.R. (1980). *A kinematic and kinetic analysis of selected parameters during the support phase of running*. Unpublished doctoral dissertation, University of Oregon, Eugene.

Miller, D.I., Enoka, R.M., McCulloch, R.G., Burgess, E.M., & Frankel, V.H. (1981). Vertical ground reaction force-time histories of lower extremity amputee runners. In A. Morecki, K. Fidelus, K. Kedzior, & A. Wit (Eds.), *Biomechanics VI-A* (pp. 453-460). Baltimore: University Park Press.

Munro, C.F. (1984). *Utilization of ground reaction force standards in the analysis of below-knee amputee running*. Unpublished master of science thesis, University of Washington, Seattle.

Munro, C.F., Miller, D.I., & Fuglevand, A.J. (1987). Ground reaction forces in running: A reexamination. *Journal of Biomechanics*, **20**, 147-155.

Nakhla, S.S., & King, A.I. (1985). Ground reaction patterns of normal gait. In D.A. Winter, R.W. Norman, R.P. Wells, K.C. Hayes, & A.E. Patla (Eds.), *Biomechanics IX-A* (pp. 401-405). Champaign, IL: Human Kinetics.

Nigg, B.M. (1983). External force measurements with sport shoe and playing surfaces. In B.M. Nigg & B.A. Kerr (Eds.), *Proceedings of the International Symposium on Biomechanical Aspects of Sport Shoes and Playing Surfaces* (pp. 11-32). Calgary, AB: University of Calgary.

Nigg, B.M. (Ed). (1986). *Biomechanics of running shoes*. Champaign, IL: Human Kinetics.

Nigg, B.M., & Bahlsen, H.A. (1988). Influence of heel flare and midsole construction on pronation, supination, and impact forces for heel-toe running. *International Journal of Sport Biomechanics*, **4**, 205-219.

Nigg, B.M., Bahlsen, H.A., Luethi, L.M., & Stokes, S. (1987). The influence of running velocity and midsole hardness on external impact forces in heel-toe running. *Journal of Biomechanics*, **20**, 951-959.

Nigg, B.M., Denoth, J., & Neukomm, P.A. (1983). Quantifying the load on the human body: Problems and some possible solutions. In H. Matsui & K. Kobayashi (Eds.), *Biomechanics VIII-B* (pp. 88-99). Champaign, IL: Human Kinetics.

Nigg, B.M., Hertzog, W., & Read, L.J. (1988). Effects of viscoelastic shoe insoles on vertical impact forces in heel-toe running. *The American Journal of Sports Medicine*, **16**, 70-76.

Payne, A.H. (1968). The use of force platforms for the study of physical activity. In J. Wartenweiler, E. Jokl, & M. Hebbelinck (Eds.), *Biomechanics* (pp. 83-86). Basel: S. Karger.

Payne, A.H. (1983). Foot to ground contact forces of elite runners. In H. Matsui & K. Kobayashi (Eds.), *Biomechanics VIII-B* (pp. 746-753). Champaign, IL: Human Kinetics.

Payne, A.H., Slater, W.J., & Telford, T. (1968). The use of a force platform in the study of athletic activities. A preliminary investigation. *Ergonomics, 11,* 123-143.

Roy, B. (1982). Charactéristiques biomechaniques de la course d'endurance [Biomechanical characteristics of endurance running]. *Canadian Journal of Applied Sports Sciences, 7,* 104-115.

Snel, J.G., Delleman, N.J., Heerkens, Y.F., & van Ingen Schenau, G.J. (1985). Shock-absorbing characteristics of running shoes during actual running. In D.A. Winter, R.W. Norman, R.P. Wells, K.C. Hayes, & A.E. Patla (Eds.), *Biomechanics IX-B* (pp. 133-138). Champaign, IL: Human Kinetics.

Spolek, G.A., Day, B.E., Lippert, F.G., & Kirkpatrick, G.S. (1975, July). Ambulatory-force measurement using an instrumented-shoe system. *Experimental Mechanics, 15,* 271-274.

Stacoff, A., Denoth, J., Kaelin, X., & Stuessi, E. (1988). Running injuries and shoe construction: Some possible relationships. *International Journal of Sport Biomechanics, 4,* 342-357.

Ward-Smith, A.J. (1985a). A mathematical theory of running, based on the first law of thermodynamics, and its application to the performance of world-class athletes. *Journal of Biomechanics, 18,* 337-349.

Ward-Smith, A.J. (1985b). A mathematical analysis of the influence of adverse and favourable winds on sprinting. *Journal of Biomechanics, 18,* 351-357.

Wyss, C.R. (1984). Planning a computerized measurement system. *Byte, 9,* 114-123.

Chapter 9

Transmission and Attenuation of Heelstrike Accelerations

Gordon A. Valiant

Typical distance running speeds result in more than 300 footstrikes per leg per kilometer, often on very hard surfaces. The foot striking the ground is an example of a collision or an impact. This impact results in a transmission of a vertical shock through the body and carries with it the potential for injury.

What is Acceleration?

One of the characteristics of the foot impacting the running surface is a rapid change in the velocity of the foot. The rearfoot typically strikes the ground with velocities exceeding 1.0 m • s⁻¹ (Cavanagh, Valiant, & Misevich, 1984), and this velocity decreases to zero in a short period of time. The time required for the velocity to decrease to zero is related to the hardness of the interacting surfaces. With harder surfaces, the compression or deflection of the surface is smaller, and hence the velocity goes to zero in a shorter time. By quantifying this rate of change of velocity, the nature of the impact of the foot with the running surface can be characterized.

The time rate of change of velocity is acceleration, expressed in meters/second/second (m • s⁻²). It is often normalized to the acceleration due to gravity (g), which is approximately 9.81 m • s⁻². Impact accelerations are measured by transducers called accelerometers. These devices consist of a small mass that is supported by a stiff spring element. Whenever there is movement of the transducer, the mass is displaced and exerts a force against a sensing element. This mechanical stress results in an electrical output that is proportional to the force acting on the sensing element. The inertial mass is a known constant; therefore the electrical output is proportional to the acceleration of the mechanical linkage.

Peak accelerations during impacts often exceed several gs. For example, when an automobile is involved in a head-on crash and comes to a sudden stop, passengers inside the vehicle retain their original velocity and move forward with respect to the automobile with accelerations that can exceed 40 g. It is not uncommon in simulated crashes for the heads of unrestrained dummies to strike the windshield with recorded peak impact accelerations exceeding 100 g (Cichowski, 1963). This chapter will focus on the types of accelerations experienced by running athletes when their feet impact on the ground. The measurement of acceleration reveals various characteristics about the transmission of shock through the body, and because this shock could possibly injure the athlete, its description is important. With an increased understanding of impact forces, ways of modifying them can be developed such that the risk of injury is reduced.

Measurement of Accelerations at the Shank

Bone is the primary structure responsible for the transmission through the body of the shock that occurs when the foot strikes the ground during running. The best way to accurately quantify the transmission of shock to the leg would be to attach an accelerometer directly to the tibia, but of course this is not practical in most cases. Usually, measurements are made by externally attaching an accelerometer to the leg. This is demonstrated by Figure 9.1, which shows a lightweight (4.4 grams) accelerometer attached very tightly to the leg with an elastic strap. The sensitive axis of the unidirectional accelerometer is aligned with the long axis of the tibia. The site of attachment is the medial distal surface of the lower leg, or shank. At this site there is very little soft tissue between the rigid bone and the outer skin, and the triangular cross-sectional shape of the tibia provides a fairly flat surface against which to secure an accelerometer.

The limitations of using externally attached accelerometers to measure the transmission of an impact shock through the leg were examined by Hennig and Lafortune (1988). These researchers undertook the task of attaching an accelerometer to the free end of a Steinmann pin inserted 5 to 7 cm into the proximal tibia of a volunteer. In addition, a second 6.0-gram accelerometer was encased in balsa wood and externally strapped to the shank in a fashion similar to that shown in Figure 9.1. The axial accelerations from each transducer were recorded simultaneously during overground running. The peak acceleration measured externally by the skin-mounted accelerometer was, on average, 50% greater than the peak acceleration measured internally by the bone-mounted accelerometer. The acceleration measured by the bone-mounted accelerometer is considered the more accurate. The authors observed that immediately after heelstrike, the externally measured acceleration increased less rapidly than the internally measured acceleration. They attributed this initial delay to relative motion between skin and bone. Shortly

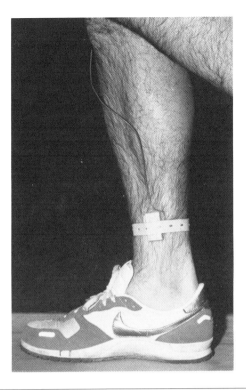

Figure 9.1.   Lightweight accelerometer encased in balsa wood and strapped to the anteromedial distal aspect of the lower leg with an elastic strap.

afterward, however, the rate of increase of the externally measured acceleration increased dramatically and resulted in a 50% overshoot in the peak acceleration. It was hypothesized that the stretched skin stiffened and an elastic recoil accounted for the acceleration overshoot.

It is not possible to routinely insert traction pins into the bones of runners, and most investigators must rely upon skin-mounted accelerometers to measure transmitted force. The 50% error reported by Hennig and Lafortune (1988) occurred when using a skin-mounted accelerometer package with a mass that exceeded 6 grams. They concluded that smaller masses would reduce the error in the externally measured acceleration. In fact, during plantar foot impacts, Valiant, McMahon, and Frederick (1987) modeled the link between the tibia and an accelerometer with a linear spring and dashpot and approximated a smaller 17% to 24% overestimation of peak acceleration when using the 4.4-gram skin-mounted accelerometer shown in Figure 9.1. Also, during barefoot vertical jump landings, Gross and Nelson (1988), using the same accelerometer/tibia model, calculated an 8% overestimation of peak tibial acceleration measured with a 1.0-gram accelerometer glued to the distal anteromedial tibia.

It is thus apparent that the signals from skin-mounted accelerometer attachments are most accurate when the mass of the accelerometer is kept small. The accuracy is also improved when the tension in the strap holding the accelerometer against the leg is as high as can be tolerated by subjects (Valiant et al., 1987). If the accelerometer is securely attached prior to any data collection and does not relocate during any trials, then within-subject comparisons can confidently be made. Between-subject comparisons cannot validly be made if the repeatability of accelerometer attachment is at all suspect.

Shank Accelerations During Barefoot Walking and Running

Figure 9.2a shows the signal from an accelerometer externally attached to the leg of a subject walking barefoot at a speed of 1.53 m • s⁻¹ on a treadmill. The signal is referenced to the accelerometer output during erect standing, which is simply +1 g. The axial acceleration rises to a peak value of 2.3 g in the first 20 ms after heelstrike. This means that the combined foot and lower leg, which have a weight of mg, are subjected to a force that is approximately equal to 2.3 mg very soon after the heel strikes the ground while walking. The acceleration signal decays to 0, and then increases toward the end of the support period as the foot is pushing against the ground prior to toe-off. This signal can be compared to the signal in Figure 9.2b, which was collected for the same person running barefoot on the treadmill at 3.83 m • s⁻¹. Initial contact with the treadmill belt was made with the rear part of the foot. The peak value of shank acceleration is considerably greater while running: 8.2 g. The greater magnitude is primarily due to the increased velocity of the foot at the time of footstrike.

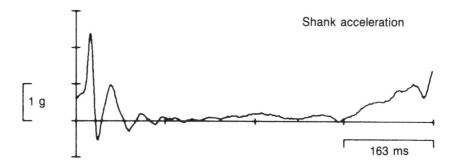

Figure 9.2a. Shank acceleration measured for the entire support period while walking barefoot on a treadmill at 1.53 m • s⁻¹. The horizontal axis represents the acceleration measured during erect standing (1 g).

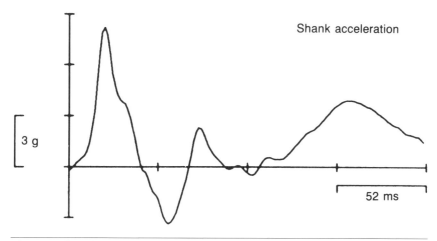

Figure 9.2b. Shank acceleration measured for one support period during barefoot running at 3.83 m • s⁻¹.

This signal also reaches peak acceleration in a much shorter time after footstrike, indicating a more rapid rate of loading during running. Magnitudes of peak shank accelerations during running tend to be an individual characteristic. A peak magnitude of 8 g falls within the range that is typical of runners who make initial contact with the running surface with their rearfoot. Thus, running with a rearfoot strike results in relatively high axial accelerations at the shank, and when compared to walking, the need for adequate cushioning is greater and should not be overlooked.

Shank Accelerations When Running in Shoes

The signal in Figure 9.3 shows the shank acceleration for the same subject running at the same speed on a treadmill while wearing shoes. Running in shoes with a well-cushioned midsole can result in a small reduction in the peak magnitude of shank acceleration. In these trials, the peak magnitude averaged over six footstrikes is reduced from 8.8 g to 7.9 g, but the shocks are still more than three times higher than those for walking. If running speed were to be increased further, the amount of shock transmitted to the shank would increase. Clarke, Cooper, Clark, and Hamill (1985) report that the peak value of the shank acceleration for a group of 10 subjects increases by a mean of 68% when running speed is increased from 3.35 m • s⁻¹ to 5.36 m • s⁻¹. Also, there is increased shock when running downhill and decreased shock when running uphill (Hamill, Clarke, Frederick, Goodyear, & Howley, 1984). This is demonstrated in Figure 9.4, where the mean peak axial acceleration for a group of 10 subjects running at 3.83 m • s⁻¹ on a treadmill are shown. Average shank

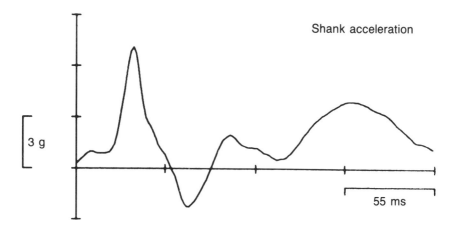

Figure 9.3. Shank acceleration measured during treadmill running at 3.83 m • s⁻¹ while wearing running shoes with a soft midsole.

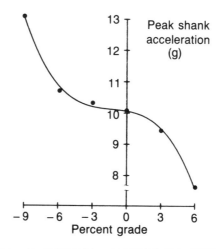

Figure 9.4. Mean values of the peak shank acceleration measured while running on a treadmill at different slopes. Treadmill speed = 3.83 m • s⁻¹ (n = 10).

acceleration increases from 10.1 g during level running to 13.1 g when running on a −9% grade, and decreases to 7.6 g when the grade is +6%.

Measurement of Shank Accelerations in Three Dimensions

Previous examples have discussed shank accelerations in an axial direction only. Lafortune and Hennig (1988), using lightweight triaxial accelerometers attached to the end of a Steinmann pin inserted into the proximal tibia of a volunteer, were able to make measurements of shank accelerations along mediolateral and anteroposterior axes as well as in the axial direction. These researchers reported that for treadmill running, the greatest tibial acceleration component was that measured in an anteroposterior direction. The peak value of this component equaled 7.6 g for running at 4.7 m $\cdot$ s^{-1}. This compares with a smaller 5.0-g peak in the axial direction. The 4.5 peak g recorded in the mediolateral direction was the smallest of the three acceleration components. Peak axial and anteroposterior components were less at a slower running velocity (3.5 m $\cdot$ s^{-1}), and the mediolateral peak was unchanged.

Lafortune (1987) and Lafortune and Hennig (1988) are the first investigators to measure and report anteroposterior and mediolateral components of shank acceleration. The relatively high magnitudes in relation to axial accelerations, especially in the anteroposterior direction, are somewhat surprising. However, knowledge of these acceleration components will be useful in the study of the etiology of running injuries and in the development of running footwear. In fact, the existence of a high anteroposterior acceleration component provides some support to the hypothesis proposed by MacLellan (1984) implicating a horizontally transmitted shock wave generated shortly after heelstrike, which has been observed in high-speed films of the distal leg, in the development of chronic Achilles tendinitis.

These examples show that the amount of transmitted shock to the leg during running can be high. The next section will present some indirect evidence from which some researchers draw the potential implication relating this high shock transmission to higher incidence of injury. However, if one accepts the possibility that these high shocks may be related to the development of a running injury of the overuse type, one should be careful *not* to assume that running is a dangerous sport for competitive runners or a dangerous exercise activity. Clearly, the potential physiological and psychological benefits of running outweigh the assumed risk (Morgan, 1985).

Implied Relationship Between Acceleration and Injury

An early animal study implied that repetitive impact loading is a mechanical factor that can lead to osteoarthritis through microfracture of osseous tissue, a resultant stiffening of the remodeled bone, and finally a wearing out of the shock-absorbing capacity of the joints (Radin, Paul, & Rose, 1972). Though this has never been established in humans through either experimental or epidemiological studies, several investigators have anecdotally linked the repetitive loading of the skeletal system during normal gait with the progression of osteoarthritis.

Light, McLellan, and Klenerman (1980) have measured the magnitude of the acceleration transmitted to the shank while walking on a hard surface by attaching a lightweight accelerometer to two Kirschner pins drilled directly into the tibia of a volunteer. They compared the signals measured while walking in hard leather-soled shoes with those measured when walking in shoes that contained a highly viscous shock-absorbing heel insert. The heel insert was found to reduce the peak shank acceleration from 5 g to 2.5 g, and to also reduce the rate of rise of the shank acceleration. The investigators suggest that these shocks measured at the tibia play a part in the development of osteoarthritic changes and conclude that the use of viscoelastic shoe inserts attenuate the shock transmitted to the tibia, and therefore moderate or even prevent this osteoarthritic degeneration (Light & McLellan, 1977). To support this conclusion, MacLellan and Vyvyan (1981) offer clinical evidence that suggests that the introduction in sports shoes of these same viscoelastic inserts resulted in the elimination of symptoms after a 3-month period in patients presenting with pain beneath the heel or with Achilles tendinitis.

Voloshin and Wosk (1982) also recognized that during walking, heel-strike generates a shock wave that propagates from the heel though the whole body to the head, carrying with it a potential for musculoskeletal damage. These investigators have externally attached lightweight accelerometers to the tibial tuberosity, the femoral condyle, and the forehead of large groups of healthy and pathological subjects. They identified a relationship between reduced shock-absorbing capacity between the femur and the head with the presence of low back pain. They also report a relationship between reduced shock absorption across the knee joint with knee joint pathology (Voloshin & Wosk, 1980). These authors conclude that if a person has joint pathology, the shock-absorbing capacity of that joint is reduced. The shock wave that is generated at heelstrike overloads the shock absorber proximal to the pathological joint, leading to the development of osteoarthritic degeneration (Voloshin, Wosk, & Brull, 1981).

If the heelstrike shock wave generated during walking is linked with joint degeneration, then the force transmission during running is likely implicated in skeletal damage. There is some, albeit indirect, evidence relating injury to the high-frequency components of the force transmitted to the shank upon landing on the ground. For example, there are suggested relationships between incidence of injury and the hardness of a playing surface (Nigg, Denoth, Luethi, & Stacoff, 1983), and between the hardness of a running surface and the magnitude of peak shank acceleration upon landing (Nigg, 1983). In light of these relationships, cushioning systems that do not jeopardize other aspects of an athlete's performance ought to play a major role in protecting the distance runner from potential injury.

Role Played by the Heel Pad for Cushioning

Evolution has provided the human foot with a built-in shock absorber—the fat pad beneath the calcaneus. The function of the human heel pad is to protect the heel from impact-generated shocks and from excessive pressures. The dissected heel region shown in Figure 9.5 shows the anatomy of the fat pad beneath the calcaneus. The heel pad is perhaps more expansive than one would think. It is about 20 mm thick, extending from the bottom surface of the calcaneus to the outer surface of the skin. In a frontal section, the heel pad is crescent-shaped, surrounding the calcaneus up to a level just exceeding the mid-height on both medial and

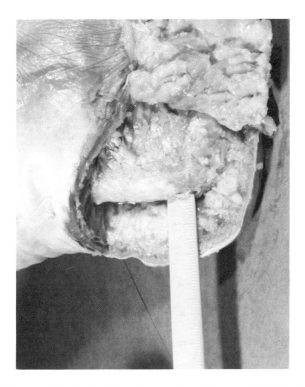

Figure 9.5. Sagittal view of a partially dissected right heel pad showing approximately 21 mm of soft tissue between the inferior calcaneal surface and the plantar skin surface. *Note:* From "Biological Aspects of Modeling Shoe/Foot Interaction During Running" by P.R. Cavanagh, G.A. Valiant, and K.W. Misevich, 1984, in E. Frederick (Ed.), *Sport Shoes and Playing Surfaces*, (p. 33), Champaign, IL: Human Kinetics. Copyright 1984 by NIKE, Inc. Reprinted by permission.

lateral sides. It is composed of two main constituents, fat and a tough fibrous tissue. The fat is contained in small compartments, which are formed by the fibrous tissue. In sagittal sections, these fat compartments are oriented vertically with the more caudal chambers being plumper. The compartments are all securely tied together by connective tissue bands. There is a greater density of fibrous septa on the lateral side of the heel pad (Tietze, 1921/1982; Blechschmidt, 1933/1982). It is the viscous nature and the orientation of these compartments of fat in the heel region that allow the heel pad to absorb a high percentage of the energy imparted to it during an impact.

The impacts to which a human heel would be subjected during running were simulated by impacting the heels of a group of 24 subjects with a free-swinging pendulum (Valiant & Cavanagh, 1985). Standing on their left legs, subjects rested their right legs on a platform and with knees flexed beyond 90 degrees firmly pressed the femoral condyles against a rigid wall. Trials were performed with the 1.92-kg pendulum striking the heel pad at three different velocities, the range of which encompassed the mean velocity of heelstrike, $1.1 \text{ m} \cdot \text{s}^{-1}$, while running at a speed of $3.6 \text{ m} \cdot \text{s}^{-1}$ (Cavanagh et al., 1984).

The mechanical characteristics of the heel pad were investigated by instrumenting the pendulum with an accelerometer and measuring the acceleration of the pendulum during its collision with the heel. The force of the collision was determined by simply multiplying the measured acceleration by the mass of the pendulum. The data plotted in Figure 9.6 are an example of the average impact response of the heel pad for a group of 10 subjects. The difference in the impact response when the velocity

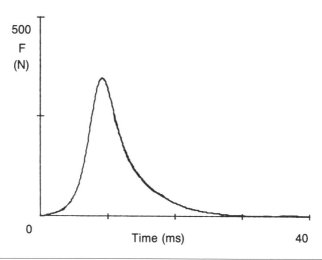

Figure 9.6. Average impact response of the human heel pad when impacted at a velocity of $1.0 \text{ m} \cdot \text{s}^{-1}$ by a pendulum that has a mass of 1.92 kg.

of the impacting pendulum was varied is shown in Figure 9.7. The peak force is closely related to the square of the impact velocity. Thus, given that increased running speed leads to increased heelstrike velocity, it can be assumed that the runner, or at least the heel pad of the runner, is subjected to increased energy levels when running faster. This implication is supported by the increased shank acceleration that is measured at increased running speeds. Of course, this applies only to those runners who make initial contact with the ground with the rearfoot and do not change their footstrike patterns at the increased running speeds.

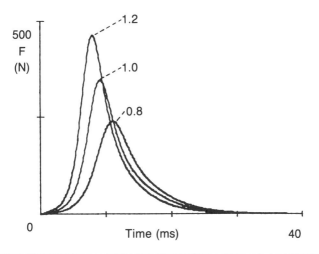

Figure 9.7. The effect of impact velocity on the response of the human heel pad to pendulum impacting. Impact velocities are expressed in m • s^{-1}.

When the average acceleration-time data is integrated twice to obtain the displacement of the pendulum and then plotted against the force of the collision, a hysteresis curve like the one in Figure 9.8 is obtained. This curve reveals an energy loss of 88.5%. This energy loss is primarily confined to the heel pad, although it is recognized that during the pendulum impacting, it is not possible to totally isolate the heel pad. Other structures such as the calcaneus, the shank, and the surfaces of the ankle and knee joints may be involved to a small extent in the absorption of energy. The energy absorption of the heel pad for the entire group of 24 subjects ranged from 84% to 99%, thus indicating that the heel pad is a highly shock-absorbing structure.

The heel pad is quite incompressible, and if one were to palpate one's own heel pad, one would find that it hardens very quickly. The hardness that is felt is due to the rapidly increasing resistance to compression as the tissue is deflected and demonstrates the stiffness characteristics of the heel pad tissue, which can be determined from the hysteresis curve

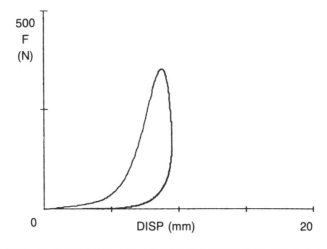

Figure 9.8.   Force-deformation characteristics of the human heel pad, exhibiting a relatively low stiffness for small displacements but a higher stiffness for displacements greater than 4 or 5 mm.

in Figure 9.8. The first portion of the hysteresis curve is characterized by a small increase in force with a relatively large increase in displacement. However, even small increases beyond about 5 mm of deformation result in very large increases in force. It is this higher stiffness that is felt during palpation of the heel pad. One also would notice that when the load is removed from the heel pad, the tissue takes some time to return to its original shape. This is a demonstration of the viscoelastic nature of the tissue in the heel pad region.

The heel pad was flattened considerably against the face of the pendulum during the impact (Valiant & Cavanagh, 1985). In addition to a simple one-dimensional compression, the tissue of the heel pad can also be displaced laterally and medially, and in anterior and posterior directions. Three-dimensional stroboscopic pictures revealed that skin markers on the distal area of the posterior heel pad (i.e., the region directly below the insertion of the Achilles tendon) were displaced an average of 2.0 mm and 2.1 mm in lateral and medial directions, respectively, during 1.0 m • s^{-1} impacts of the pendulum (Valiant, 1984). There is no difference in the kinematic response of medial and lateral sides of the heel pad tissue in spite of slight differences in anatomy. It is conceivable that most of this medial/lateral displacement occurs during the first 4 or 5 mm of forward movement of the pendulum against the heel pad. The freedom of the heel pad to change its shape during the impact contributes to the rather low initial stiffness of the tissue. This mechanism is responsible for part of the energy absorption by the heel pad. However, the tissue of the heel pad is firmly bound to the underlying bone structure. Therefore, as the impacting pendulum continues to advance forward against the heel pad, the skin becomes stretched and taut and contributes more to the overall stiffness of the heel pad. The resistance to the compression of the tissue increases, which results in a dramatic increase in the reaction force. This limited compression is demonstrated in the latter part of the hysteresis curve in Figure 9.8, and is characterized by a stiffness coefficient that is more than an order of magnitude larger than the initial stiffness of the tissue.

The 24 subjects in the pendulum-impacting experiments (Valiant & Cavanagh, 1985) were equally divided between runners and nonrunners. It was observed that not all subjects' heel pads exhibited the same response to pendulum impacting; however, these differences were not related to the subject grouping. There is clinical evidence suggesting that a loss of the integrity of the heel pad leads to injuries such as diffuse heel pain (Marr & Pod, 1980), plantar fasciitis (Sewall, Black, Chapman, Statham, Hughes, & Lavander, 1980), and Achilles tendinitis (Jorgensen, 1985), but there is no definitive evidence suggesting that a regime of distance running specifically leads to a deterioration in the protective makeup of the heel pad.

It has often been implied, however, that the transmission of high acceleration at high repetitive rates during running is related to overuse injuries. Thus, additional cushioning systems are warranted when the lower limb is subjected to the accelerations that occur while running. There has thus been a major emphasis on the development of well-cushioned running shoes in the last decade and a half.

Comparison Between Shank Accelerations and Head Accelerations

Up to now, mention has been made only of the shock measured at the shank, but during running most parts of the body are subjected to shock. Figure 9.9 shows a runner biting on a stiff nylon bar that has an accelerometer mounted on it. This bite bar provides a direct attachment of an accelerometer to the skeletal system via the teeth, and measurements of the shock transmitted to the head can be made. Figure 9.10 shows an example of the accelerations measured with the shank accelerometer and with the bite bar accelerometer while running on a treadmill at 3.83 m • s⁻¹. By the time the impact shock reaches the level of the head, the peak acceleration is greatly reduced and the high-frequency components that are present in the shank signal disappear. This is because there is a considerable amount of damping and compliance in the body between the shank and the head.

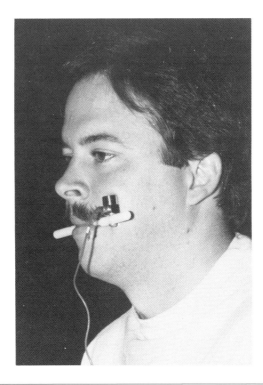

Figure 9.9. By biting down on keys in the bite bar, the nylon and the teeth provide a relatively direct attachment between the skeletal system and an accelerometer ($f_n > 230$ Hz) mounted close to the face on the nylon rod.

The bite bar acceleration signal bears a strong resemblance to a vertical ground reaction force signal measured while running across a force platform, indicating that the vertical response at the skull is similar to the response of the runner's center of gravity. This is primarily because the mass to which the bite bar accelerometer responds, the mass of the whole body, is the same as the mass to which the force platform responds. If the body were modeled as a single lumped mass, this mass would directly relate the vertical acceleration (measured by the bite bar) to the vertical ground reaction force (measured by the force platform). However, to relate the measured vertical ground reaction forces and measured axial accelerations at the shank by Newton's second law, the accelerated mass must be considerably smaller than total body mass. Denoth (1986) has therefore developed a concept termed *effective mass*. Using this concept, the vertical ground reaction force developed by a runner during the first 10 to 20 ms can be modeled by a single rigid body with effective mass m^* and an axial acceleration equivalent to the sum of measured shank acceleration and the acceleration due to gravity.

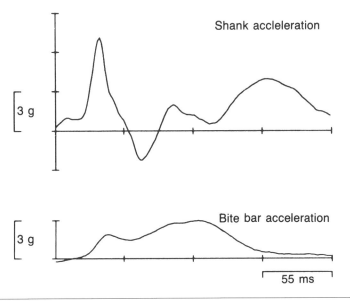

Figure 9.10. Examples of acceleration measures at the shank and the head while running in shoes at 3.83 m · s⁻¹. Baseline values represent accelerations measured during erect standing.

The Concept of Effective Mass

Effective mass is defined to be that part of the mass of the human body that has an effect on the forces that are developed passively within the knee joint during the impact phase. These forces are estimated by multiplying the difference between the effective mass and the mass of the distal segment of a rigid body model by the sum of gravitational acceleration and the measured axial shank acceleration. This calculation assumes that there is a muscle latency period of at least 30 ms after ground contact. Because there is no muscular influence on the forces at the knee joint during this time, the resultant force is only that component developed as a result of impact forces.

One can thus think of effective mass as a representation of the combination of tibia mass and knee joint reaction force. For example, if the effective mass were simply the mass of the tibia, the predicted force in the knee caused by the impact would be zero. Because impact forces do result in a transmission of forces through the knee joint, the effective mass must have some value greater than the mass of the lower leg. Consider the concept of effective mass as being similar to the unsprung weight of a vehicle. Decreasing a vehicle's unsprung weight results in a smoother ride for the passengers. Similarly, a smaller effective mass is related to greater shock absorption. Denoth (1986) showed that a greater knee flexion angle results in smaller effective mass, and it will be shown in the next section that a greater knee flexion angle reduces the effective spring stiffness of a runner, leading to greater shock absorption.

Effective mass is not a true inertial quantity, however, because its value is not a constant. It is dependent upon variables such as knee angle, ankle angle, limb masses, limb lengths, the frequency content of the shank acceleration signal, impact velocity, and rate of flexion at the knee joint. Also, the shock attenuation by soft tissue structures distal to the knee joint is neglected. As a result, Denoth and colleagues modified the model of the impacting athlete to take into account relative motion between the rigid mass and the soft tissue mass to yield better approximations of the resultant forces transmitted through the joints (Denoth, Gruber, Ruder, & Keepler, 1985).

Vertical Compliance of the Body

Many factors account for the change in the acceleration signal as the shock wave travels upward through the body from the shank to the head. There are many joints in the body between these two points. The joints and the intervertebral discs in the back serve to absorb much of the energy and attenuate the traveling shock wave. The hyaline cartilage lining the joints in the body accounts for some shock attenuation. The movement of pronation at the subtalar joint after footstrike also absorbs energy and attenuates the propagating shock wave that results from the foot striking the ground (Clarke, Frederick, & Hamill, 1983; Denoth, 1986). Even the long bones in the leg, although considered rigid, may bend slightly when subjected to the loads of running and can therefore be responsible for a slight attenuation of the shock.

Groucho Running.   The knee joint can also account for some attenuation of the traveling shock wave by acting like a soft spring after footstrike. In an effort to determine the effect that knee flexion has on shock attenuation, a group of runners were trained to run with increased amounts of knee flexion (McMahon, Valiant, & Frederick, 1987). Running with increased amounts of knee flexion was termed Groucho running, as this style of running resembled the style of walking adopted by Groucho Marx in many of his films.

The supported runner was modeled as a simple mass and spring striking the ground with a downward velocity that is dependent on the runner's forward running speed, and then rebounding off the ground with the same vertical velocity. For any given impact velocity, the length of time required for the model to rebound from the ground is determined by the stiffness of the spring. Knowing impact velocity, total contact time, and the mass of a runner, an effective spring stiffness for the runner can be calculated. In the increased knee flexion experiments, at a constant running speed the ground contact time was observed to progressively increase as knee flexion was increased, thereby showing that the effective spring stiffness was related to the amount of knee flexion.

A dimensionless parameter, $u\omega_0/g$ was introduced to relate the ground reaction force normalized to body weight, F/mg, to the phase of resonant vibration of the impacting mass-spring model, $\omega_0 t$, by

$$\frac{F}{mg} = \left(\frac{u\omega_0}{g}\right) \ \sin(\omega_0 t) + 1 - \cos(\omega_0 t).$$

In this parameter, u represents the vertical impact velocity, ω_0 represents the natural frequency of the mass-spring system that models the runner, and g is gravitational acceleration. The parameter $u\omega_0/g$ is directly related to the peak ground reaction force and is useful for describing the nature of the impact. Its magnitude is small for soft landings and large for hard landings.

The knee joints of the subjects in this study flexed through a range of 20 degrees during the support phase when running normally. When runners purposely increase the maximum amount of knee flexion during the support phase there is a reduction in the peak impact shock measured at the head, as illustrated in Figure 9.11. The ratio of bite bar acceleration to shank acceleration decreases from 38% to 17% as mean midstance knee

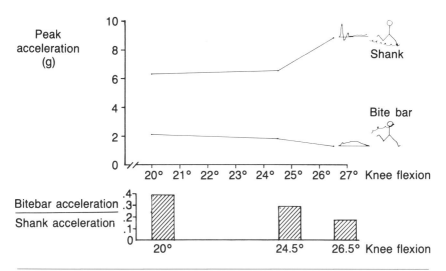

Figure 9.11.   Mean peak accelerations and the ratio of accelerations measured at the shank and at the head when running with varying amounts of maximum midstance knee flexion. *Note*: From "Groucho Running" by T.A. McMahon, G.A. Valiant, and E.C. Frederick, 1987, *Journal of Applied Physiology*, **62**, p. 2333. Copyright by the American Physiological Society. Adapted by permission.

flexion increases from 20 to 26.5 degrees. This decrease is not due to a decreased acceleration at the shank because the measured accelerations at the shank actually increase with the increased knee flexion running style. It can be argued that by modifying the amount of knee flexion, the vertical stiffness of the runner is regulated, thus regulating the shock that is transmitted through the body.

There is, however, a simple reason why runners do not voluntarily select this style of running as a means of modifying impact force transmission. The penalty to pay is an increase of about 25% in the oxygen cost of running for each 5-degree increase in midstance knee angle.

Cost of Cushioning.   There is further evidence that suggests that runners use the action of knee flexion as a means of regulating the transmission of shock through the body. Clarke, Frederick, and Cooper (1983) found that the rate of knee flexion after touchdown increased with increasing hardness of the shoe. The shoes that were tested differed only in the cushioning characteristics of the midsole material. A correlation coefficient of 0.81 related rate of knee flexion to shoe cushioning (cushioning was determined by a standardized impact test). It is possible that the runners were able to perceive the increased hardness of the shoes and went about reducing the level of transmitted shock by mechanically increasing the compliance of the spring at their knee joint. The increased rate of knee flexion in the harder shoes, however, required adjusted muscular action to control the knee joint, and the result was an increase in the submaximal oxygen cost of running with increased rates of knee flexion (Frederick, Clarke, Larsen, & Cooper, 1983). Thus, questions related to kinematic adaptations to transmitted shock will be good research topics in future studies of the transmission and attenuation of heelstrike accelerations.

Characterization of Accelerations in the Frequency Domain

Rather than just measuring the magnitude and the time course of shank and head accelerations, it is also useful to examine the frequency content of these signals. To date, very few researchers have conducted spectral analyses of accelerations. From data collected by a lightweight accelerometer strapped to the tibial tuberosity of a normal subject, Voloshin, Burger, Wosk, and Arcan (1985) presented shank acceleration results during overground walking in both the time and frequency domains. Their results showed the magnitude of the peak acceleration at the tibial tuberosity typically measured 3.8 g. The peak frequency content occurred between 18 and 20 Hz. Harmonics above 30 Hz were essentially zero.

Shorten, Valiant, and Cooper (1986) made similar measurements during treadmill running and recorded equivalent results. Shank and bite bar accelerations were measured for a group of eight subjects running at 3.83 m • s⁻¹ barefoot and in shoes with varying midsole hardness. The power spectrums of each signal were determined for the frequency range 3 to 45 Hz. The mean peak frequency components of the shank acceleration occurred between 20 and 25 Hz. The slight difference in range of peak amplitudes measured by Voloshin et al. (1985) can be attributed to factors such as different accelerometer locations, or differences in accelerometer fixations. The mean peak frequency components of the bite bar accelerations occurred at 3.5 Hz, which is close to twice the stride frequency. This finding simply reflects the low frequency motion of the head imposed by the running stride and further verifies that much of the high-frequency content of the shank acceleration signal is attenuated a great deal during transmission to the head.

It was not possible to measure any differences in the power of shank accelerations between the different cushioning conditions (Shorten et al., 1986). However, the power of the bite bar acceleration, which was greatest for barefoot running, was shown to decrease for frequency components above 20 Hz as midsole hardness decreased. The differences in power in the bite bar acceleration for frequency components less than 20 Hz between the different midsoles were very small. It was concluded that cushioning in a running shoe midsole was capable of attenuating the higher frequency components of the shock transmitted to the head but did not directly influence frequency components below about 20 Hz.

Based on these findings, it would be logical to expect increased shoe cushioning to attenuate the high-frequency components transmitted to the shank. However, the effects of shoe cushioning on the shock transmitted to the shank were not as clearly evident, particularly because the natural frequency of the shank accelerometer attachment was limited to 51 to 61 Hz (Valiant et al., 1987).

It is conceivable that certain frequency components will carry a greater potential for injury than other frequency components. Thus, future characterization of shank and head accelerations in the frequency domain will be important. If it becomes possible to identify the potentially more harmful frequency components, then a characterization of the frequency content of the shocks generated at heelstrike will lead to a greater understanding of how to filter or dampen them.

Summary

This chapter focused primarily on the nature of the skeletal accelerations that occur as a result of the foot striking the ground during running, and the transmission of these accelerations through the body. Peak accelerations measured at the shank exceed 8 g, and occur within 17 to 25 ms after first contact with the ground. The accelerations increase when running down a slope and when running at higher speeds. There are suggestions in the literature that repeated exposure to shocks such as these can, in part, be responsible for overuse-type injuries such as Achilles tendinitis and diffuse heel pain, and can even lead to osteoarthritic degeneration. The fat pad beneath the heel plays a very important role in the cushioning of the skeletal system from the shocks that are generated as a result of heelstrike. Also, wearing well-cushioned running shoes results in modest reductions in these acceleration signals. There is a damping and an attenuation of the shock wave as it is propagated through the skeletal system. Attenuation is most evident for frequency components greater than 20 Hz. The action of knee flexion has a direct effect on the transmission of the shock waves and hence on the magnitude of the signal at the level of the head.

Acknowledgments

The author wishes to express his gratitude to S.P. Moore, M.R. Shorten, L.B. Cooper, and T. McGuirk for their constructive ideas and comments during the preparation of this manuscript.

References

Blechschmidt, E. (1982). The structure of the calcaneal padding. *Foot and Ankle*, **2**, 260-283. (Reprinted from *Gegenbaurs Morphologisches Jahrbuch*, 1933, **72**, 20-68)

Cavanagh, P.R., Valiant, G.A., & Misevich, K.W. (1984). Biological aspects of modeling shoe/foot interaction during running. In E. Frederick (Ed.), *Sport shoes and playing surfaces* (pp. 24-46). Champaign, IL: Human Kinetics.

Cichowski, W.G. (1963). *A new laboratory device for passenger car safety studies* (Technical Paper No. 663A). Warrendale, PA: Society of Automotive Engineers, Inc.

Clarke, T.E., Cooper, L.B., Clark, D.E., & Hamill, C.L. (1985). The effect of increased running speed upon peak shank deceleration during ground contact. In D. Winter, R. Norman, R. Wells, K. Hayes, & A. Patla (Eds.), *Biomechanics IX-B* (pp. 101-105). Champaign, IL: Human Kinetics.

Clarke, T.E., Frederick, E.C., & Cooper, L.B. (1983). Biomechanical measurement of running shoe cushioning properties. In B. Nigg & B. Kerr (Eds.), *Biomechanical aspects of sport shoes and playing surfaces* (pp. 25-33). Calgary, AB: University of Calgary.

Clarke, T.E., Frederick, E.C., & Hamill, C.L. (1983). The effects of shoe design parameters on rearfoot control in running. *Medicine and Science in Sports and Exercise*, **15**, 376-381.

Denoth, J. (1986). Load on the locomotor system and modelling. In B. Nigg (Ed.), *Biomechanics of running shoes* (pp. 63-116). Champaign, IL: Human Kinetics.

Denoth, J., Gruber, K., Ruder, H., & Keepler, M. (1985). Forces and torques during sports activities with high accelerations. In S. Perren & E. Schneider (Eds.), *Biomechanics: Principles and applications* (pp. 663-668). Boston: Martinus Nijhoff.

Frederick, E.C., Clarke, T.E., Larsen, J.L., & Cooper, L.B. (1983). The effects of shoe cushioning on the oxygen demands of running. In B. Nigg & B. Kerr (Eds.), *Biomechanical aspects of sport shoes and playing surfaces* (pp. 107-114). Calgary, AB: University of Calgary.

Gross, T.S., & Nelson, R.C. (1988). The shock attenuation role of the ankle during landing from a vertical jump. *Medicine and Science in Sports and Exercise*, **20**, 506-514.

Hamill, C.L., Clarke, T.E., Frederick, E.C., Goodyear, L.J., & Howley, E.T. (1984). Effects of grade running on kinematics and impact force. *Medicine and Science in Sports and Exercise*, **16**, 185.

Hennig, E.M., & Lafortune, M.A. (1988). Tibial bone and skin accelerations during running. *Proceedings of the Fifth Biennial Conference and Human Locomotion Symposium of the Canadian Society for Biomechanics* (pp. 74-75). London, ON: Spodym.

Jorgensen, U. (1985). Achillodynia and loss of heel pad shock absorbency. *American Journal of Sports Medicine*, **13**, 128-132.

Lafortune, M.A. (1987). Impact loading of the lower limbs. *Abstracts of the XIth International Congress of Biomechanics* (p. 179). Amsterdam: Free University Press.

Lafortune, M.A., & Hennig, E.M. (1988). Effects of velocity and uphill slope on tibial shock during running. *Proceedings of the Fifth Biennial Conference and Human Locomotion Symposium of the Canadian Society for Biomechanics* (pp. 94-95). London, ON: Spodym.

Light, L.H., & McLellan, G.E. (1977). Skeletal transients associated with heel strike. *Journal of Physiology*, **272**, 9P-10P.

Light, L.H., McLellan, G.E., & Klenerman, L. (1980). Skeletal transients on heel strike in normal walking with different footwear. *Journal of Biomechanics*, **13**, 477-480.

MacLellan, G.E. (1984). Skeletal heel strike transients, measurement, implications, and modification by footwear. In E. Frederick (Ed.), *Sport shoes and playing surfaces* (pp. 76-86). Champaign, IL: Human Kinetics.

MacLellan, G.E., & Vyvyan, B. (1981). Management of pain beneath the heel and achilles tendinitis with visco-elastic heel inserts. *British Journal of Sports Medicine*, **15**, 117-121.

Marr, S.J., & Pod, F.A. (1980). The use of heel posting orthotic techniques for relief of heel pain. *Archives of Orthopaedic and Traumatic Surgery*, **96**, 73-74.

McMahon, T.A., Valiant, G.A., & Frederick, E.C. (1987). Groucho running. *Journal of Applied Physiology*, **62**, 2326-2337.

Morgan, W.P. (1985). Affective beneficence of vigorous physical activity. *Medicine and Science in Sports and Exercise*, **17**, 94-100.

Nigg, B.M. (1983). The load on the lower extremities in selected sports activities. In G. Wood (Ed.), *Collected papers on sports biomechanics* (pp. 62-73). Nedlands: University of Western Australia.

Nigg, B.M., Denoth, J., Luethi, S., & Stacoff, A. (1983). Methodological aspects of sport shoe and sport surface analysis. In H. Matsui & K. Kobayashi (Eds.), *Biomechanics VIII-B* (pp. 1041-1052). Champaign, IL: Human Kinetics.

Radin, E.L., Paul, I.L., & Rose, R.M. (1972). Role of mechanical factors in pathogenesis of primary osteoarthritis. *Lancet*, **7749**, 519-522.

Sewell, J.R., Black, C.M., Chapman, A.H., Statham, J., Hughes, G.R.V., & Lavender, J.P. (1980). Quantitative scintigraphy in diagnosis and management of plantar fasciitis (calcaneal periostitis): Concise communication. *Journal of Nuclear Medicine*, **21**, 633-636.

Shorten, M.R., Valiant, G.A., & Cooper, L.B. (1986). Frequency analysis of the effects of shoe cushioning on dynamic shock in running. *Medicine and Science in Sports and Exercise*, **18**(Suppl.), S80-S81.

Tietze, A. (1982). Concerning the architectural structure of the connective tissue in the human sole. *Foot and Ankle*, **2**, 252-259. (Reprinted from *Beitrage Klinischen Chirurgie*, 1921, **123**, 493-506)

Valiant, G.A. (1984). *A determination of the mechanical characteristics of the human heel pad in vivo*. Unpublished doctoral dissertation, The Pennsylvania State University, University Park.

Valiant, G.A., & Cavanagh, P.R. (1985). An *in vivo* determination of the mechanical characteristics of the human heel pad. *Journal of Biomechanics*, **18**, 242.

Valiant, G.A., McMahon, T.A., & Frederick, E.C. (1987). A new test to evaluate the cushioning properties of athletic shoes. In B. Jonsson (Ed.), *Biomechanics X-B* (pp. 937-941). Champaign, IL: Human Kinetics.

Voloshin, A.S., Burger, C.P., Wosk, J., & Arcan, M. (1985). An *in vivo* evaluation of the leg's shock-absorbing capacity. In D. Winter, R. Norman, R. Wells, K. Hayes, & A. Patla (Eds.), *Biomechanics IX-B* (pp. 112-116). Champaign, IL: Human Kinetics.

Voloshin, A.S., & Wosk, J. (1980). Shock absorbing capacity of the human knee (*in vivo* properties). In *Human Locomotion I: Pathological Gait to the Elite Athlete* (Proceedings of the Special Conference of the Canadian Society for Biomechanics, pp. 104-105). Ottawa, ON: The organizing committee.

Voloshin, A.S., & Wosk, J. (1982). An *in vivo* study of low back pain and shock absorption in the human locomotor system. *Journal of Biomechanics, 15*, 21-27.

Voloshin, A.S., Wosk, J., & Brull, M. (1981). Force wave transmission through the human locomotor system. *Transactions of the American Society of Mechanical Engineers Journal of Biomechanical Engineering, 103*, 48-50.

Chapter 10

Biomechanical Aspects of Distance Running Injuries

Stanley L. James
Donald C. Jones

Runners are a unique group of people in our population. First of all, they comprise a very healthy segment that would most likely not have any difficulty performing many other types of athletic endeavors besides the one they have chosen. Second, contemporary runners are logging extremely high mileage, and this serves to magnify the deleterious effect of any basic anatomic variation that probably could be tolerated in most sporting events or activities. It is not unusual for a runner to complain of a problem associated only with distance running, whereas other activities such as tennis, squash, skiing, and basketball are tolerated well by the same individual. It appears that this is the result of accumulated impact loading (Brand, 1976; James, Bates, & Osternig, 1978; Nigg, 1985; Perry, 1983) of the lower extremities encountered uniquely in long-distance running.

In 1978, James presented the results of a study of 180 injured runners. It was the expectation of this study that there would be a very definite correlation between anatomic malalignment or variations in the lower extremity and specific pathologic entities generally classified as overuse syndromes, but no definite correlation was found. The results of this study listed the etiology of the various problems under three categories:

- Training errors, which constituted almost two thirds of the causes of injury
- Anatomic factors generally involving abnormal biomechanics or malalignments of the lower extremities
- Shoes and surfaces

In our experience, training errors still remain the predominant factor in producing runners' injuries. Originally high mileage, intense workouts, and changes in surfaces or terrain were listed as the most significant factors, but as the years have gone by, the bottom line seems to be a rapid change in the program, whether it is sudden increased mileage, increased intensity, or some other factor that has been inserted into the program with inappropriate time for the body to accommodate to the new forces.

In the group of 180 runners presenting with 232 conditions, 71% fell into six categories. Knee pain was 34%, posterior tibial syndrome (now often referred to as medial tibial stress syndrome) 13%, Achilles tendinitis 11%, plantar fasciitis 7%, and stress fractures 6%. The remaining 29% were miscellaneous problems.

Knee problems, which constituted 34% of the injuries, were divided into the following categories:

• Patellofemoral joint pathology (25%)
• Knee pain—no diagnosis (20%)
• Iliotibial tract tendinitis (17%)
• Peripatellar pain (15%)
• Patellar tendon tendinitis (7%)
• Medial retinacular discomfort (6%)
• Miscellaneous (10%)

Recently McKenzie, Clement, and Taunton (1985) reported an increasing incidence of knee injuries in runners when today's statistics are compared with those of a few years ago. It is felt that the improved construction of running shoes has probably played a role in reducing the number of lower leg and foot injuries in runners but has not been as effective in resolving the problem more proximally at the knee.

Few studies directly correlate abnormal anatomic factors with a specific type of injury in a reliable fashion, but McKenzie et al. (1985) postulate that a runner with excessive or compensatory pronation of the foot is predisposed to injuries that more frequently occur on the medial aspect of the lower extremities, such as medial tibial stress syndrome and posterior tibial tendinitis, Achilles tendinitis, and plantar fasciitis. Compensatory pronation with increased internal tibial rotation places additional stress upon the foot, ankle, knee, hip, or lower back, probably due to increased transverse plane motion in the lower extremity. On the contrary, an individual with a cavus-type foot, which is more rigid and represents a poor shock-absorbing mechanism due to inadequate foot flexibility to dissipate forces, is more likely to have injuries on the lateral side of the foot and knee, creating problems such as iliotibial band friction syndrome, trochanteric bursitis, peroneal tendinitis, plantar fasciitis, stress fracture, Achilles tendinitis, gastrocnemius-soleus muscle strain, and metatarsalgia.

Ross and Schuster (1983) present a method of predicting injuries in distance runners based on measurement of a total varus value, which was the summation of tibia varum measured in stance with rearfoot varus and forefoot varus measured non–weight bearing. Runners with less than 8 degrees total varus had a low incidence of injury. Those with a total varus between 10 and 16 degrees had a moderate increase in injuries, and runners with over 18 degrees total varus appeared to have a marked increase in running-related injuries. Using this method in 63 runners, the authors had a predictability rate of 75% over a 5-year period. Although not specifically discussed, it is implied that increased total varus would require compensatory measures with a high potential for injury.

However, to date we simply have not been able to correlate specific anatomic abnormalities and abnormal biomechanics of the lower extremity with specific injuries on a predictable basis. A comprehensive lower extremity examination can certainly suggest potential problems and can be useful in a preventive fashion for prescribing shoe modifications, orthotics, and perhaps most importantly providing training parameters for injury prevention.

Through the years, however, we have come to the realization that there are certain people who probably should not be running long distance for aerobic fitness and would be better off in other sports. Fortunately, there has recently been a change in attitude toward total fitness, and people are much more willing to pursue other types of physical activity. The popularity of triathlons has attracted many people to the alternatives of swimming and cycling in addition to running, and there is also a tendency toward fitness centers where various types of exercise machines and aerobic-type indoor activities are offered. There seems little doubt that a broader-based fitness program, even though running may be the foundation of the program, is more injury-free than a fitness program that consists of distance running alone.

Relationship of Biomechancial Abnormalities With Running Injuries

Biomechanics

At the risk of being redundant, some review of the biomechanical function of the lower extremity from an orthopedic standpoint should be briefly discussed. This is a very complex subject and will be covered rather superficially primarily as a quick review rather than to add any new information. Much of the examination and assessment of foot function is in relation to motion and position of the subtalar joint. Alignment of the leg-heel and heel-forefoot are evaluated based upon the assumption that the foot functions best with the subtalar joint at or near its neutral position. Two complex motions take place about the subtalar joint: pronation and supination. With pronation, the foot is everted, abducted, and dorsiflexed; and with supination it is inverted, adducted, and plantar flexed.

Total range of motion of the subtalar joint based upon James, Bates, and Osternig (1978) was 31 degrees, with inversion being approximately 23 degrees and eversion 8 degrees, thus basically a 3:1 ratio with inversion being greater. During pronation, the foot becomes a more adaptable, flexible mechanism to accommodate to various types of surfaces during weight bearing. Supination, on the other hand, positions the foot as a more rigid device to propel the body forward during the propulsive phase of running.

Just before footstrike, which in most runners is generally on the lateral side of the heel (although runners do assume different positions at footstrike, with some landing essentially flatfoot and others on the ball of the foot) (Nigg, 1985; Slocum & James, 1968), there is initially a very rapid period of pronation. Pronation continues for approximately 70% of the weight-bearing phase with maximum pronation occurring at about 40% through the support phase, which is approximately when the center of gravity passes over the weight-bearing foot and the patellae cross. During this period of time, as mentioned, the midtarsal-subtalar joint complex is unlocked, allowing the foot to become more flexible to accommodate to the underlying surface. After maximum pronation, the subtalar joint gradually supinates, passing from pronation into supination at about 70% of the support phase, stabilizing the midtarsal-subtalar joint complex and creating a more rigid lever for pushoff.

The problem arises with excessive or prolonged pronation during the support phase, resulting in increased forces being applied to the supporting structures of the foot and leg by requiring additional effort of the intrinsic and extrinsic muscles in an attempt to stabilize the foot during pushoff. The tibia internally rotates with pronation and obligatorily externally rotates during supination. Transverse plane rotations occur at the knee secondary to the obligatory tibial rotation during pronation and supination. If internal tibial rotation is increased or prolonged with excessive pronation, which is compensatory to an anatomic malalignment of the foot or leg, then more transverse rotation must be absorbed at the knee. It has been assumed that under these circumstances the normal tibial-femoral rotation relationship at the knee is quite likely disturbed and may well account for much of the high incidence of knee problems in runners. Cavanagh, Valiant, and Misevich (1984) conducted studies that do not verify this concept, but it must be studied further. Excessive pronation of the foot seems to be a compensatory mechanism generally for one or more of the following anatomic conditions (Kuhns, 1949): (a) tibia varum—excess of 10 degrees is abnormal, (b) functional equinus with a tight triceps surae, (c) subtalar varus, and/or (d) forefoot supination (varus) (James et al., 1978).

Bates, Osternig, Mason, and James (1979a) have indicated that the degree of pronation is increased with barefoot running, begins sooner, and ends later, whereas a shoe diminished the period of pronation. The maximum point of pronation occurred significantly later into the pronation phase under the barefoot condition than it did with the simple addition of a running shoe. Another interesting phenomenon was the fact that the angle of the leg relative to the running surface was less when running faster, thus increasing leg varus in relation to the running surface and requiring more pronation for the foot to be placed plantigrade on the running surface.

Perry (1983) discussed muscular function associated with subtalar motion during the support phase of walking and running activities. The anterior tibial, tibialis posterior, and soleus muscles are the primary muscular restraints against subtalar eversion (pronation). The tibialis anterior is the primary muscle at initial heel contact, but once the forefoot comes into contact with the ground, the tibialis posterior activity replaces that of the tibialis anterior, and as the body weight moves forward of the ankle axis, the soleus begins to contract. The tibialis posterior has the longest inversion lever with the greatest mechanical advantage, and although the soleus has a shorter lever arm, its size and strength overcome the effect of its shorter lever arm with the resultant inversion torque on the calcaneus being twice that of the tibialis posterior. However, even with maximum muscular participation, the ability of these muscles to meet the tremendous valgus torques (pronation) imposed on the foot during the support phase necessitates careful selection of footwear and/or the use of orthoses to help control excessive eversion forces.

Considerable attention has been given to the configuration of the weight-bearing foot with the thought than an individual with compensatory pronation would be more likely to have a given syndrome of overuse and the individual with a more rigid cavus-type foot another type of injury. An assessment of the arch can be simply whether the arch is high, low, or normal. In our study of 180 runners (James et al., 1978) we found that 58% of the group had a low arch, 20% had a high arch, and 22% were considered normal. In the low arch group of 58%, the following conditions were found:

- Posterior tibial syndrome or medial tibial stress syndrome (15%)
- Plantar fasciitis (15%)
- Achilles tendinitis (12%)
- Knee pain (18%)

Unfortunately, none of the foot types could be associated with a single predominant diagnosis.

Although in reading the literature one might assume that pronation is an abnormal or undesirable form of motion of the subtalar joint, this could not be further from the truth. A normal amount of pronation is an extremely important part of the impact-absorbing mechanism at foot-strike and coordinates with other events more proximal in the extremity, such as ankle dorsiflexion, knee flexion, and hip motion, all of which are necessary for proper dissipation of the forces imparted to the lower extremity in running. Thus, although there is a great deal of emphasis placed on the function of the subtalar joint, the remainder of the lower extremity cannot be ignored.

The function of the entire lower extremity, as pointed out by Mann, Baxter, and Lutter (1981) is dynamic, and as the speed of gait increases, the magnitude of the forces involved may increase manifold (Bates et al., 1979a; Cavanagh & Lafortune, 1980; Clarke, Cooper, Clark, & Hamill, 1985) and must be dissipated efficiently to prevent overuse injury to structures of the lower extremity.

Functions of the support extremity are to (a) absorb the impact of foot-strike, (b) support the body's weight, (c) maintain forward motion, and (d) accelerate the body's center of gravity against internal and external resistances. Thus the segmental articulation of the lower extremity, along with the mobile lumbar spine-pelvic unit, functions as a stable, adjustable strut to assure the body's center of gravity follows a smooth, undulating path in the sagittal plane, with the low point occurring during the midsupport phase and the high point just after takeoff. This necessitates a relative shortening of the extremity during midsupport and relative lengthening during takeoff (James & Brubaker, 1973; Slocum & James, 1968). Relative shortening is accomplished by downward pelvic tilt, knee flexion, ankle dorsiflexion, and subtalar pronation. Relative lengthening is a function of upward pelvic tilt, posterior pelvic rotation, hip extension, knee extension, ankle plantar flexion, and resupination of the subtalar joint, creating a rigid level for pushoff.

Bates's 1970s studies revealed that there is a definite segmental coordination relationship between pronation/supination and knee flexion/extension. Pronation and knee flexion are accompanied by internal tibial rotation, and supination and knee extension relate to external tibial rotation. If these particular motions are not synchronous, as might be in the situation with prolonged compensatory pronation and internal tibial rotation, asynchrony will develop, which can disrupt normal tibial-femoral transverse plane rotation. This can result in soft tissue stress about the knee or perhaps a patellofemoral malalignment syndrome. Maximum dorsiflexion of the ankle was noted to occur at a later point during the support phase than did maximum pronation. The body's center of gravity had passed over the foot, which had entered into the propulsive phase. Maximum dorsiflexion of the ankle at this point placed the gastrocsoleus group under stretch immediately before it was called upon to contract during the propulsive phase, thus enhancing its force of contraction by the stretch mechanism and also placing great tension on the Achilles tendon.

During takeoff, a powerful extensor thrust is created, which initiates in the stronger, slower muscles of the lumbar spine–pelvic unit and migrates to the distal, but faster and weaker, muscles of the leg and foot. This sequential muscular activity provides a summation of forces to propel the body into its airborne trajectory. A disruption of the normal sequence can overload one of the more distal, weaker musculotendinous units, resulting in injury (e.g., Achilles tendinitis).

Nigg (1985) emphasizes that forces acting on the body during running have a magnitude, a point of application, and a direction, all of which must be considered in load analysis. He points out that sports injuries are generally secondary to an overload of forces on a specific region of the body, with the local forces or stresses exceeding the critical limits. Overload can occur due to a single force being above a critical limit, inducing an acute injury such as one might expect in a contact sport; or overload can occur secondary to cyclic forces below the critical limit, but producing an accumulation of forces with eventual fatigue and injury.

Nigg (1985, p. 379) suggests that the most effective biomechanical strategies to reduce load and stress on the locomotor system are as follows: "(A) the movement; such as a change in running style; (B) the surface; running on a soft surface as opposed to a hard surface; (C) the shoe; and (D) diminish the frequency of repetition of a movement such as reducing mileage in running."

Shoes

In 1978, James et al. postulated that if an adequately designed shoe were available, many problems attendant to long-distance running, short of training errors, could be prevented. This has probably been substantiated through the years with fewer distal leg and foot problems occurring. There are numerous well-conducted studies on the effect of shoes on the biomechanics of foot and leg function. Great emphasis has particularly been placed on rearfoot control.

Cavanagh, Valiant, and Misevich (1984) stated that in general shoes with harder midsole materials, a stiffer counter, and a wider heel base at the outsole do better on rearfoot control studies. Clarke, Frederick, and Hamill (1984) derived the following conclusions from shoes of similar design features:

- Shoes having a midsole softer than 35 durometer will allow significantly more maximum pronation and total rearfoot movement.
- Shoes having less rearfoot flare will allow more maximum pronation and total rearfoot movement.
- Heel height has no significant effect on the amount or rate of pronation that occurs.
- Time to maximum pronation velocity increases as the shoe becomes softer and has less flare and a lower heel lift.

Maximum pronation averages about 9.4 degrees with a standard deviation around 3.5 degrees, and maximum pronation greater than 13 degrees is excessive.

Frederick, Clarke, and Hamill (1984) summarize their recommendations for shoes as follows:

> When choosing a shoe with more cushioning, picking a thicker soled shoe is a wiser choice than a softer shoe. Thicker soles have no effect on maximum pronation, but increased thickness can increase cushioning significantly. Choosing a softer rather than a thicker sole, however, would mean significantly less rearfoot control. Optimum combination of rearfoot control and cushioning occurs in thickly soled shoes with a 35 durometer midsole/wedge, flared to 15 degrees. There are many shoes on the market today which meet these criteria. (p. 197)

This hypothesis is further supported by Denoth (1985), who indicated that shoes with a rounded lateral heel tend to produce higher impact forces, and shoes with a flared heel tend to reduce impact forces. Nigg (1985) also found that an experimental shoe with a rounded lateral heel and a midsole with increasing density toward the medial side produced smaller moments to initiate pronation, and increasing density resisted pronation. The rounded heel, however, was associated with increased impact peak, whereas the flared lateral heel helped reduce lateral impact but enhanced the rapidity of pronation if the durometer was too high.

Cavanagh et al. (1984) have pointed out the importance of the heel pad property in impact absorption. This study found that there is approximately the same thickness of uncompressed adipose tissue under the calcaneus as there is foam material in the rear part of most running shoes. The heel pad must, therefore, be considered as part of the impact-absorbing mechanism in running, but its degree of use or overuse may be modified by appropriate heel construction. Heel pad problems are often difficult to differentiate from plantar fascial problems and nerve entrapments about the heel (Murphy & Baxter, 1985).

Frederick and Clarke (1982) studied the effect of body size on the biomechanical consequences for runners. Taller runners were found to weigh more, and vertical ground reaction forces were disproportionately greater than in shorter runners. Although the taller runners had larger feet, on the average their feet were not large enough to compensate for the increased forces at footstrike. Conventional running shoe construction generally has a given thickness for the midsole and heel wedge throughout all sizes, which means that what might be adequate cushioning in midsize ranges will be inadequate in the larger shoe sizes for taller, heavier runners. Conversely, in the smaller shoe sizes the cushioning will be relatively great, with possible instability and a relative lack of flexibility. This is based on the concept that the load on the foot is proportional to foot length to the 2.5 power, and because the surface area under the foot is proportional only to foot length squared, the pressure that the foot is applying to the shoe should be greater, on the average, for larger, taller runners and generally lesser for shorter, lighter runners. This implies that the cushioning for larger shoes will have a lower shock-absorbing capacity unless the cushioning is enhanced with a thicker midsole or some other means of increasing its shock-absorbing capability.

Bates, James, and Osternig (1978a) conducted a study on a comparison between walking and running shoes, studying several parameters. They concluded that in spite of a multitude of tests that can be conducted on shoes and shoe materials, one must keep in mind that once the shoe is placed upon the foot, it becomes a dynamic, functional element. Much of the testing is done under static conditions, thus not truly reflecting the effect the shoe will have on the foot during use. At best today, we can say that shoes have definitely improved due to design features based upon what appear to be sound biomechanical principles, but there is no best shoe for every runner, and in the extreme instances, runners may even require customized shoes.

Orthotics

Studies (Bates, Osternig, Mason, & James, 1979b) have definitely shown that orthotic devices have a significant effect on the amount of maximum pronation, the time to maximum pronation, the maximum pronation velocity, the period of pronation, and movement of rearfoot angle in the first 10 degrees of foot contact. In our 1978 study (James et al.) 70% of the cases treated with orthotics were resolved. Even a soft orthotic can affect the degree of pronation and rapidity of pronation. A heel lift of soft materials alone reduces the amount of ankle dorsiflexion and thus the required pronation, as dorsiflexion is a component of pronation, and Bates et al. (1978a) found the effect of just wearing a shoe with a heel reduced pronation.

Nigg (1985) and coworkers did a study that linked various leg ailments to patterns of rearfoot movement in which the average pronation during heel contact was significantly greater in subjects complaining of Achilles tendon problems than in a group who ran pain-free. This same study showed the addition of a medial support to help control the amount of pronation in the subjects with symptoms reduced the pronation below that of the symptom-free subjects with uncorrected shoes.

Smith, Clarke, Hamill, and Santopietro (1983) studied the effect of soft versus semirigid orthotic devices on leg and foot motions in well-trained runners. Significant differences were noted only between controls and semirigid orthotics where the maximal calcaneal eversion was reduced to some degree but a relatively large decrease in the rate of maximal calcaneal eversion was noted. This may play a role in reducing injury.

Shoe modifications and orthotic devices should not be designed to completely eliminate pronation, because it is a required function in the dissipation of normal stresses and accommodation of the foot to the running surface.

Although we can describe certain elements of biomechanical function of the lower extremity, particularly in the sagittal plane, we still have difficulty in total evaluation, particularly in transverse plane rotational abnormalities about the knee. Most of the biomechanical studies today that have implicated various force patterns within the knee joint are implied from other known parameters such as force plate studies, sagittal plane joint motion, and other forms of mechanical analysis. We still do not have an accurate and practical way for the clinician to thoroughly assess the transverse plane rotations of the extremity objectively before treatment to elucidate the diagnosis and then to determine the efficacy of treatment.

Alignment Problems in Runners

Malalignment of the lower extremity may occur at any of several locations from the pelvis to the foot. It is incumbent upon the examiner to do a thorough lower extremity examination as a prerequisite to appropriate diagnosis and treatment. It must be emphasized that the cause of injury may be remote from the site of injury. Emphasizing the site of injury alone in the examination quite likely will not resolve the condition.

A complete examination initially involves the standing alignment of the lower extremity. Excessive genu valgum or genu varum will not only affect foot placement along the line of progression but will rather dramatically change the loading of the medial and lateral compartments of the knee. There are very few genu valgum runners in our experience. Most good runners have a mild amount of genu varum. Excessive amounts of genu varum through the years can cause potential problems. Generally speaking, the medial compartment of the knee joint has 60% of the contact area and carries 60% of the load. Excessive genu varum will increase the amount of loading on the medial compartment, which through the years may result in degenerative changes, particularly if the individual has a medial meniscectomy or some other injury to the joint, disrupting its normal architecture. We think it is incumbent upon the practitioner to warn people, particularly with this type of situation, that running may enhance deterioration of the joint.

When the patient is standing, patellar alignment should be positioned directly anteriorly. Squinting patella, in which the patellae are inclined toward the midline, are often indicative of femoral and/or tibial malrotation. This brings to mind a group of people with a condition termed the *miserable malalignment syndrome* (James et al., 1978) who, in our estimation, would probably be better off seeking alternative sports to running. Multiple malalignment variations in the lower extremity are well exemplified by this group, who are generally seen with the complaint of anterior knee pain associated with running. Examination reveals a femoral neck anteversion with a relatively greater degree of internal rotation of the hip than external, a genu varum that is sometimes enhanced by hyperextension, squinting patellae that are inclined toward the midline, an excessive Q angle (Brattstrom, 1964), tibia varum, functional equinus secondary to tightness of the gastrocsoleus muscle group, and, frequently, compensatory foot pronation. Upon inquiring into their past history, it is not unusual to find that as children they wore corrective shoes, braces, or both. Through the years, these people also typically avoided sports in school because they were rather awkward. They were also instructed to walk with their feet straight ahead, which they did eventually accomplish, but at the expense of many rotational abnormalities from hip to foot.

Torsional and angular malalignments of the lower extremity have a significant influence on the patellofemoral joint mechanics, which may result in *anterior knee pain*. Much of the anterior knee pain formerly referred to as chondromalacia patellae is actually secondary to a patellofemoral joint incongruency with the patella maltracking in the sulcus, creating abnormal patellofemoral joint compression forces. This is not unlike driving one's car with the front wheels out of alignment, creating wear on the tires. Persistent anteversion of the femoral neck, or at least a clinically apparent anteversion of the femoral neck, with greater internal femoral rotation than external rotation often goes unrecognized. This is frequently associated with a series of compensatory growth disturbances throughout the limb. Somerville (1957) pointed out what he called "persistent fetal alignment of the hip" (p. 107) with femoral neck anteversion that is normally molded away with growth. In this situation, although the hip joint develops normally, the arc of rotation is abnormal with internal rotation being greater than external rotation, which creates an in-toed, somewhat clumsy gait in children. In an attempt to correct this, the feet are voluntarily placed in the frontal plane, but this is accomplished with eventual secondary changes in the form of external torsion of the tibia and frequently compensatory pronation of the foot.

Fabry and McEwen (1973) indicated a similar syndrome in 30% of subjects studied who had a femoral torsion. Practically none of these patients complained of knee pain, which was felt to be a potential problem secondary to possible malalignment of the patella, or compensatory changes at the knee joint. Kleiger (1968) has described the *anteversion syndrome*. This syndrome may not be caused by a true femoral neck anteversion but may include such elements as soft tissue contractures about the hips limiting external rotation, asymmetrical placement of the head of the femur onto the neck, or abnormal shape, location, or rotation of the acetabulum with respect to the pelvis and hip joint. Persistent anteversion of the hip can result in compensatory growth deformities distally.

Sikorski, Peters, and Watt (1979) studied the importance of femoral rotation as an etiological factor in patellofemoral pain. Their study found that in normal children there was internal or medial femoral rotation with the onset of muscular activity in the quadriceps, but in children with patellofemoral joint pain, the femur externally rotated with muscular activity. It was postulated that with internal femoral rotation, a large lateral condyle was presented to the lateral facet, and the patella was able to track without rotation or angulation of the long axis. With abnormal external femoral rotation, the medial femoral condyle becomes more prominent, possibly abutting on the medial facet of the patella, resulting in patellofemoral pathology.

Subtle maltracking of the patella, most apparent on a tangential view of the patella on X ray, may not cause problems for most daily activities but, with the accumulated stress of running, can create serious problems in the patellofemoral joint due to an accumulation of stress with the repetitive act of running. This relates to load application with magnitude and geometry (point of application and direction) as described by Nigg (1985). Examination of the patella may reveal instability with lateral pressure on the patella while the knee is flexed at about 30 degrees. This condition is frequently associated with a maltracking patella resulting in abnormal patellofemoral joint forces. Significant genu varum or valgum deformities likewise can affect patellofemoral joint function, as well as medial and lateral compartments of the knee due to excessive stresses on the articular surfaces.

The varus angle of the leg to the floor is directly related to the amount of compensatory foot pronation. A varus angle of the leg in excess of 10 degrees is abnormal, requiring compensatory foot pronation. As earlier pointed out in the Bates et al. (1978a) study, increased running speed will also dynamically increase the leg varus alignment to the running surface, and this, in addition to perhaps an already excessive varus angle, may certainly lead to a problem commensurate with those associated with compensatory foot pronation.

The angle of gait can also give some clue to malalignments that may indicate tibial torsion problems that are generally congenital, although in some instances secondary to trauma. In our experience, we have often encountered runners who had a natural toeing-out gait and developed subsequent knee problems when they were coached to run with their feet straight ahead. Toeing out represents a normal alignment for their lower extremity and may well stem from their femoral neck angle. Generally speaking, the abductors of the hip place the hip joint in a position where the hip abductors can function most efficiently. In some people with a reduced femoral neck angle, this may result in external rotation of the entire lower extremity, and an attempt to run with the feet straight ahead will place the abductors at a mechanical disadvantage and quite likely induce excessive transverse plane rotations at the knee, which is caught between the foot and the hip. They may look better running but be more inefficient and develop problems. Thus, not only can local conditions affect patellofemoral joint forces, but structural malalignment proximal and distal to the knee can also be a cause of problems.

The treatment of malalignment of the lower extremity, particularly proximal to the foot, presents a rather complex challenge to the clinician. We feel that the miserable malalignment group, as described, can be seen very early on in the developmental stages of childhood, but once the age of 3 or 4 has been attained, probably very little can be done. This suggests that we should pay more attention to malalignment in young children when corrections can still be made conservatively.

Modalities that can be used in the treatment of malalignment problems once they have occurred in the mature runner are flexibility exercises; muscle strengthening to restore muscle balance; orthotics; shoe modification; and, in the more severe, persistent cases, surgical correction, particularly of malrotations. We have, in some instances with severe symptoms, performed high tibial derotation osteotomies to correct a rather severe out-toeing condition that had been causing knee pain. This, of course, is dramatic surgery and is done only on rare occasion. Most conditions in regard to the patellofemoral joint can be resolved by an intensive rehabilitation program to restore muscle balance, endurance, and flexibility (particularly hamstrings).

General Problems

Patellar tendinitis or ''jumper's knee'' is a very common problem related to excessive stress at the tendon insertion on the patella, resulting in microruptures and localized areas of tendon degeneration. The symptom is pain at the insertion site of the tendon on the distal pole of the patella. This is characterized by aching after exercise that often becomes progressively disabling, preventing running in many instances. This type of individual may also have an increased Q angle or a Q angle that dramatically changes dynamically during pronation/supination of the foot with subsequent internal-external tibial rotation, causing an eccentric traction on the tendon.

Patellar tendinitis, as with other anterior knee problems, generally responds to conservative treatment involving medication, relative rest or reduction in mileage, functional orthotics, or shoe modifications. Intractable cases may require surgical exploration. Patellar tendon ruptures can occur but are rather rare and are usually preceded by a prolonged period of chronic patellar tendon tendinitis with perhaps steroid injections into the tendon.

Other common problems about the knee are the iliotibial tract friction syndrome, popliteal tendon tenosynovitis, meniscal lesions, and plica syndromes.

The *iliotibial tract friction syndrome* is the most common problem involving the lateral aspect of the knee. It presents with a spontaneous onset of pain, usually initiated by an extra-long run, and is characterized by lateral pain that usually starts a mile or two into the run and becomes worse for the duration of the run. Findings are tenderness over the distal iliotibial tract and lateral femoral condylar region with occasional swelling and sometime crepitation. Clinically, this type of individual will frequently present with an increased genu varum or tibial varum in conjunction with a heel and forefoot varus and shoes that show excessive lateral heel wear. This type of alignment theoretically causes increased friction between the iliotibial band and the lateral femoral condyle, due to greater tension on the lateral structures.

Popliteal tendon tenosynovitis is not as frequent as iliotibial tract friction syndrome. It involves the popliteal tendon posterolaterally and may vary from a transient condition to one that becomes chronic.

The pathomechanics of popliteal tendinitis according to Mayfield (1977) is that in downhill running there is an increased vector to displace the weight-bearing femur forward on the relatively fixed tibia as the knee is increasingly flexed. Electromyographic studies (Mann & Hagy, 1980) indicate that the popliteus muscle is active during this particular weight-bearing phase of gait and may act to retard the femur from forward displacement on the tibia in conjunction with the quadriceps. More specifically, it helps retard the lateral femoral condyle from rotating forward in relation to the lateral tibial plateau. Downhill running, therefore, may cause increased stress on the popliteus muscle-tendon unit in an effort to decelerate body weight against the altered angle of gravitational pull with resultant tenosynovitis and symptoms.

Treatment for the iliotibial tract friction syndrome and popliteal tendon tenosynovitis are similar. One must thoroughly evaluate the training program, look for anatomic variations in the lower extremity as described, rest or reduce mileage, change or modify shoes, consider the use of functional orthotics, contrast heat and cold over the area of involvement, and consider anti-inflammatory medication and possibly steroid injection during the acute/subacute phase, along with stretching exercises, particularly for the iliotibial tract.

Another source of lateral knee pain, particularly in the middle-aged and older runner, is a *meniscal lesion* involving the lateral meniscus of the knee. This is usually an attrition-type injury and not very commonly encountered, but it must be considered with lateral knee pain. Meniscal lesions are actually more frequent on the medial side of the knee and are usually a degenerative type of lesion rather than the acute tear that one might encounter in a younger athlete involved in sports requiring rapid change in direction. A genu varum deformity of the knee in a runner will enhance the forces transmitted across the medial aspect of the joint and may result in degenerative tears, particularly involving the posterior horn of the medial meniscus.

Another condition frequently encountered in runners and one that has been diagnosed primarily since the advent of arthroscopy is that of the *plica syndrome* (Hardaker, Whipple, & Bassett, 1980). The plica may be described as an atavistic fold of synovial tissue within the joint that may become irritated with chronic or repetitive activities such as running. It has not necessarily been associated with any specific biomechanical abnormality in running. The most common locations are in the suprapatellar pouch area and along the medial retinacular region. The medial parapatellar plica is the more common variety encountered and may mimic other conditions affecting the knee. It is often a diagnosis of suspicion.

The soleus and posterior tibial muscles have been implicated in the *medial tibial stress syndrome*. Michael and Holder (1985) feel that the medial tibial stress syndrome is actually secondary to inflammation of the fascial attachments of the soleus muscle rather than the posterior tibial muscle. They point out that the fascial attachments of the soleus muscle attach along the medial border of the tibia distally where the pain is frequently located, and the posterior tibial muscle takes origin more proximally. The tibialis posterior muscle actually rises from the upper two thirds of the interosseous membrane, the medial fibula and the lateral tibial, and its origin is not contiguous with the posterior medial border of the tibia as is that of the soleus. Electromyographic studies indicated that the medial half of the soleus muscle was not only a strong plantar flexor of the foot but also an inverter of the heel and thus is also involved in resisting compensatory pronation or eversion of the heel during the support phase. Biomechanically, it is felt that the soleus, which inserts more medially on the calcaneus, is eccentrically stressed with excessive heel eversion.

Viitasalo and Kvist (1983) compared the alignment of the lower leg and heel while standing, the passive range of mobility of the subtalar joint, and angular displacement between the calcaneus and midline of the lower leg in a group of patients with shin splints. Their statistics showed significant change in the Achilles tendon angle with values being greater in the shin splint group. The athletes with shin splints also had significantly greater angular displacement values in inversion and eversion, and during running, the Achilles tendon angle in this group was found to be significantly greater than with the asymptomatic group. (The authors pointed out the term *shin splints* is also termed *medial tibial stress syndrome*.) They found the increased pronation of the subtalar joint to be due to the causes listed elsewhere such as forefoot varus, rearfoot varus deformity, and/or a functional ankle joint equinus with a tight triceps surae.

Achilles tendon disorders have been biomechanically associated with a greater Achilles tendon angle in conditions usually associated with compensatory pronation of the foot. Clement, Taunton, and Smart (1984), by the use of slow-motion/high-speed cinematography, noted prolonged pronation producing a whipping action of the Achilles tendon, which could potentially initiate microruptures or tears within the substance. Smart, Taunton, and Clement (1980) speculate that torsional forces transmitted through the Achilles tendon in situations with prolonged excessive pronation may cause vascular impairment, resulting in degenerative changes of the Achilles tendon. Most Achilles tendon involvement is located 2 to 5 cm proximal to the insertion into the calcaneus, which is an area of decreased vascularity. Thus any condition resulting in increased torsional forces in this area might further hamper blood supply.

Biomechanical considerations of shoes must also be taken into account with Achilles tendon disorders. An Achilles pad or heel counter that applies pressure to the Achilles tendon may result in a chronic source of tendinitis or peritendinitis. A shoe with an inadequate medial heel wedge allowing for excessive pronation, particularly one with a durometer that is too soft to support the heel or control hindfoot motion, may lead to increased Achilles tendon angulation and subsequent problems. The sole of the shoe should also be flexible. A stiff-soled running shoe creates a longer anterior lever from the ankle, thus creating greater Achilles tendon tension during heel rise. Likewise, a tight gastrocsoleus muscle group limiting ankle dorsiflexion will cause excessive tension on the Achilles tendon.

A *retrocalcaneal bursitis* and combined tendinitis are not an unusual source for hindfoot pain (Jones & James, 1984; Leach, James, & Wasilewski, 1981). Biomechanical and anatomical factors to consider in this condition are compensatory pronation, cavus-type feet, shoe construction, training surface, and training errors.

Plantar heel pain is most commonly caused by *plantar fasciitis*, which is secondary to microruptures at the insertion of the calcaneus. Not infrequently, treatment is directed at the insertion site with injections, heel pads, heel cups, and, in some instances, removal of secondary bone spurs. All of these are usually to no avail as the problem is not at the site of injury but secondary to abnormal foot biomechanics such as compensatory pronation or a rigid cavus-type foot, both conditions placing excessive tension on the insertion of the plantar fascia. Higher values occur during heel rise when body weight is concentrated on the forefoot and the triceps surae exerts traction on the calcaneus. In addition to this, the downward acceleration of the body increases ground reaction force by 20% (Perry, 1983). These factors combine to place tremendous tension on the plantar fascia during running and in conjunction with abnormal foot biomechanics, such as compensatory pronation or a cavus foot, may result in microtrauma to the plantar fascia due to repetitive overloading or accumulation of stress. Until the abnormal biomechanical function is corrected, local treatment for the plantar fasciitis will probably fail.

Murphy and Baxter (1985) describe a condition often associated with plantar fasciitis in which the small nerve to the abductor digiti quinti muscle, passing deep to the insertion of the plantar fascia, may become involved. Decompression of this nerve by partial fascial release will frequently relieve heel pain. Again, biomechanical abnormalities of the foot are the basic etiology, and initial treatment should be directed at full biomechanical analysis of foot function.

Stress fractures (McBryde, 1982) in runners most commonly involve the tibia, fibula, and metatarsals with the femur and pelvis occasionally involved. Femoral stress fractures are the most *critical* and require protective weight bearing until healed. Bone is a dynamic tissue capable of remodeling its structure in response to stress. If bone has sufficient time to remodel, it occurs uneventfully as a natural physiological process. The problem arises when the stress is applied too rapidly. The normal remodeling process requires a balance between bone resorption and replacement, but if the resorption process proceeds faster than the replacement due to continued repetitive stress application (running), microfractures occur, resulting in the clinical picture of a stress fracture. Etiological factors include training errors (rapid change), extremity malalignment problems, muscular imbalance, and compensatory gait alterations often secondary to another injury.

Benefits of Combining Clinical Evaluation and Laboratory Studies

The benefits of combining a clinical evaluation with laboratory studies is typified by the case of a 34-year-old world-class distance runner complaining of chronic, recurrent left Achilles tendinitis and a sense that his right lower extremity was not functioning normally. The onset of his problems began some two years earlier when he trained for and won the New York Marathon, which was a new distance for him, and he felt the increased training intensity had a deleterious effect on his body.

Examination revealed a marked pelvic tilt to the right. The left Achilles tendon was enlarged and he had cavus-type feet. Leg length X rays revealed a 2.4-cm shortening of the *right* lower extremity. Cybex testing showed the left gastrocsoleus muscles to be stronger.

It was felt that his problems related to the leg length discrepancy that he had accommodated to through the years. However, training for the marathon constituted a dramatic change and his body was now unable to accommodate.

Biomechanical studies were done on a treadmill with lightweight accelerometers on the lower tibia to measure peak deceleration at footstrike as a measure of the "shock" being transmitted to the foot and lower leg. Also, rearfoot movements were analyzed. The left leg was experiencing on the average 19% more shock at footstrike than the shorter right leg. The left foot contact was on the heel whereas on the right it was midfoot. Rearfoot motion analysis revealed more supination on the right at touchdown. Maximum pronation was greater on the left. Insertion of orthotics generally increased supination and decreased pronation on both feet. There was consistently longer contact time on the left foot and a longer flight time when pushing off with the right foot. He also exhibited an exceptionally large distance between the right and left foot placement. These studies revealed significant asymmetry requiring considerable compensation on the left, resulting in increased stress on the Achilles tendon.

The treatment program was to modify his training program to an acceptable level and gradually modify the right shoe midsole height to accommodate to the leg length discrepancy. The initial effect has been positive. Unfortunately, this type of comprehensive approach is rarely available in the usual clinical setting.

Conclusion

Conclusions drawn in 1978 (James, et al.) have remained true even today for the most part and they are as follows:

- The vast majority of injuries among long-distance runners are the result of improper training.
- No specific anatomical factor correlates consistently with any specific injury.
- A wide variety of overuse syndromes or injuries may respond to one specific modality of treatment.
- Most problems in distance runners can be resolved with adequate rest but may recur with resumption of running if the etiology has not been determined.

- Present methods of applying foot and leg biomechanics from a practical standpoint are still somewhat inexact clinically in an office environment but nevertheless are often very effective in diagnosis and treatment.

In summary, an appropriate approach to the analysis and treatment of running injuries consists of the following:

1. Evaluate the training program, particularly looking for a rapid change.
2. Look for anatomical variations in the lower extremity from hip to foot.
3. Consider reduced mileage and/or rest.
4. Consider shoe change or modification.
5. Consider orthotics.
6. Consider alternative exercise.
7. Consider a structured, progressive return to running after the injury has been resolved.

Even with apparent lower extremity malalignment or abnormal biomechanical conditions, most runners will do well with an appropriately designed training program. The body is a tremendously adaptable mechanism and, if given time to accommodate to stress, will usually respond favorably.

References

Bates, B.T., James, S.L., & Osternig, L.R. (1978a). Foot function during the support phase of running. *Running*, **3**, 24-29.

Bates, B.T., James, S.L., & Osternig, L.R. (1978b). Foot function during the support phase of running. In P. Asmussen & K. Jorgensen (Eds.), *Biomechanics VI*. Baltimore: University Park.

Bates, B.T., Osternig, L.R., Mason, B.R., & James, S.L. (1979a). Functional variability of the lower extremity during the support phase of running. *Medicine and Science in Sports and Exercise*, **11**, 328-331.

Bates, B.T., Osternig, L.R., Mason, B.R., & James, S.L. (1979b). Foot orthotic devices to modify selected aspects of lower extremity mechanics. *American Journal of Sports Medicine*, **7** (Suppl 6), 338-342.

Brand, P.W. (1976). Pressure sores—the problem. In R. Kenedi & J. Cowden (Eds.), *Bedsore biomechanics* (pp. 19-23). Baltimore: University Press.

Brattstrom, H. (1964). Shape of intercondylar groove normally and in recurrent dislocation of the patella. *Acta Orthopedica Scandinavica*, (Suppl. 68), 5-198.

Cavanagh, P.R., & Lafortune, M.A. (1980). Ground reaction in distance running. *Journal of Biomechanics*, **13**, 393-406.

Cavanagh, P.R., Valiant, G.A., & Misevich, K.W. (1984). Biological aspects of modeling shoe/foot interaction during running. In E. Frederick (Ed.), *Sport shoes and playing surfaces: Biomechanical properties* (pp. 24-46). Champaign, IL: Human Kinetics.

Clarke, T.E., Cooper, L.B., Clark, D.E., & Hamill, C.L. (1985). The effect of increased running speed upon peak shank deceleration during ground contact. In D. Winter, R. Norman, R. Wells, K. Hayes, & A. Patla (Eds.), *Biomechanics IX-B* (pp. 101-105). Champaign, IL: Human Kinetics.

Clarke, T.E., Cooper, L.B., Clark, D.E., & Hamill, C.L. (1985). The effect of increased running speed upon peak shank deceleration during ground contact. In D. Winter, R. Norman, R. Wells, K. Hayes, & A. Patla (Eds.), *Biomechanics IX-B* (pp. 101-105). Champaign, IL: Human Kinetics.

Clarke, T.E., Frederick, E.C., & Hamill, C. (1984). The study of rearfoot movement in running. In E. Frederick (Ed.), *Sport shoes and playing surfaces: Biomechanical properties* (pp. 166-189). Champaign, IL: Human Kinetics.

Clement, D.B., Taunton, J.E., & Smart, G.W. (1984). Achilles tendinitis and peritendinitis: Etiology and treatment. *American Journal of Sports Medicine*, **12**, 179-184.

Denoth, J. (1985). Load on the musculoskeletal system and modelling. In B. Nigg (Ed.), *Biomechanics of running shoes* (pp. 103-172). Champaign, IL: Human Kinetics.

Fabry, G., & McEwen, G.C. (1973). A follow-up study in normal and abnormal conditions. *Journal of Bone and Joint Surgery*, **55-A**, 1726-1738.

Frederick, E.C., and Clarke, T.E. (1982). Body size and biomechanical consequences. In R.C. Cantu & W.J. Gillespie (Eds.), *Sports medicine, sports science: Bridging the gap* (pp. 47-57). Lexington, MA: Collamore.

Frederick, E.C., Clarke, T.E., & Hamill, C.L. (1984). The effect of running shoe design on shock attenuation. In E. Frederick (Ed.), *Sport shoes and playing surfaces: Biomechanical properties* (pp. 190-198). Champaign, IL: Human Kinetics.

James, S.L., & Brubaker, C.E. (1973). Biomechanics of running. *Orthopedic Clinics of North America*, **4**, 605-615.

Jones, D.C., & James, S.L. (1984). Partial calcaneal ostectomy for retrocalcaneal bursitis. *American Journal of Sports Medicine*, **12**, 72-76.

Kleiger, B. (1968). The anteversion syndrome. *Bulletin of Hospital Disease*, **29**, 22-37.

Kuhns, J.G. (1949). Changes in elastic adipose tissue. *Journal of Bone and Joint Surgery*, **31-A**, 541-545.

Leach, R.E., James, S.L., & Wasilewski, S. (1981). Achilles tendinitis. *American Journal of Sports Medicine*, **9**, 93-98.

Mann, R.A., Baxter, D.E., & Lutter, L.D. (1981). Running symposium. *Foot and Ankle*, **1**, 190-224.

Mann, R.A., & Hagy, J.L. (1980). Running, jogging and walking. A comparative electromyographic and biomechanical study. In J. Bateman & A. Trott (Eds.), *The foot and ankle* (pp. 161-165). New York: Brian C. Decker.

Mayfield, G.W. (1977). Popliteus tendon tenosynovitis. *American Journal of Sports Medicine*, **5**, 31-36.

McBryde, A.M., Jr. (1982). Stress fractures in runners. In R. D'Ambrosia & D. Drez (Eds.), *Prevention and treatment of running injuries* (pp. 21-42). Thorofare, NJ: Charles B. Slack.

McKenzie, D.C., Clement, D.B., & Taunton, J.D. (1985). Running shoes, orthotics and injuries. *Sports Medicine*, **2**, 334-337.

Michael, R.H., & Holder, L.E. (1985). The soleus syndrome. A cause of medial tibial stress (shin splints). *American Journal of Sports Medicine*, **13**, 87-94.

Murphy, P.C., & Baxter, D.E. (1985). Nerve entrapment of the foot and ankle in runners. *Clinics in Sports Medicine*, **4**, 753-763.

Nigg, B.M. (1985). Biomechanics, load analysis and sports injuries in the lower extremities. *Sports Medicine*, **2**, 367-379.

Perry, J. (1983). Anatomy and biomechanics of the hindfoot. *Clinical Orthopedics*, **177**, 9-15.

Ross, C.F., & Schuster, R.O. (1983). A preliminary report in predicting injuries in distance runners. *Podiatric Sports Medicine*, **73**, 275-277.

Sikorski, J.M., Peters, J., & Watt, I. (1979). The importance of femoral rotation in chondromalacia patellae as shown by serial radiography. *Journal of Bone and Joint Surgery*, **61-B**, 435-442.

Slocum, D.B., & James, S.L. (1968). Biomechanics of running. *Journal of the American Medical Association*, **205**, 720-728.

Smart, G.W., Taunton, J.E., & Clement, D.B. (1980). Achilles tendon disorders in runners—a review. *Medicine and Science in Sports and Exercise*, **12**, 231-243.

Smith, L., Clarke, T., Hamill, C., & Santopietro, A. (1983). *The effect of soft and semirigid orthoses upon rearfoot movement in running*. Unpublished presentation to the American College of Sports Medicine. Nike Research Laboratory, Beaverton, Oregon.

Somerville, E.W. (1957). Persistent frontal alignment of the hip. *Journal of Bone and Joint Surgery*, **39-B**, 106-113.

Viitasalo, J.T., & Kvist, M. (1983). Some biomechanical aspects of the foot and ankle in athletes with and without shin splints. *American Journal of Sports Medicine*, **11**, 125-130.

Chapter 11

Relationships Between Distance Running Biomechanics and Running Economy

Keith R. Williams

Through casual observation it is obvious that there is considerable variability in the running mechanics of different individuals. It is of interest to know how this variation is related to the performance ability of runners. Are there aspects of running mechanics that can give a runner an advantage in performance, or conversely, patterns of movement that might adversely affect performance? Though performance measures might be the most appropriate criteria for investigations into relationships between performance and running mechanics, experimental studies involving distance running performance are very difficult to control because so many factors affect performance. An alternative criterion commonly used is running economy, submaximal metabolic energy expenditure ($\dot{V}O_2$submax). As will be discussed below, running economy can be logically linked to performance and has often been used as a dependent variable in biomechanical studies of distance running.

A large variation in running economy is usually found within a group of runners with similar performance abilities. It is typical to find a range in $\dot{V}O_2$submax of between 15% and 30% (Cureton & Sparling, 1980; Daniels, Scardina, & Foley, 1984; Mayhew, 1977; Williams & Cavanagh, 1987). Figure 11.1 illustrates such a distribution for a group of elite male runners. From the biomechanical standpoint we are interested in how much of this variability in $\dot{V}O_2$submax between individuals can be accounted for by differences in running mechanics. Though there are certain to be physiological factors (e.g., muscle fiber composition, state of training) and psychogenic factors (e.g., hypnosis, meditation, relaxation) that affect running economy, running mechanics are frequently cited as being important to the economy or efficiency of running (Cavanagh & Kram, 1985; Hagan, Strathman, Strathman, & Gettman, 1980). It appears reasonable to expect that mechanics of movement do have a substantial influence

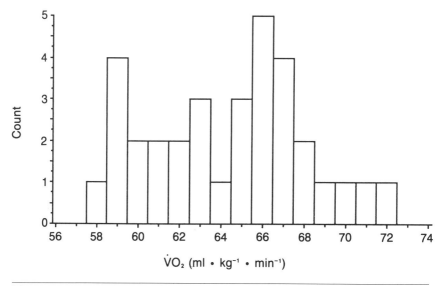

Figure 11.1. Distribution of net oxygen consumption among a group of subjects running at 5.36 m • s⁻¹.

on the metabolic energy costs involved in running. Daniels et al. (1984) compared running economy for a group of subjects with other measures of exercise economy ($\dot{V}O_2$submax during arm cranking, stepping, cycling, and grade walking), and concluded that economy in one mode of exercise did not insure economy in another mode. They hypothesized that variations in economy might be due to differences in genetic factors (which cannot be changed through practice), motor unit recruitment, anatomical mechanical advantage, or movement skill. The purpose of this chapter is to examine the evidence that is available relating running mechanics and running economy.

Running Economy and Performance

Although having a low $\dot{V}O_2$submax during running at a given speed does not necessarily correlate with a good performance, there is an important relationship between running economy and performance. Many studies have shown a strong relationship between maximal oxygen consumption ($\dot{V}O_2$max) and running performance (Daniels, 1985), but there are also clear indications that running economy can be an important factor (Conley & Krahenbuhl, 1980; Daniels, 1985; Sjödin & Shele, 1982). Conley and Krahenbuhl (1980) found correlations between $\dot{V}O_2$submax and running performance of 0.79, 0.82, and 0.83 for three different submaximal speeds using a group of runners of very similar abilities. A more complete discussion of the relationships between running economy, $\dot{V}O_2$max, and performance has been presented by Daniels (1985). Though it has been the measure most frequently used, submaximal oxygen consumption is not the only physiological measurement that could be used to investigate relationships between the mechanical and physiological aspects of running (Dotan, Rotstein, Dlin, Inbar, Kofman, & Kaplansky, 1983). For example, Komi, Ito, Sjödin, Wallenstein, and Karlsson (1981) reported significant relationships between mechanical power during distance running and the running speed at which blood lactate begins to accumulate, muscle fiber composition, and capillary density.

Perhaps most convincing is a simple argument relating the effect of a change in $\dot{V}O_2$submax on performance. If an individual's running mechanics were altered, and this resulted in reduced energy costs throughout a range of submaximal speeds of running, it is logical to assume that the individual's maximal running performance would improve. For example, assume that originally an individual could sustain some maximal average speed of running over a given distance. Both aerobic and anaerobic metabolism would be involved at this speed, and performance would ultimately be limited by one or more factors, such as lactate accumulation, glycogen depletion, psychological stress due to fatigue, and so forth. Assuming that changes in running mechanics could be effected that lowered metabolic costs, energy demands at the original maximal average speed would be lowered, and the factors limiting performance would be at subcritical levels. The individual would then be able to increase running speed by an amount that would bring these factors back to the point where they would again limit performance. This rationale is illustrated in Figure 11.2.

It is important to put the magnitude of potential changes or differences in $\dot{V}O_2$submax into perspective. Frederick (1983a) reasoned that even what seem to be relatively small changes in $\dot{V}O_2$submax can be important ones. We might rationalize that a 2% decrease in $\dot{V}O_2$submax could lead to a 2% improvement in performance time. Though a 2% change at first glance seems small, when put into the context of a distance running time it can easily be the difference between individual performances, as illustrated in Table 11.1. At world-record 10K pace a 2% difference would amount to approximately 32 seconds, and at a world-record marathon pace the difference would be more than 2 minutes. The potential importance of small differences in $\dot{V}O_2$submax creates some problems for experimental research in this area. Although a 2% difference might be meaningful to the performer, it is much more difficult to statistically validate differences this small due to the measurement errors inherent in the analysis procedures employed. More definitive identification of relationships between running economy and running mechanics may necessitate either improved experimental procedures and measurement techniques or a different approach to the evaluation of statistical significance.

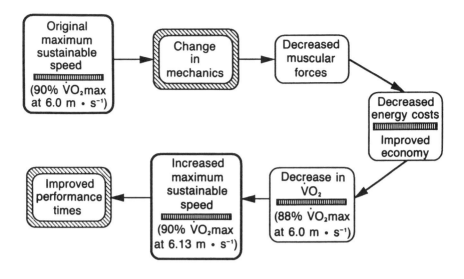

Figure 11.2. A proposed mechanism for improved performance times based on alterations in running mechanics and their effects on running economy.

Table 11.1 Comparisons of Best Performances for the Marathon in 1985 for Men and Women

	Men		Women	
Rank	Time (hr.min:sec)	Difference (%)	Time (hr.min:sec)	Difference (%)
1	2.07:12	100.0	2.21:06	100.0
2	2.07:13	99.9	2.21:21	99.8
3	2.08:08	99.3	2.23:29	98.3
4	2.08:09	99.3	2.27:50	95.2
10	2.09:05	98.5	2.28:38	94.7
20	2.10:23	97.6	2.32:23	92.0

Gaining an understanding of the process of optimization of movement patterns is important if we are to truly understand how running mechanics influence running economy. An implicit assumption is usually made that the human body will self-optimize a performance task over a period of time, probably subconsciously (Cavanagh & Williams, 1982). Though some contradictions exist in the literature, Daniels (1985) deems it justifiable to accept the premise that training will generally have the effect of reducing $\dot{V}O_2$submax. Bransford and Howley (1977) showed that trained men and women on average have lower metabolic energy costs at a given speed of running when compared to their untrained counterparts. In contrast, Petray and Krahenbuhl (1985) monitored changes in economy in groups of 10-year-old boys who received either instruction on the technique of running, running training, or a combination of instruction and running training over an 11-week period, but no differences in economy were found between any of the groups. Instruction, which focused on reducing vertical oscillation and being aware of stride length and stride rate, may have been too minimal to cause marked changes in economy. How completely an individual can self-optimize, or by how much individuals vary in their ability to optimize, is poorly understood. In a study that did not involve running but instead a simpler leg movement, Hatze (1976) demonstrated a remarkable improvement in performance when kinematic feedback was provided after a period of initial self-optimization. The extent to which the addition of feedback not normally present might affect running is currently not known.

Measurement Considerations

The precision of measuring systems can be an important factor affecting studies of the biomechanics of running in relation to running economy. The greater the variability resulting from the methodology employed, the more difficult it will be to identify clear relationships. This is confounded even more when trying to draw conclusions from several studies where different investigators have used different procedures. Because it is unlikely that there will soon be any adoption of standard methods for either biomechanical or physiological data collection, the best that can be done is to try to minimize experimental error as much as possible.

There are several aspects of data collection that are particularly relevant to the task of identifying relationships between running economy and running mechanics, and these will be discussed briefly.

Planar Versus Three-Dimensional Analyses

During the last decade there has been a great deal of work done in the area of three-dimensional (3D) kinematic data collection. Much of the research has emphasized the techniques of data collection, and relatively few studies have involved the 3D kinematics of running. Given the added complexity and increased costs associated with 3D methodology, the necessity of such techniques is a very relevant question. Certainly there are many measures of running mechanics that cannot be obtained using traditional planar methods. An assessment of the contributions of the arms to running mechanics (Hinrichs, Cavanagh, & Williams, 1983; Hinrichs, Cavanagh, & Williams, 1987) would be very limited without three-dimensional data, as would measures of the pattern of hip and shoulder rotation during running (Williams, 1982). For these types of studies, 3D methods are a necessity.

Of particular interest, however, is the question of whether important information is being missed when a variable having three-dimensional components is measured using planar techniques. Very little data is available to answer this question. Though running certainly has three-dimensional components, the greatest proportion of segmental movements takes place in a sagittal plane, and 2D measures may be representative of the more complex 3D movements. There is evidence available showing that for some types of variables 2D estimates are very close to results obtained using 3D methodology (Williams, 1985a). Whether 3D methods are necessary may depend on the objectives of a particular study. If variables are used that have major out-of-the-plane components, or if a high degree of precision is necessary to identify subtle differences, 3D methods might be warranted. Otherwise, it may be safe to use 2D methods and keep in mind that the data may lack some precision.

Treadmill Versus Overground Running

Another important consideration in analyzing the precise relationships between running economy and mechanics of running concerns the differences between treadmill and overground running. A number of studies have examined differences in mechanics between overground and treadmill running, with some disagreement over the nature and extent of the differences (Williams, 1985b). Of particular interest would be the question of whether there are differences in energy costs between overground and treadmill running that might be related to differences in mechanics. It has generally been found that overground running does incur greater metabolic energy costs compared to treadmill running, particularly at faster speeds of running (Daniels, 1985), and at least part of these differences can be attributed to air resistance (Daniels, 1985; Pugh, 1970). At the present time the extent that other differences in running mechanics between overground and treadmill running affect oxygen uptake is not known.

Generalizing Results of Experimental Studies

Samples of runners used as subjects in scientific studies of distance running have ranged from novice untrained runners to highly conditioned elite athletes. Because the physiological capabilities of disparate groups of subjects can be very different, it is possible that relationships between the mechanics of running and energy expenditure will also be affected. Though there may be fundamental relationships between running mechanics and metabolic energy costs that are common to all levels of running, there may also be some interactions that are influenced by factors such as speed of running or state of training. If this is the case, it may be inappropriate to apply the conclusions from a study involving elite runners to a group of collegiate or high school runners, and vice versa.

Another important question concerns the appropriateness of applying mean results for a group in a particular situation to individual runners, or to different running conditions. Although the results cited for a group of runners as a whole might suggest general relationships between running economy and biomechanical measures, they do not necessarily provide guidelines for shaping the running style of individual athletes. It has often been found that some individuals are physiologically economical but have some biomechanical measures that would be predicted to make them inefficient (Williams & Cavanagh, 1987; Williams, Cavanagh, & Ziff, 1987). At least two possibilities exist to explain this apparent inconsistency. First, it is possible that the overall economy associated with an individual runner depends on the influences of a large number of mechanical variables, as illustrated in Figure 11.3. For a given individual some variables might represent economical movement patterns whereas others are uneconomical, and it would be a weighted sum of the various influences that determines overall economy. Second, even though an individual might show mechanical characteristics that seem to be uneconomical based on group results, there may be reasons why those patterns

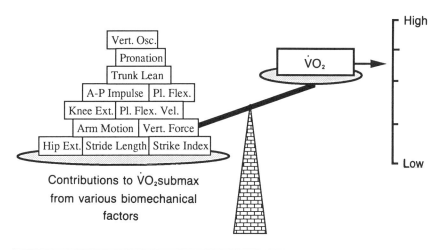

Figure 11.3. A conceptual model that suggests that oxygen consumption at a given speed of running is the result of combined influences from a variety of biomechanical factors.

are, in fact, economical for them. For example, there might be structural reasons why a particular running motion is best for a given individual, even though it results in a nontraditional running style.

Economy Versus Running Speed

A question relevant to those engaged in distance running research concerns how running economy changes over a range of speeds. Is a runner who is economical at one speed also going to be economical at a faster speed? Does the speed at which a runner trains most frequently become the most efficient speed for the runner? If one assumes that improved economy comes at least in part from more efficient mechanics, it seems logical that the more a runner trains at a given speed, the more finely tuned the associated mechanical movements might become. In a summary of the literature concerning running economy, Daniels (1985) stated that there was a linear or very nearly linear relationship between running speed and $\dot{V}O_2$submax, though some studies have cited curvilinear relationships as being the best fit for data (Hagan et al., 1980). Correlations between $\dot{V}O_2$submax and speed are typically in the range of 0.9 (Hagan et al.). A runner who is economical at a given speed of running will usually be economical at other speeds, as shown in Figure 11.4a. There are indications, however, that some runners become more (or less) efficient across a speed range relative to other runners. Figure 11.4b shows $\dot{V}O_2$submax for two runners across a range of speeds. Though one runner has lower energy costs at the slower speeds, that runner becomes less economical, compared to the other runner, at the faster speeds.

This presents a nagging problem to the researcher. Often it is desirable to study a runner at a speed different from his or her typical race pace.

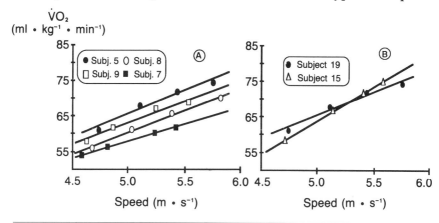

Figure 11.4. Changes in oxygen consumption during distance running over a range of speeds.

This is particularly true when a group of runners is being evaluated and it is desired to evaluate all runners at the same speed. Though it seems appropriate to assume that economy is a general characteristic of a runner across a range of speeds, the possibility exists that a given runner might be considered economical at one speed and uneconomical at another. The best procedure would be to test subjects across a range of speeds and use the results of a regression analysis to rank runners in terms of economy.

Variables Selected for Biomechanical Analyses

Several approaches have been used to investigate the relationships between running mechanics and running economy. One has been to measure $\dot{V}O_2$submax in a group of runners and, using correlational and regression techniques, identify those biomechanical variables that best relate to running economy. Ideally one could identify a set of mechanical variables that accounts for a major portion of the variance in $\dot{V}O_2$submax. A great deal of effort and expense is required to obtain a comprehensive set of biomechanical variables, and often a large number of subjects are needed before results can be statistically meaningful. Only a limited number of comprehensive studies have been done. An alternative approach has been to manipulate one or more specific aspects of running mechanics and observe the effect on submaximal oxygen consumption. Each of these methods has the potential to provide insight into the general relationship between running mechanics and running economy.

A multitude of variables might be used to describe the running motion, and it is difficult to know a priori which variables are the most appropriate to use. Should one collect as many variables as possible, or limit the analysis to a selected set? When a large number of variables are obtained, two problems exist: (a) Often there are high correlations between variables; and (b) the probability of identifying statistically significant relationships by chance alone is high. A problem with using only a small set of variables for analyses is that the most meaningful measures might be unknowingly left out. Often the variables chosen are ones that have been used in previous studies or that are easy to obtain, even though there are no definitive indications that they are the most appropriate.

One solution to some of these problems is to collect data for a large number of variables and then use a factor analysis procedure to identify independent sets of variables (Williams & Cavanagh, 1987; Williams et al., 1987). Table 11.2 shows a number of independent sets of variables that are highly correlated among themselves, obtained from sagittal view kinematic analysis of 12 collegiate runners. By using either a representative variable from each set, or more preferably a new factor variable that is a weighted sum influenced most by the highly related variables within

Table 11.2   Variables Included in Each Factor Identified by Factor Analyses for Side-View Kinematics During Distance Running at Speeds Between 3.6 and 5.5 m · s^{-1}

Factor 1
Support time
Nonsupport time
Thigh flexion
Thigh angle @ TO
Knee flexion in swing
Dorsiflexion
Plantar flexion
Ankle @ FS
Ankle @ TO
Heel horizontal velocity @ FS

Factor 2
Cycle time*
Thigh extension*
Thigh @ TO*
Thigh flexion velocity
Thigh extension velocity
Max knee linear velocity
Step length

Factor 3
Thigh @ FS
Vertical oscillation*
Knee @ FS*
Knee flexion velocity
Thigh extension*

Factor 4
Thigh extension*
Knee extension
Knee @ TO

Factor 5
Plantar flexion velocity
Heel vertical velocity @ FS

Factor 6
Trunk angle
Vertical oscillation*
Vertical oscillation asymmetry

Factor 7
Cycle time*
Knee extension velocity

Note. References to lower extremity joints indicate joint angles or angular velocity. TO = toe-off; FS = footstrike.
*Indicates variables with moderate loading weights for more than one factor.

a factor, the large set of initial variables can be reduced to a much smaller set of independent ones. Such an analysis should be done for each experimental situation, because the factors identified differ somewhat from study to study (Williams & Cavanagh, 1986; Williams & Cavanagh, 1987; Williams et al., 1987).

In reviewing studies that have examined the relationship between specific measures of running style and $\dot{V}O_2$submax it is important to consider the relevance of measures selected for analysis. Some measures are very specific to a particular time or interval during the running cycle, such as stride length or maximal knee angle during support, whereas others are more global in nature, such as total mechanical power during a running cycle. Relationships involving a single biomechanical measure and running economy must be kept in perspective. For example, assume that a study is done manipulating stride length, and relationships between varying stride length and $\dot{V}O_2$submax are identified. Much more is involved with a change in stride length than is evident from measuring the distance between successive footstrikes. Differences in stride length require adjustments in overall patterns of segmental movements. Lower extremity angles will be altered, vertical oscillation will be affected, and so forth. It would be more appropriate to say that variations in $\dot{V}O_2$submax are due to differences in the overall patterns of movement rather than to the specific measure of stride length. Thus, although investigators might concentrate on a limited number of specific measures, it is the more general changes in running patterns that will be directly related to energy costs.

Mechanisms Relating Running Economy and Running Mechanics

Why should there be a relationship between running economy and running mechanics? The metabolic energy costs above resting levels associated with running are primarily the result of increased muscular activity in the muscles directly involved with running movements or stabilization, with additional costs associated with increases in supportive metabolic functions (respiration, etc.). Oxygen carried in the blood is directly involved in aerobic metabolism, being a necessary component of the chain of chemical reactions that provide adenosine triphosphate (ATP) for the contractile process in muscle. Any alteration in movement patterns will be associated with changes in muscle activation, and this opens the possibility of either increasing or decreasing associated energy costs. Changes could occur either in the number of muscle fibers activated during a given movement or in the level of activation of the involved fibers. An example of such changes in muscle use can be found in a study by Komi, Golihofer, Schmidtbleicher, and Frick (1987) that presented preliminary data indicating that footwear having different shock-absorption characteristics can alter the activation patterns of lower extremity muscles. As someone becomes more skilled at running it is likely that metabolic energy costs will decrease. One simple explanation for this would be that the runner learns to activate only the necessary muscles. Inefficient movement patterns, or unnecessary co-contractions, are likely to increase metabolic costs without contributing to the forward progress of the runner.

Although alterations in muscular activation levels and firing patterns are likely to be associated with the process of becoming more economical, the relationship between a change in movement patterns and changes in energy cost are more complex than might first appear. There are a number of sources for the mechanical power involved in running at a given submaximal speed. The most obvious source is muscular contraction, but several other factors can also be major contributors (Williams, 1985c; Williams & Cavanagh, 1983; Winter, 1978). The stretch-shortening cycle, involving the elastic components of the musculotendinous tissues and often referred to as storage and return of elastic strain energy, has often been cited as a major contributor to the positive power generated during running (Cavagna, 1978; Cavagna, Komarekko, & Mazzolen, 1971). Potentiation of the contractile mechanisms has also been attributed to the stretch-shortening cycle (Bosco, Tarkka, & Komi, 1982; Cavagna, 1978). Transfer of energy between segments has more recently been proposed as an important power source, and some energy changes might also be attributed to passive sources, such as limitation to joint range of motion by bony or ligamentous restraints (Williams, 1985c).

Metabolic energy costs might decrease during running at a given speed if one of two changes were to occur: (a) a reduction in energy costs due to a decrease in total muscle use (which would be consistent with the idea of reducing unnecessary movements and muscle contractions); or (b) a reduction in the energy costs–associated segmental movements while maintaining a level of muscle use, such as might occur as the result of increased contributions from the stretch-shortening cycle or from transfer of energy. A cyclic activity such as running involves the same amount of positive and negative *mechanical* work over a running cycle (Cavagna, 1978). This does not mean, however, that contributions to *metabolic* work from positive and negative muscle actions are in equal proportion. Both the stretch-shortening cycle and between-segment energy transfer involve eccentric muscular work. These contractions are used to slow down a segment's movement, decreasing its kinetic energy, providing the opportunity for either the return of strain energy during subsequent movements or the immediate transfer of that energy to another segment. Both of these mechanisms would contribute to the positive work done but involve eccentric rather than concentric muscular contractions.

Certainly a great deal of concentric muscular work is involved in the running cycle, but concentric work is spared to the extent that strain energy and transfer of energy can be utilized. Positive work is done by the release of elastic strain energy or when energy is transferred from one segment to another, and though this work is not metabolically free, it is more closely associated with the metabolic cost of eccentric rather than concentric muscular contractions. The energy costs associated with eccentric contractions have been shown to be anywhere from three to five times lower than concentric costs (Nagle, Balke, & Naughton, 1965). Thus, any changes to segmental movements that can enhance the use of elastic energy or transfer of energy are likely to result in increased economy.

Relationships Between Economy and Selected Biomechanical Variables

Stride Length and Frequency

Stride length is an obvious candidate as a variable that, when altered, would affect metabolic energy costs. If someone runs at a given speed but experiments with varying stride lengths, there would be obvious differences in muscle firing patterns and probably also in the forces exerted on the ground during the running cycle for each stride length. Clarke, Cooper, Clarke, and Hamill (1983) have shown that longer strides cause greater shock to the legs following footstrike as measured using accelerometers attached to the lower leg. Because of these differences in mechanical and muscular factors, variations in stride length could be expected to affect running economy.

Stride length, the length from one foot contact to the next contact of the same foot, is often reported in both absolute and relative terms, with relative measures using either body height or leg length for scaling. Step length will be used here to define the length between successive footstrikes of different feet. In general there has not been evidence to show that a longer or shorter absolute or relative stride is associated with better or worse running economy. Cavanagh, Pollock, and Landa (1977) found a group of elite-level runners to take shorter absolute and relative (to leg length) steps than a group of good runners, though differences just missed being significant at the 0.05 level. Van der Walt and Wyndham (1973) found a correlation between relative step length and $\dot{V}O_2$submax of 0.62, again indicating that shorter relative strides were associated with lower energy consumption, but they concluded that variations in step length contributed less than 1% of the variance in $\dot{V}O_2$submax. Hagan et al. (1980) concluded that natural choice of stride length did not significantly influence oxygen uptake during treadmill running.

Williams and Cavanagh (1986) found a correlation between absolute step length and $\dot{V}O_2$submax of -0.47, but this was found to be more appropriately associated with body size than with step length as only very low correlations were found when stride length relative to leg length was used. In other studies (Williams & Cavanagh, 1987; Williams, Cavanagh, & Ziff, 1987) even lower correlations between step length and $\dot{V}O_2$submax have been found.

The lack of evidence for a general relationship between stride length and economy certainly does not mean that stride length does not have an important effect on energy expenditure. Hogberg (1952) measured $\dot{V}O_2$submax in a single individual during a series of runs where step length was varied both shorter and longer than the subject's freely chosen length. He found that both increases and decreases in step length resulted in increased energy costs and concluded that a well-trained subject will adopt a freely chosen step length that is most economical. Based on his data for the single subject he also concluded that steps longer than the one freely chosen resulted in greater increases in energy consumption than did shorter steps. In a study with a somewhat different design, Knuttgen (1961) also showed that running with shorter-than-normal strides at various speeds resulted in increased energy costs.

Hogberg's study was replicated by Cavanagh and Williams (1982) using ten well-trained subjects. Results showed that, in general, the freely chosen step length was the most economical. Eight of the ten subjects showed chosen step lengths that were associated with energy costs within 0.2 ml • kg^{-1} • min^{-1} of $\dot{V}O_2$submax values predicted for their optimal stride length. Results concerning the relative influence of overstriding versus understriding were in contrast to those from Hogberg's study. Although some subjects did show relatively greater increases in $\dot{V}O_2$submax at steps longer than chosen compared to the response to shorter steps, others showed greater increases at steps shorter than chosen—the opposite trend. Subjects whose chosen step length as a percentage of leg length (SL[%LL]) was low showed greater increases in energy costs when they ran at shorter-than-chosen strides, whereas those who ran at relatively long SL(%LL) were affected most by step lengths longer than chosen.

In a study with a somewhat different design, Kaneko, Matsumoto, Ito, and Fuchimoto (1987) measured energy consumption during running at varying step frequencies while speed was kept constant at either 2.5, 3.5, or 4.5 m • s^{-1}. Energy was minimized for all three speeds at approximately the same step length (about 2.9 steps per second), and this was very close to the freely chosen step rate (2.8 steps per second).

Vertical Oscillation

It seems logical that excessive vertical oscillation would be adversely related to energy consumption. Energy "wasted" in vertical motion might be better channeled into contributions to horizontal running speed. Within a group of subjects at a given speed there is a high degree of variability in vertical oscillation, illustrated in Figure 11.5. However, correlations between measures of $\dot{V}O_2$submax have generally been low, though there have been exceptions (Cavanagh et al., 1977; Williams & Cavanagh, 1987). It would seem that for many individuals it is possible to run economically despite having a relatively high vertical oscillation.

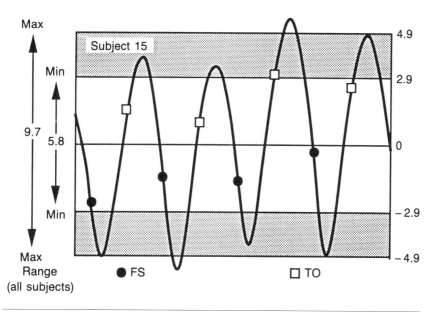

Figure 11.5. Vertical oscillation of a marker on the head during running at 5.96 m • s⁻¹ for two complete cycles of running.

There are at least two possibilities why this could be so. The first explanation is similar to one presented earlier. No one pattern of motion is likely to account for a major portion of the total energy costs. Instead, each is likely to have some weighted contribution. It is possible that high vertical oscillation does in fact incur greater metabolic energy costs, but that this is only one of many contributing variables. The inefficiency of high vertical oscillation might be offset by other more efficient patterns. The second reason could be that high vertical oscillation causes the body to adopt a sequence of movement patterns that are, in fact, more economical. For example, if a high vertical oscillation were linked to an increased stride length, perhaps the increased time to get the trail leg through to the next footstrike actually reduces the energy costs associated with the swing phase, making the entire movement pattern more efficient.

Mechanical Power and Stretch-Shortening Energetics

Because mechanical power is a global measure of the mechanical power expended during running, and $\dot{V}O_2$submax is a global measure of aerobic metabolic costs, one would assume that the two would be directly related, with more economical runners showing the lowest power output. There is little research that directly links submaximal energy costs with measures of mechanical power. Most studies have dealt with a measure of efficiency (mechanical power/metabolic energy costs) rather than economy. Williams and Cavanagh (1983) showed trends toward the expected relationship for groups of runners divided on the basis of submaximal oxygen consumption and cited energy transfer between segments of the body as a major reason for differences in mechanical power between the groups. Norman, Sharratt, Pezzack, and Noble (1976), in contrast, found no direct relationship for the three subjects studied. In fact, the subject who had the highest mechanical power was the most economical. Kaneko et al. (1987) examined the influence of varying step frequency on both mechanical power and energy expenditure. From data for a limited number of subjects at three speeds, they found that both mechanical power and metabolic energy expenditure were minimized at approximately 2.9 steps per second for each of the speeds studied.

It has been argued that metabolic energy costs are greatly influenced by contributions to mechanical power coming from events involving the stretch-shortening cycle of lower extremity muscles during activities such as running (Aura & Komi, 1986; Bosco et al., 1982; Ito, Komi, Sjödin, Bosco, & Karlsson, 1983; Williams, 1985c; Williams & Cavanagh, 1983). There appears to be considerable variability between individuals in the ability to utilize elastic energy (Aura & Komi, 1986; Ito et al., 1983; Williams & Cavanagh, 1983), suggesting that this could be one source of differences in metabolic energy costs associated with running at a given speed. To the extent that elastic strain energy can be recovered or the contractile mechanism potentiated, contributions to mechanical power from concentric muscular actions should be reduced. Cavagna et al. (1971; Cavagna, Saibene, & Margaria, 1964) estimated that oxygen consumption during running was reduced by 30% to 40% due to contributions from elastic storage and return of energy. Ito et al. (1983) reported that the efficiency of positive work remained the same across a range of speeds and attributed this ability to maintain efficiency to increased utilization of elastic energy.

Ground Reaction Forces

One can use logic to implicate high ground reaction forces being associated with increased energy costs. Greater forces in any of the ground reaction components might result in the need for more intense muscular contributions to control segmental movements and stabilize body position during the support phase. This could result in greater metabolic demand from the involved muscles. Across a range of distance running speeds ground reaction force peaks increase with increased running speed (Munro, Miller, & Fuglevand, 1987), as do measures of submaximal oxygen consumption. During treadmill running at four speeds Bhattacharya, McCutcheon, Schvartzx, and Greenleaf (1980) measured oxygen uptake and shock to the lower extremity using an accelerometer attached to the ankle. With increased speed they found a linear relationship between increased acceleration at the ankle at footstrike and increased $\dot{V}O_2$submax. Because increased force is only one of many changes that occur with speed, a causal relationship between increased force levels and increased energy expenditure cannot be determined. At a given speed of running some support for a relationship between forces and metabolic energy costs has been shown by Williams and Cavanagh (1987), where a correlation of 0.56 was found between the first peak of the vertical ground reaction force (FZ1) and $\dot{V}O_2$submax for running at 3.6 m · s^{-1}. Higher oxygen consumption values were related to greater vertical force peaks, as illustrated in Figure 11.6. In contrast, data collected on elite runners (Williams & Cavanagh, 1987; Williams, Cavanagh, & Ziff, 1987) showed only very low correlations between ground reaction force variables and $\dot{V}O_2$submax. It is evident that the complete nature of this relationship with have to await further study.

Footstrike Position

A variable that has received increased consideration over the last few years in relation to energy consumption is the position of the foot at initial contact with the ground. One method used to indicate footstrike position has involved the center of pressure pattern available from force platforms. Cavanagh (1982) described a measure called strike index (SI) to provide a quantitative assessment of the point of initial contact with the ground. Of interest is whether there is a more economical way to strike the ground. Do midfoot strikers have an advantage over rearfoot strikers, or vice versa?

In a cross-sectional study, Kerr, Beauchamp, Fisher, and Neil (1983) reported that in a marathon race the majority of the runners would be classified at rearfoot strikers, based on side-view films of ankle position. However, it was noted that the faster runners included a greater proportion of midfoot strikers than did the slower runners. It is commonly

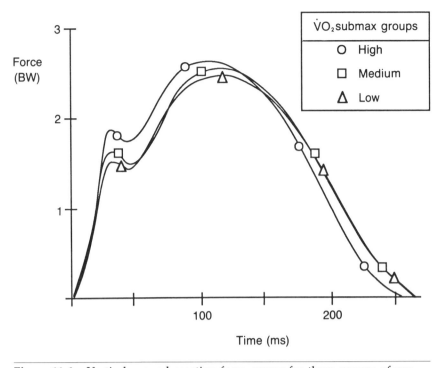

Figure 11.6.   Vertical ground reaction force curves for three groups of runners separated on the basis of submaximal oxygen consumption during running at 3.57 m • s⁻¹. *Note*: From "Relationship Between Distance Running Mechanics, Running Economy, and Performance" by K.R. Williams and P.R. Cavanagh, 1987, *Journal of Applied Physiology*, **63**, p. 1241. Copyright by the American Physiological Society. Reprinted by permission.

assumed that strike index increases (a landing further forward on the foot) as speed increases, but it should not be assumed that initial contact further forward on the ground is necessarily linked to more economical running. Figure 11.7 shows a group of center of pressure patterns taken from elite runners moving at a speed of 5.96 m • s⁻¹. Even at this fast 10-km speed there was a wide variety of foot contact patterns (Williams & Cavanagh, 1986; Williams, Cavanagh, & Ziff, 1987). Strike index correlated poorly with V̇O₂submax in this group of runners (Williams & Cavanagh, 1987). Nilsson and Thorstensson (1987) reported no significant differences in running economy between groups of six forefoot and six rearfoot strikers, nor were there changes when the subjects voluntarily changed from fore to rear strike patterns and vice versa. In contrast to this, Williams and Cavanagh (1987) found trends indicating that striking more to the rear of the foot was more economical. There does not appear to be a clearly demonstrated advantage to any particular landing position in terms of

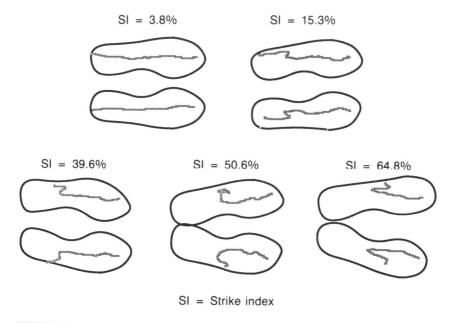

SI = Strike index

Figure 11.7. Variations in foot contact position from a group of 13 elite runners running at 5.96 m • s⁻¹ as determined from center of pressure patterns.

economy. It is likely that specific footstrike patterns are influenced by an individual's anatomical structure, flexibility, muscular strength, and training. The pattern that is efficient for one runner may not be efficient for another. Repeated training with a particular footstrike pattern would be expected to optimize energy costs compared to those associated with a different pattern.

Anatomical Structure and Gender Differences in Relation to Oxygen Consumption

There are conflicting reports in the literature concerning whether there is a gender difference in running economy (Bhambhani & Singh, 1985), and little information is available that evaluates differences in biomechanical measures between the sexes in relation to differences in $\dot{V}O_2$submax. Although Williams, Cavanagh, and Ziff (1987) did report some differences between males and females in the relationships between biomechanical variables and running economy, they concluded that there was insufficient evidence that the differences were gender-based. Howley and Glover (1974) and Bransford and Howley (1977) suggested that the increased metabolic energy costs they found for female runners compared to male runners resulted from increased vertical displacement of the body, but they had no kinematic measurements to support their hypothesis. In a study where subjects ran at self-selected speeds, Bhambhani and Singh reported a significantly higher $\dot{V}O_2$submax for female runners compared to males, despite the females having an average running speed that was 7% slower than that chosen by the males. Though the females were found to have shorter step lengths (by 6.1 cm) than males and greater vertical lift per stride (8.89 vs. 8.80 cm), differences were nonsignificant. Cureton and Sparling (1980) investigated the question of whether differences in percent body fat would account for the differences seen in $\dot{V}O_2$submax in groups of male and female runners matched for ability. By adding weights to the male runners to simulate the additional body fat carried by the females, they showed that approximately 32% of the difference in $\dot{V}O_2$submax (as measured relative to fat-free weight) could be accounted for by sex-specific body fat. Although it is possible that some of the remaining differences in $\dot{V}O_2$submax between the male and female groups could be due to biomechanical factors, no quantitative information to support such an idea was available.

External Factors Affecting Running Economy

Air Resistance.   The effects of the wind, or air resistance, on the economy of running is very relevant to the runner. Not only will it have an effect on energy consumption on a windy day, but the effects of air resistance during drafting, where one runner blocks the wind for another, can also have important energy consequences. Factors such as running velocity, wind velocity, projected cross-sectional area of the runner, air density, and a drag coefficient all affect the air resistance encountered by a runner (Shanebrook & Jaszczak, 1976).

On a day with no wind, where the runner's speed determines relative wind velocity, the energy costs involved in overcoming air resistance have been reported to be proportional to both the cube (Pugh, 1970; Pugh, 1971) and the square (Davies, 1980) of running speed. Davies estimated the cost of overcoming air resistance to be a 4% increase in $\dot{V}O_2$submax for running at 6 m $\cdot$ s^{-1} on a calm day and a 2% increase for speeds of 5 m $\cdot$ s^{-1}. Running against a wind brings increased energy costs proportional to the speed of the wind squared (Davies, 1980; Pugh, 1970). The energy costs associated with overcoming air resistance have been estimated at 7% to 8% of the total metabolic costs for middle-distance running speeds, and 13% to 16% for maximal sprint speeds. Davies reported that a tail wind reduced energy costs compared to running in the absence of wind, but that the relative effects were about 50% of those associated with running into the head wind of the same speed. Hatsell (1975) mathematically derived a model for predicting an optimal strategy for running on a circular track on a windy day. For an assumed 1,000-m lap track, he estimated a reduction of 0.6 seconds per lap if a strategy of accelerating when the wind was blowing into the face and decelerating when the wind was at the back were used instead of a constant velocity strategy.

Pugh (1971) estimated that a runner traveling 1 m behind another would have air resistance decreased by 80%, causing a reduction in energy costs of approximately 6%. Kyle (1979) estimated that a 2-m spacing would reduce resistance by 40%, lessen energy costs by 3%, and result in a 1.42 seconds/lap reduction in time for a typical 400-m track.

Footwear and Shoe Inserts.   Although footwear is most often associated with injury prevention, there is convincing evidence that shoes have an effect on energy consumption in two different ways. Frederick (1983a, 1983b) and Frederick, Clarke, Larsen, and Cooper (1983) showed how both the weight of the shoes and their cushioning properties can affect energy expenditure during running. Most obvious is the effect that the weight of the shoe has on energy costs. The heavier the shoe, the greater the metabolic costs to run in it (Catlin & Dressendorfer, 1979; Frederick, Daniels, & Hayes, 1984), though some studies have not shown significant effects (Fukuda, Ohmichi, & Miyashita, 1983). Frederick et al. (1984) added weights to running shoes during treadmill running at speeds from 3.8 to 4.9 m • s^{-1} and reported significant increases in $\dot{V}O_2$submax for added weight as low as 75 g per shoe. $\dot{V}O_2$submax increased 1.2% for each 100 g per shoe added for running at a speed of 3.83 m • s^{-1}, with effects diminishing somewhat at the higher speeds. Although it might be hypothesized that the use of foot orthotic devices could actually decrease metabolic energy costs by favorably altering running mechanics and reducing unnecessary muscular activity, experimental data generally show a slight additional energy cost proportional to the weight of the inserts (Burkett, Kohrt, & Buchbinder, 1985; Clement, Taunton, Wiley, Smart, & McNicol, 1984; Hayes, Smity, & Santopietro, 1983).

The cushioning properties of shoes also seem to be of importance to metabolic costs. As much as a 2.8% lower oxygen cost has been attributed to shoes with better cushioning properties than standard ethyl vinyl acetate (EVA) shoes (Frederick, Howley, & Powers, 1986), though other similar studies have shown smaller differences (0.4 ml • kg^{-1} • min^{-1}, Frederick et al., 1983). Berg and Sady (1985) added shock-absorbing inserts weighing 100 g to running shoes and found slight, but nonsignificant, increases in $\dot{V}O_2$submax. It may be that tendencies for reduced metabolic costs due to added cushioning were balanced by additional energy costs resulting from the added weight.

Added Weight.   In recent years there has been increasing interest in the addition of weight to various locations on the body during both walking and running. Although Stevens (1983) concluded that the weight of clothing worn during running had minimal effect on $\dot{V}O_2$submax, higher weights or different distributions of weight have been shown to affect energy costs. Cureton and Sparling (1980) reported that the addition of 7.5% body weight to the trunk during running significantly increased oxygen consumption by 0.16 L • min^{-1}. When this increase was expressed as ml • kg^{-1} • min^{-1}, however, $\dot{V}O_2$submax was shown to significantly decrease. The fact that added weight can increase the absolute level of oxygen consumption is relevant to the runner because this will put energy expenditure at a higher level relative to $\dot{V}O_2$max and could adversely affect performance. Epstein, Stroschein, and Pandolf (1987) developed a model for predicting metabolic energy costs during running over a range of speeds from 2.2 to 3.2 m • s^{-1}, with external loads from 0 to 30 kg, and up inclines of 2% and 4%. The model is limited in that estimates of energy costs during running are based on predicted oxygen consumption values for walking.

Martin (1985) investigated the influence of added weight not only on $\dot{V}O_2$submax but also on a number of measures of running mechanics. Weights totaling either 0.25 or 0.5 kg were added to each leg on either the thigh or the foot, and changes compared to no added weight were observed. Increases in $\dot{V}O_2$submax were generally linear and similar in magnitude to those of other investigations, averaging 3.5% for 0.5 kg added to each thigh, and 7.2% for 0.5 kg added to each foot. No changes in a set of temporal and kinematic variables were observed for weight added to the thigh, but with 0.5 kg added to the feet significant changes were found indicating an increased stride length, longer swing and flight times, and a slower peak linear velocity of the ankle. Increased mechanical work was done by the limb segment to which the weight was added. The changes in mechanics that occurred were considered to be small, and the increases in energy costs were primarily attributed to the increased mechanical work required of the musculature in response to the increased inertia of the leg with the added weight.

Surfaces. The most celebrated attempt at using a running surface to enhance performance has been through the development of an indoor track surface that attempts to match the spring characteristics of the track with the natural stiffness of the lower extremity of the runner (McMahon & Greene, 1979). In theory, the surface will store energy during the initial stages of support and return some of this energy during toe-off. Presumably one benefit of the system would be that the energy returned by the track would then lessen the work required of the muscles, resulting in a lower energy cost to the runner. This would presumably allow the runner to run faster for the same relative metabolic work load and thus improve performance. Though there have been observations of markedly better performances from such a track installed at Harvard University (McMahon, 1984), no statistical evidence is available. Perhaps as important as possible improvements in performance is the belief that the potential for injury is lessened during running and prolonged training on the track. The compliance of the surface could serve to lessen the stresses on the lower extremities of the runners.

The metabolic cost of running on different types of surfaces has not been explored in detail. Bonen, Gass, Kachadorian, and Johnson (1974) reported no differences in $\dot{V}O_2$submax for two subjects running on a treadmill, a cement surface, a Tartan track surface, and a cinder track. Pugh (1970) also reported no difference for comparisons of metabolic energy costs during running on an all-weather track and a cinder surface.

Downhill Running. Metabolic costs during running downhill are markedly lower than the costs associated with level running, with differences of 17.8 ml • kg^{-1} • min^{-1} reported by Dick and Cavanagh (1987) for treadmill running at 3.83 m • s^{-1}. Of particular interest to these investigators was the 2.8 ml • kg^{-1} • min^{-1} upward drift in $\dot{V}O_2$submax associated with prolonged downhill running but not level running. Based on increased integrated electromyographic measures for the quadriceps muscles and delayed onset of muscle soreness they hypothesized that a combination of local muscle fatigue and damage to muscle and connective tissues during downhill running caused increased motor unit recruitment within the eccentrically acting muscles. No significant changes in stride length were found during the prolonged runs, so that the increased metabolic costs could not be explained by alterations in stride length. Because no other biomechanical measures were obtained, it is not possible to say whether there were any other changes in running mechanics that might have been associated with the drift in $\dot{V}O_2$submax. Pierrynowski, Tüdus, and Plyley (1987) also examined changes in $\dot{V}O_2$submax during downhill running, using subjects previously trained in either uphill or downhill running. They found stride frequency (and thus stride length) to be unchanged throughout a downhill run, except during the second day of downhill training where stride frequency increased and stride length shortened. They suggested that muscle soreness was responsible for the change in stride length on the second training day, but the differences were not large enough to significantly alter $\dot{V}O_2$submax.

Relationships With Economy Derived From Comprehensive Sets of Variables

In 1977 Cavanagh et al. reported on a variety of biomechanical variables for a group of 14 elite and 8 good runners. Although an analysis between running mechanics and running economy was not specifically done, a simultaneous study of the same athletes by Pollock (1977) did show the group of elite-level runners to have lower $\dot{V}O_2$submax than the group of good runners at 4.5- and 5.4-m • s^{-1} paces. In the comparisons between the running mechanics of the good and elite groups that follow, it should be remembered that relationships between mechanical variables and running economy are inferred here and were not statistically examined in the original study.

The elite runners in this study, compared to the good runners, took shorter absolute (1.56 vs. 1.64 m) and relative (165% vs. 172% leg length) strides, though results were not statistically significant (Cavanagh et al., 1977). The elite runners had slightly lower vertical oscillation of the center of gravity during the running cycle than did the good group (7.6 vs. 8.0 cm), though again results were nonsignificant, and the elite athletes were significantly more symmetrical in vertical oscillation between right and left sides compared to the good runners. Significant differences were found between groups in maximal plantar flexion angle at toe-off, with the elite runners plantar flexing an average of 10° less than the good runners. Angular kinematics of the lower extremity were very similar between the groups, with no significant differences found for maximal knee flexion during support and swing, maximal hip flexion during swing, or the position of the leg segments at footstrike. Knee joint torques estimated during the swing phase also showed no significant differences between groups.

It is apparent that only minor differences were found in running mechanics between the two levels of runners. A number of methodological considerations may have limited the analysis, as pointed out by the authors. For example, there were variations in experience among the subjects in treadmill running, and the analysis was limited to relatively few variables describing sagittal plane motion.

A study by Williams (1987) was designed specifically to investigate the relationship between a wide variety of biomechanical measures and running economy. Thirty-one subjects ran at a pace of 3.6 m $\cdot$ s^{-1} and were divided into three significantly different groups based on $\dot{V}O_2$submax. Comparisons were then made between these groups to identify those mechanical variables most closely related to running economy. To provide as diverse a set of measures as possible, the data collected included ground reaction forces and center of pressure patterns; linear and angular kinematics derived from three-dimensional cinematography; measures of mechanical power using a segmental energy analysis; anthropometric measures; and physiological measures including $\dot{V}O_2$submax, $\dot{V}O_2$max, stretch-shortening cycle elastic contributions, and muscle fiber composition. A factor analysis procedure was used to identify independent sets of biomechanical variables, and representative variables were chosen from each of these factors for comparison with oxygen consumption measures. These variables were entered into a multiple regression analysis, and the five variables included in the regression gave an R^2 value of 0.67, indicating that a considerable portion of the variance in oxygen consumption could be explained by mechanical variables. Significant differences between the running economy groups, illustrated for some of the variables in Figure 11.8, showed the runners with lower $\dot{V}O_2$submax to have a lower initial peak in the vertical force, less plantar flexion at toe-off, greater forward lean of the trunk and a greater angle of the shank with the vertical at footstrike, greater energy transfer involving the legs, and lower minimum knee velocity during support. Consistent but nonsignificant trends between groups showed the lower $\dot{V}O_2$submax group to have lower mechanical power, greater energy transfer between segments, smaller peak anteroposterior and vertical forces, footstrike more toward the heel, greater knee flexion during support, less arm movement, and a smaller range of oscillation of the center of mass.

More recently, Williams and Cavanagh (1986) and Williams, Cavanagh, and Ziff (1987) have collected similar, though somewhat less extensive, sets of data, one on elite males and the other on elite females. In both studies, factor analysis procedures were also used to identify sets of independent variables, with an appropriate variable from each factor chosen for analysis with running economy measures. These analyses resulted in somewhat different sets of variables showing strong correlations with running economy compared to those just described. In the study involving elite males (Williams & Cavanagh, 1986), correlations indicated that lower oxygen consumption was associated with longer support time, greater maximal angle of the thigh with the vertical during hip extension, less knee extension at toe-off, greater maximal plantar flexion velocity, and greater horizontal heel velocity at foot contact. Although correlations indicated that longer step lengths were associated with lower $\dot{V}O_2$submax, the relationship disappeared when step length was measured relative to leg length. This dependency on size was consistent with relationships that linked larger individuals, as measured by leg length, pelvic width, and body mass, with better economy.

In the study of elite female runners (Williams, Cavanagh, & Ziff, 1987), better economy was associated with less maximal extension of the thigh and slower extension velocity, less extension of the knee during toe-off, less rapid knee flexion velocity during swing, and greater dorsiflexion and faster dorsiflexion velocity during support. Lower oxygen consumption was also associated with greater body weight, but unlike the study involving males, no relationship was evident with measures of segmental lengths.

Among the four studies described there does not appear to be one set of measures that is consistently related to better economy. A number of variables showed a significant relationship for one or two of the studies, but not for all. Although relationships for a smaller knee angle at toe-off and heavier body weight are consistent between the studies involving elite athletes (Williams & Cavanagh, 1986; Williams, Cavanagh, & Ziff,

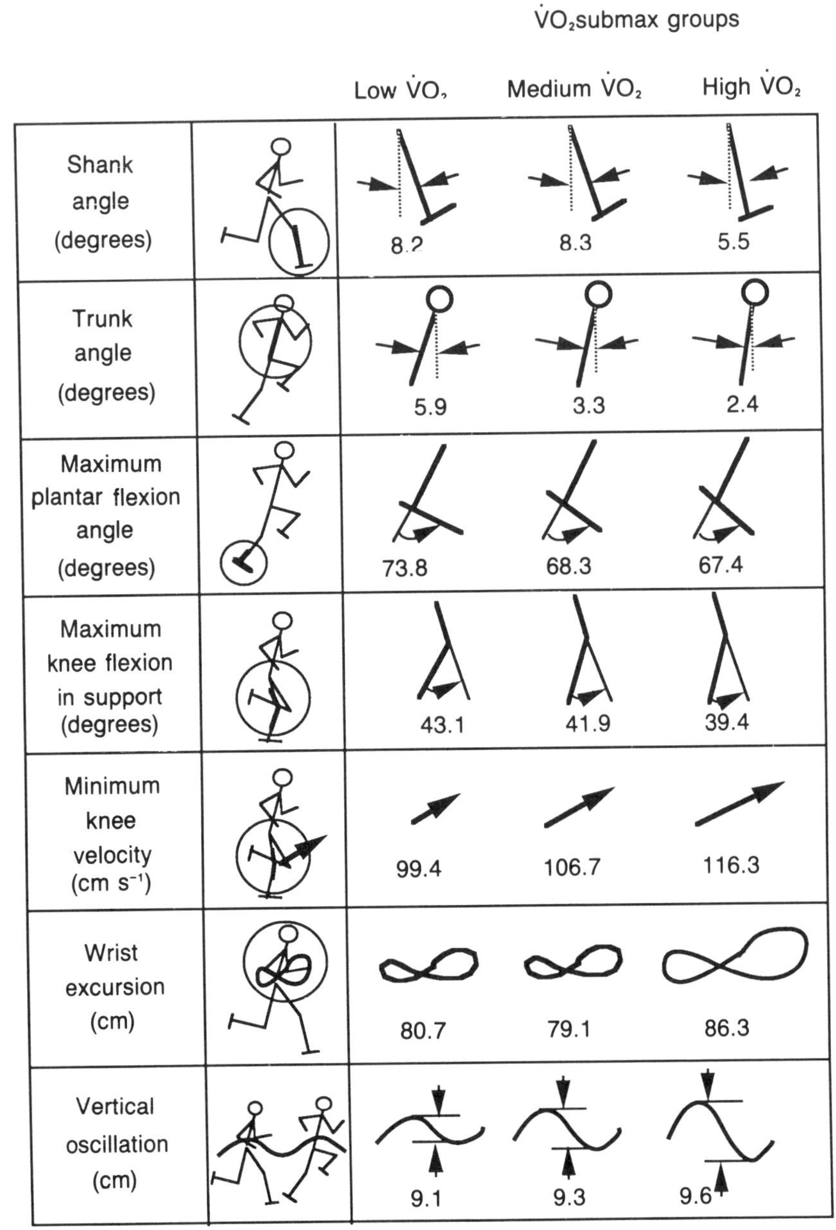

Figure 11.8. Differences in a number of biomechanical measures of running style between groups of runners separated on the basis of submaximal energy costs. *Note*: From "Relationship Between Distance Running Mechanics, Running Economy, and Performance" by K.R. Williams and P.R. Cavanagh, 1987, *Journal of Applied Physiology*, **63**, p. 1242. Copyright by the American Physiological Society. Reprinted by permission.

1987), maximal thigh extension angle relationships were in the opposite directions. Two studies showed longer support times to be associated with lower economy (Williams & Cavanagh, 1986; Williams & Cavanagh, 1987), but the same was not true for the other two studies. Similarly, two studies showed trends toward an association of higher vertical oscillation with better economy, but no relationship was identified for the other studies. It is possible that the lack of consistent relationships between the studies is the result of differences in ability level of the runners, gender, or experimental procedures. The relatively small number of subjects (14 or fewer) in three of the studies (Cavanagh et al., 1977; Williams & Cavanagh, 1986; Williams & Cavanagh, 1987) suggests that larger data bases will be needed before definitive relationships can be established. Until such time, all of these results should be considered preliminary.

Implications for Future Research

Based on the literature reviewed here it is evident that relationships between running mechanics and running economy do exist. However, much remains to be explained about the exact nature and subtlety of the relationships. One goal of research in this area might be to better understand the extent to which it is possible to alter a runner's energy expenditure by changing running style. In this case we need to understand more fully the physiological and mechanical consequences of alterations to running mechanics and how associated changes might vary between individuals. It is not clear at this time whether it is feasible to markedly enhance running economy, or distance running performance, by making alterations to the running style of individuals. There is a need for studies that identify the sensitivity of $\dot{V}O_2$ to changes in mechanics and that demonstrate that it is possible to lower energy costs and enhance performance by changing mechanics. It should be remembered that although lowering energy costs is conceptually an admirable goal, it is possible that there could be some associated changes that would be detrimental to the runner. If changes in style increased the likelihood of certain types of injuries, the improved performance coming from better economy might be negated by a debilitating injury. More information is needed concerning how changes in running mechanics will influence factors other than running economy.

A general goal for investigations into the relationships between running mechanics and running economy might be to better understand the optimization process that the body goes through as economical movement patterns develop during running training. We need to understand how and why an individual adopts specific movement patterns. We might hypothesize that runners use a self-optimization process to develop segmental movement patterns that minimize energy costs and/or minimize the stresses that might be related to the development of an injury. To understand the optimization process we need to know more about how a variety of factors influence running mechanics. General factors such as speed of running, state of training, fatigue, running surface, and gender will have an effect on the specific running patterns used, and we need to know how these factors affect mechanics before we can fully understand how mechanics influence energy costs.

Similarly, a variety of factors more specific to the individual, such as body structure, strength, and flexibility, are certain to influence running mechanics and, indirectly, economy. For example, an individual might show a movement pattern that is extreme when considered against a group mean. It is possible that this pattern is the result of poor mechanics, and a change would be beneficial to the runner. It is also possible that the pattern is an adaptation that the runner has made to some structural or functional anomaly, and in fact the runner runs more economically with this pattern of movement than with a more typical style. There is a need for research that will help identify aspects of running style that are in fact poor mechanics and not the result of a positive adaptation to some other factor. To do this there is a need for more information relevant to the individual. Although studies of groups of runners will aid in the identification of more general relationships between running mechanics and running economy, it is necessary to understand these relationships as they apply to the individual, not just to group data.

Several different experimental approaches might be used to provide the information necessary to give insight into the questions identified. A comprehensive approach would include the collection and analysis of data for a wide variety of variables. There is a general need for the development of a data base for a large number of runners of different ability levels from each gender for which multidisciplinary data is available so that generalized relationships can be identified. This would form the basis for performance models and could be used to evaluate the accuracy of theoretical models. Because this approach is pointed toward identifying interactions between different types of variables, it is important to collect data of sufficient diversity. The identification of general relationships will aid in the identification of variables that might be selected for further in-depth study on individuals.

A second approach involves the direct manipulation of one or more biomechanical measures and an evaluation of the physiological and mechanical consequences of the changes made. This would constitute a type of sensitivity analysis that could identify those variables that seem to have the greatest influence on economy. A potentially confounding factor, however, is that when one factor is manipulated, such as increasing shoe mass, there are likely to be a myriad of other changes in running mechanics that occur, and it may be difficult to identify which changes are most responsible for any alterations in economy. A variation of this approach might entail the identification of a feature or features of running mechanics that are highly variable among a group of runners. Once identified, these features could be manipulated for individual subjects to see how they affect economy for each individual. For example, maximal hip extension near toe-off has been shown to correlate well with economy, with greater extension associated with lower oxygen consumption. An individual who shows a lesser amount of extension might be involved in a study that manipulates this variable to see if learning to extend the leg to a greater degree leads to better economy. This method would focus on existing differences in mechanics between individuals rather than artificially introducing such differences.

A third approach might involve the direct training of runners using some form of augmented feedback. The mechanics of an individual runner could be evaluated on a real-time basis and put in relation to measures of economy. Various manipulations of running style could be effected and influences on economy documented. Over a training period it would ideally be possible to fine-tune the individual's mechanics to minimize energy costs. The need for relatively immediate mechanical feedback would necessitate the use of one of the automated video or opto-electronic kinematic systems that are becoming more readily available.

Summary

Running economy is one of the many factors important to running performance, and changes in economy during running at a given speed are likely to lead to changes in performance. Physiological studies have shown a wide variation in submaximal energy consumption during running at a given speed, and some portion of this variability can be accounted for by differences in running mechanics. It is of interest to understand how various aspects of the mechanics of running influence metabolic energy consumption.

Studies that investigate the relationships between running mechanics and submaximal metabolic energy costs should recognize limitations inherent in the experimental methods employed that might affect the relationships identified. These might include factors such as the use of 2D versus 3D kinematic methods, the use of treadmill versus overground running, or the influence that the use of a specific speed, population, or set of variables chosen for analysis might have on the generalization of results.

The direct manipulation of biomechanical variables, such as stride length or shoe weight, has demonstrated that oxygen consumption can be markedly influenced by factors related to running mechanics. Although studies that have manipulated stride length indicate that most runners naturally adopt a stride length that is most economical for them at a given speed, there is insufficient evidence to indicate how pervasive this is among the variety of mechanical variables that can be used to describe the running motion.

A variety of factors have been shown to affect oxygen consumption when they are considered individually. It has been easily demonstrated that factors such as shoe weight, midsole composition, wind velocity, and slope of the running surface will affect energy costs, but identification of features of running style that have predictable relationships with economy is more difficult. Altering stride length from that freely chosen will increase metabolic costs in most individuals, but it is not clear what relationship exists between economy and factors such as ground reaction force magnitudes, vertical oscillation, or footstrike position. Though some general relationships with economy have been established in studies that have examined a variety of mechanical variables for a group of runners, these relationships do not appear to be general for all groups of runners. At the present time it is not known how relationships identified for a group as a whole apply to an individual. Certainly there are economical runners who show some patterns of movement that would be considered uneconomical based on results for the group as a whole. It is possible that (a) the pattern of movement is in fact uneconomical for the specified factor, and changing it would result in better economy; or (b) other factors specific to the individuals, such as anatomical structure, strength, or flexibility, make a given movement pattern optimal for the individual even though it may be contrary to what is considered optimal for a group. There currently are no definitive guidelines that can be used to make recommendations to an individual concerning how changes in mechanics might make him or her a more economical runner. Such evaluations have to be considered on an individual basis and should include as broad an evaluation of running mechanics and economy as possible. Unusual movement patterns or large differences from norms should be more closely investigated to see if they are linked to poor economy, and any changes suggested to running mechanics should be moderate and closely monitored to evaluate their effect on economy, injury, and performance.

References

Aura, O., & Komi, P.V. (1986). The mechanical efficiency of locomotion in men and women with special emphasis on stretch-shortening cycle exercises. *European Journal of Applied Physiology*, **55**, 37-43.

Berg, K., & Sady, S. (1985). Oxygen cost of running at submaximal speeds while wearing shoe inserts. *Research Quarterly*, **56**(1), 86-89.

Bhambhani, Y., & Singh, M. (1985). Metabolic and cinematographic analysis of walking and running in men and women. *Medicine and Science in Sports and Exercise*, **17**(1), 131-137.

Bhattacharya, A., McCutcheon, E.P., Shvartzx, E., & Greenleaf, J.E. (1980). Body acceleration distribution and O_2 uptake in humans during running and jumping. *Journal of Applied Physiology*, **49**(5), 881-887.

Bonen, A., Gass, G.C., Kachadorian, W.A., & Johnson, R.R. (1974). The energy cost of walking and running on different surfaces. *Australian Journal of Sports Medicine*, **6**(1), 5-11.

Bosco, C., Tarkka, I., & Komi, P.V. (1982). Effect of elastic energy and myoelectric potentiation of triceps surae during stretch-shortening cycle exercise. *International Journal of Sports Medicine*, **3**, 137-140.

Bransford, D.R., & Howley, E.T. (1977). Oxygen cost of running in trained and untrained men and women. *Medicine and Science in Sports*, **9**(1), 41-44.

Burkett, L.N., Kohrt, W.M., & Buchbinder, R. (1985). Effects of shoes and foot orthotics on $\dot{V}O_2$ and selected frontal plane knee kinematics. *Medicine and Science in Sports and Exercise*, **17**(1), 158-163.

Catlin, M.E., & Dressendorfer, R.H. (1979). Effect of shoe weight on the energy cost of running. *Medicine and Science in Sports*, **11**(1), 80.

Cavagna, G.A. (1978). Storage and utilization of elastic energy in skeletal muscle. *Exercise and Sport Science Reviews*, **5**, 89-129.

Cavagna, G.A., Komarekko, L., & Mazzolen, S. (1971). The mechanics of sprint running. *Journal of Physiology*, **217**, 709-721.

Cavagna, G.A., Saibene, F.P., & Margaria, R. (1964). Mechanical work in running. *Journal of Applied Physiology*, **18**, 1-9.

Cavanagh, P.R. (1982). The shoe-ground interface in running. In R.P. Mack (Ed.), *Symposium on the foot and leg in running sports* (pp. 30-44). St. Louis: C.V Mosby.

Cavanagh, P.R., & Kram, R. (1985). Mechanical and muscular factors affecting the efficiency of human movement. *Medicine and Science in Sports and Exercise*, **17**(3), 326-331.

Cavanagh, P.R., Pollock, M.L., & Landa, J. (1977). A biomechanical comparison of elite and good distance runners. In P. Milvy (Ed.), *The marathon: Physiological, medical, epidemiological, and psychological studies* (pp. 328-345). New York: New York Academy of Sciences.

Cavanagh, P.R., & Williams, K.R. (1982). The effect of stride length variation on oxygen uptake during distance running. *Medicine and Science in Sports and Exercise*, **14**(1), 30-35.

Clarke, T.E., Cooper, L.B., Clarke, D.E., & Hamill, C.L. (1983). The effect of varied stride rate and length upon shank deceleration during ground contact in running (abstract). *Medicine and Science in Sports and Exercise*, **15**(2), 170.

Clement, D., Taunton, J., Wiley, J.P., Smart, G., & McNicol, K. (1984). The effects of corrective orthotic devices on oxygen uptake during running. In L. Prokop (Ed.), *Proceedings of the World Congress on Sports Medicine*, (pp. 648-655). Vienna: World Congress on Sports Medicine.

Conley, D.L., & Krahenbuhl, G.S. (1980). Running economy and distance running performance of highly trained athletes. *Medicine and Science in Sports and Exercise*, **12**(5), 357-360.

Cureton, K.J., & Sparling, P.B. (1980). Distance running performance and metabolic responses to running in men and women with excess weight experimentally equated. *Medicine and Science in Sports and Exercise*, **12**(4), 288-294.

Daniels, J.T. (1985). A physiologist's view of running economy. *Medicine and Science in Sports and Exercise*, **17**(3), 1-23.

Daniels, J.T., Scardina, N.J., & Foley, P. (1984). $\dot{V}O_2$ submax during five modes of exercise. In L. Prokop (Ed.), *Proceedings of the World Congress on Sports Medicine* (pp. 604-615). Vienna: World Congress on Sports Medicine.

Davies, C.T.M. (1980). Effects of wind assistance and resistance on the forward motion of a runner. *Journal of Applied Physiology*, **48**(4), 702-709.

Dick, R.W., & Cavanagh, P.R. (1987). An explanation of the upward drift in oxygen uptake during prolonged sub-maximal downhill running. *Medicine and Science in Sports and Exercise*, **19**(3), 310-317.

Dotan, R., Rotstein, A., Din, R., Inbar, O., Kofman, H., & Kaplansky, Y. (1983). Relationships of marathon running to physiological, anthropometric and training indices. *European Journal of Applied Physiology*, **51**, 281-293.

Epstein, Y., Stroschein, L.A., & Pandolf, K.B. (1987). Predicting metabolic cost of running with and without backpack loads. *European Journal of Applied Physiology*, **56**, 495-500.

Frederick, E.C. (1983a). Extrinsic biomechanical aids. In M. Williams (Ed.), *Ergogenic aids in sport* (pp. 323-339). Champaign, IL: Human Kinetics.

Frederick, E.C. (1983b). Measuring the effects of shoes and surfaces on the economy of locomotion. In B.M. Nigg & B.A. Kerr (Eds.), *Biomechanical aspects of sport shoes and playing surfaces* (pp. 93-106). Calgary, AB: University of Calgary.

Frederick, E.C., Clarke, T.E., Larsen, J.L., & Cooper, L.B. (1983). The effects of shoe cushioning on the oxygen demands of running. In B.M. Nigg & B.A. Kerr (Eds.), *Biomechanical aspects of sport shoes and playing surfaces* (pp. 107-114). Calgary, AB: University of Calgary.

Frederick, E.C., Daniels, J.T., & Hayes, J.W. (1984). The effect of shoe weight on the aerobic demands of running. In L. Prokop (Ed.), *Proceedings of the World Congress on Sports Medicine* (pp. 616-625). Vienna: World Congress on Sports Medicine.

Frederick, E.C., Howley, E.T., & Powers, S.K. (1986). Lower oxygen demands of running in soft-soled shoes. *Research Quarterly for Exercise and Sports,* **57**(2), 174-177.

Fukuda, H., Ohmichi, H., & Miyashita, M. (1983). Effects of shoe weight on oxygen uptake during submaximal running. In B.M. Nigg & B.A. Kerr (Eds.), *Biomechanical aspects of sport shoes and playing surfaces* (pp. 115-122). Calgary, AB: University of Calgary.

Hagan, R.D., Strathman, T., Strathman, L., & Gettman, L.R. (1980). Oxygen uptake and energy expenditure during horizontal treadmill running. *Journal of Applied Physiology,* **49**(4), 571-575.

Hatsell, C.P. (1975). A note on jogging on a windy day. *Institute of Electronics and Electrical Engineers Transactions on Biomedical Engineering,* **22**(5), 428-429.

Hatze, H. (1976). Biomechanical aspects of a successful motion optimization. In P.V. Komi (Ed.), *Biomechanics V-B* (pp. 5-12). Baltimore: University Park.

Hayes, J., Smity, L., & Santopietro, F. (1983). The effects of orthotics on the aerobic demands of running (abstract). *Medicine and Science in Sports and Exercise,* **15**(2), 169.

Hinrichs, R.N., Cavanagh, P.R., & Williams, K.R. (1983). Upper extremity contributions to angular momentum in running. In H. Matsui & K. Kobayashi (Eds.), *Biomechanics VIII-B* (pp. 641-647). Champaign, IL: Human Kinetics.

Hinrichs, R.N., Cavanagh, P.R., & Williams, K.R. (1987). Upper extremity function in running. I: Center of mass and propulsion considerations. *International Journal of Sport Biomechanics,* **3**(3), 222-241.

Hogberg, P. (1952). How do stride length and stride frequency influence the energy-output during running? *Arbeitsphysiologie,* **14** (Suppl), 437-441.

Howley, E.T., & Glover, M.E. (1974). The caloric costs of running and walking one mile for men and women. *Medicine and Science in Sports,* **6**(4), 235-237.

Ito, A., Komi, P.V., Sjödin, B., Bosco, C., & Karlsson, J. (1983). Mechanical efficiency of positive work in running at different speeds. *Medicine and Science in Sports and Exercise,* **15**(4), 299-308.

Kaneko, M., Matsumoto, M., Ito, A., & Fuchimoto, T. (1987). Optimum step frequency in constant speed running. In B. Jonsson (Ed.). *Biomechanics X-B* (pp. 803-897). Champaign, IL: Human Kinetics.

Kerr, B.A., Beauchamp, L., Fisher, B., & Neil, R. (1983). Footstrike patterns in distance running. In B.M. Nigg & B.A. Kerr (Eds.), *Biomechanical aspects of sport shoes and playing surfaces* (pp. 135-142). Calgary, AB: University of Calgary.

Knuttgen, H.G. (1961). Oxygen uptake and pulse rate while running with undetermined and determined stride lengths at different speeds. *Acta Physiologica Scandinavica,* **52**, 366-371.

Komi, P.V., Golihofer, A., Schmidtbleicher, D., & Frick, U. (1987). Inter-action between man and shoe in running: Considerations for a more comprehensive measurement approach. *International Journal of Sports Medicine,* **8**, 196-202.

Komi, P.V., Ito, A., Sjödin, B., Wallenstein, R., & Karlsson, J. (1981). Muscle metabolism, lactate breaking point, and biomechanical features of endurance running. *International Journal of Sports Medicine,* **2**, 148-153.

Kyle, C.R. (1979). Reduction of wind resistance and power output of racing cyclists and runners travelling in groups. *Ergonomics,* **22**(4), 387-397.

Martin, P.E. (1985). Mechanical and physiological responses to lower extremity loading during running. *Medicine and Science in Sports and Exercise,* **17**(4), 427-433.

Mayhew, J.L. (1977). Oxygen cost and energy expenditure of running in trained runners. *British Journal of Sports Medicine,* **11**(3), 116-121.

McMahon, T.A. (1984). *Muscles, reflexes, and locomotion.* Princeton, NJ: Princeton University.

McMahon, T.A., & Green, P.R. (1979). The influence of track compliance on running. *Journal of Biomechanics,* **12**, 893-904.

Munro, C.F., Miller, D.I., & Fuglevand, A.J. (1987). Ground reaction forces in running: A re-examination. *Journal of Biomechanics,* **20**(2), 147-156.

Nagle, F.J., Balke, B., & Naughton, J.P. (1965). Gradational step tests for assessing work capacity. *Journal of Applied Physiology,* **20**, 745-748.

Nilsson, J., & Thorstensson, A. (1987). Characterization of different foot strike patterns in running and their effects on ground reaction forces and running economy. In *Abstracts, XI International Congress of Biomechanics* (p. 230). Amsterdam: Free University.

Norman, R., Sharratt, M., Pezzack, J.J., & Noble, E. (1976). Re-examination of the mechanical efficiency of horizontal treadmill running. In P.V. Komi (Ed.), *Biomechanics V-B* (pp. 87-93). Baltimore: University Park.

Petray, C.K., & Krahenbuhl, G.S. (1985). Running training, instruction on running technique, and running economy in 10-year-old males. *Research Quarterly for Exercise and Sport,* **56**(3), 251-255.

Pierrynowski, M.R., Tüdus, P.M., & Plyley, M.J. (1987). Effects of down-hill and uphill training prior to a downhill run. *European Journal of Applied Physiology,* **56**, 668-672.

Pollock, M.L. (1977). Submaximal and maximal working capacity of elite distance runners. Part 1: Cardiorespiratory aspects. In P. Milvy (Ed.), *The marathon: Physiological, medical, epidemiological, and psychological studies* (pp. 310-322). New York: New York Academy of Sciences.

Pugh, L.G.C.E. (1970). Oxygen intake in track and treadmill running with observations on the effect of air resistance. *Journal of Physiology, 207,* 823-835.

Pugh, L.G.C.E. (1971). The influence of wind resistance in running and walking and the mechanical efficiency of work against horizontal or vertical forces. *Journal of Physiology, 213,* 255-276.

Shanebrook, J.R., & Jaszczak, R.D. (1976). Aerodynamic drag analysis of runners. *Medicine and Science in Sports, 8*(1), 43-45.

Sjödin, B., & Shele, R. (1982). Oxygen cost of treadmill running in long-distance runners. In P.V. Komi (Ed.), *Exercise and sport biology* (pp. 61-67). Champaign, IL: Human Kinetics.

Stevens, E.D. (1983). Effect of the weight of athletic clothing in distance running by amateur athletes. *Journal of Sports Medicine, 23,* 185-190.

Van der Walt, W.H., & Wyndham, C.H. (1973). An equation for predicion of energy expenditure of walking and running. *Journal of Applied Physiology, 34*(5), 559-563.

Williams, K.R. (1982). Non-sagittal plane movements and forces during distance running. In Abstracts, 6th Annual Conference of the American Society of Biomechanics, p. 24, Seattle: University of Washington.

Williams, K.R. (1985a). A comparison of 2D vs. 3D analyses of distance running. In D.A. Winter, R.W. Norman, R.P. Wells, K.C. Hayes, & A.E. Patla (Eds.), *Biomechanics IX-B* (pp. 331-337). Champaign, IL: Human Kinetics.

Williams, K.R. (1985b). The biomechanics of running. *Exercise and Sport Science Reviews, 13,* 389-441.

Williams, K.R. (1985c). The relationship between mechanical and physiological energy estimates. *Medicine and Science in Sports and Exercise, 17*(3), 317-325.

Williams, K.R., & Cavanagh, P.R. (1983). A model for the calculation of mechanical power during distance running. *Journal of Biomechanics, 16*(2), 115-128.

Williams, K.R., & Cavanagh, P.R. (1986). Biomechanical correlates with running economy in elite distance runners. *Proceedings of the North American Congress on Biomechanics* (pp. 287-288). Montreal: Organizing Committee.

Williams, K.R., & Cavanagh, P.R. (1987). Relationship between distance running mechanics, running economy, and performance. *Journal of Applied Physiology, 63*(3), 1236-1245.

Williams, K.R., Cavanagh, P.R., & Ziff, J.L. (1987). Biomechanical studies of elite female distance runners. *International Journal of Sports Medicine, 8*(Suppl.), 107-118.

Winter, D.A. (1978). A new definition of mechanical work done in human movement. *Exercise and Sport Science Reviews, 6,* 183-201.

Chapter 12

Scale Effects in Distance Running

E.C. Frederick

Distance runners come in an impressive array of sizes and shapes, from tall and lean to short and stout. In a survey of 1,468 marathon runners (Frederick & Clarke 1981, 1982) the average marathoner was 176.5 cm tall and had a body mass of 65.8 kg, but height ranged from 152 cm to 197 cm and body mass from 40 kg to 100 kg. This observation raises a number of questions about the potential biomechanical consequences of such dimensional variety. Is it advantageous for distance runners to be tall or short? Do taller and consequently heavier runners experience greater impact forces? Are shorter runners slower because of their shorter strides? This chapter addresses these and other questions by way of exploring the relationship between body dimensions and the kinetics and kinematics of running. The tools of static dimensional analysis are discussed here to the exclusion of dynamic scale effects. McMahon and Bonner (1983) have included some discussion of the biological significance of dynamic scaling, but these effects have not been well explored in humans and are beyond the scope of this review.

The effects of this relationship are not obvious and are sometimes counterintuitive. For example, when Swift conceived the tiny Lilliputians and giant Brobdingnagians of *Gulliver's Travels*, he was laboring under a considerable misconception about the way in which bodies scale and the consequences of that scaling. He gave his creations the same shape as the normal-sized Gulliver, but dimensional analysis teaches that, if that were the case, the 15-cm-tall Lilliputians and 18-m-tall Brobding-nagians would be burdened with seemingly insurmountable problems as a result of their relative sizes.

The Brobdingnagians, 10 times taller than average humans, would collapse under some 90 tons of body mass borne on a skeleton capable of supporting perhaps a ton of weight. Even at a height only twice that of Gulliver, human-shaped creatures would scarcely be able to take a single step without snapping bones and tendons. At the other extreme, the Lilliputians, due to an order of magnitude higher ratio of surface area to mass, would lose heat to the environment at a rate 10 times normal.

This would require the little folk to spend nearly every waking moment in search of food rather than entertaining Gulliver as Swift would have us imagine. This altered relationship of surface area to mass would also mean that some frictional forces that we usually consider negligible for full-sized people would be a major mechanical constraint for the Lilliputians. These complex problems that considerably change the nature of Swift's characters are not products of the imagination but real consequences of scaling. A number of these consequences significantly affect the biomechanics of distance running when body size is considered as a variable. The concept of body size is taken to describe the average combination of height and weight of typical "large" versus "small" people. The characteristic dimension referred to when talking about large and small people is body stature, or standing height.

Theory

In geometric (isometric) scaling, as espoused by Swift, objects maintain their shape when size varies by holding mass (m) and volume proportional to the cube of linear dimensions (l), and areas (A) proportional to the square of linear dimensions such that

$$m \propto l^3,$$
$$l \propto m^{1/3},$$
and
$$A \propto l^2;$$
therefore,
$$A \propto m^{2/3}.$$

Limitations on the strength of biomaterials, usually a function of cross-sectional area, and various physiological considerations, such as thermoregulation, mean that animals generally change in shape when one compares large versus small forms to compensate for scale effects. This explains the stout versus lean shapes we usually see in large animals like elephants when compared with small animals like mice. McMahon (1973) has shown that this *allometric* scaling of animals tends to follow a model of elastic rather than geometric similarity. McMahon argues that when animal structures scale according to elastic similarity, lengths increase as the 2/3 power of diameters (d). This influences the scaling of other dimensions, such that $l \propto m^{1/4}$, $d \propto m^{3/8}$, and $A \propto m^{5/8}$. This model of elastic similarity seems to work for animals over a wide range of sizes, but it does not work well for comparisons over limited size ranges, such as we find within the real, as opposed to the imagined, human species.

   The range of human dimensions is relatively modest when compared with the several orders of magnitude of size variation found in the animal kingdom. Adult human stature has a "possible" range of about 70 cm to 270 cm (Diagram Group, 1980), although 95% of adults fall between

147 cm and 188 cm in height. Body mass is even more variable, from as much as 490 kg, unlikely in a runner, to less than 25 kg (Diagram Group, 1980).

McMahon and Bonner (1983) point out that because of such limitations on size variation, the dimensions of individuals within the human species generally follow a model of geometric similarity. And they note that actuarial tables reporting average human heights and weights show a relationship of $m \propto l^{2.9}$. Although this exponent shows a slight allometric tendency that would make humans leaner with increasing height, a fact easily corroborated by even the casual observer, it is quite close to what would be expected with geometric similarity.

Frederick and Clarke (1982) have shown, curiously enough, that in a large sample of marathoners, $m \propto l^{2.5}$. This means that marathoners, who are certainly a highly self-selected population, are leaner and scaled even more allometrically than their nonrunning counterparts. For the purposes of this paper, however, I will assume that we are dealing with a more diverse human population than marathoners and therefore $m \propto l^{2.9}$ should prevail when the precise exponent is needed. And the model of choice for describing human scale effects is the model of geometric similarity, $m \propto l^3$. To distinguish between actual and theoretical results I will hold to the convention of using whole fractions for theoretically determined exponents and decimal fractions for experimentally determined exponents.

Hill's Predictions and Beyond

A.V. Hill gave considerable thought to the biomechanical consequences of scale effects in his classic paper, "The Dimensions of Animals and Their Muscular Dynamics" (1950). Using a model of geometric similarity, Hill made several predictions that we can apply to humans by way of developing a theoretical understanding of how scale effects should influence the biomechanics of running.

Based on the correct assumption that the speeds of the distal segments of limbs relative to the body are the same regardless of size, Hill determined that time, $t \propto l \propto m^{1/3}$. He also predicted that, based on his experiments with muscle, maximum force, $F \propto l^2$. And he further determined that maximum stride length, L, should be proportional to l and therefore $\propto m^{1/3}$; and maximum stride frequency, f, should be proportional to l^{-1} and $\propto m^{-1/3}$. The relationship between the previous two variables means that maximum speed, V_{max}, should be a constant and independent of size. Continuing with Hill's findings, maximum power $P_{max} \propto m^{2/3} \propto l^2 \cdot V_{max}$ (derived as follows: $P = F \cdot V$; $P \propto l^2 \cdot V$(a constant); therefore $P \cdot m^{2/3}$). This should also apply to metabolic power such that maximum oxygen uptake, $\dot{V}O_2max \propto m^{2/3}$. McMahon (1975) and Taylor (1977) have also pointed out that Hill's isometric model also predicts that the power required to run at any speed should be proportional to the cube of the speed. Hill further maintained that the kinetic energy of a limb KE $\propto l^3 \propto m$, and his determinations can be extended further to show that accelerations (a) connected with maximum forces are $\propto l^{-1}$ ($a = F/m$; therefore, $a \propto l^2/l^3 \propto l^{-1}$).

Scaling and Human Running

These theoretical determinations lead us to a number of predictions about the relationship between human size and the biomechanics of running that have relevance to this synthesis. The rest of this paper is devoted to examining a selection of these predictions and related issues.

Predicted Scale Effects on Human Running

1. The maximal metabolic power available for running should be $\propto m^{2/3}$. Also, the power required to run at any speed should increase as the cube of speed (McMahon, 1975).
2. Size should not affect maximum sprint speed but may affect the ability to maintain speed over a long distance because of the importance of sustained metabolic power.
3. Maximum forces (non-impact–related) should be $\propto m^{2/3} \propto l^2$, and peak forces and accelerations should also be size-influenced but not $\propto m^{2/3}$.
4. Plantar pressures due to loading of the extremities should increase with size, because weight increases at a greater rate than plantar surface area as we compare larger and larger people.
5. Because it is a function of cross-sectional area, the power required to overcome air resistance should be independent of size for runners, but other frictional forces may vary with size.

Metabolic Power

Within the human species basal metabolic rate, $\dot{V}O_2$basal $\propto m^{.67}$ (Guyton, 1966). More importantly, $\dot{V}O_2$max has also been shown to be $\propto m^{.67}$ for trained athletes ranging in body mass from 55 kg to 95 kg (Åstrand & Rodahl, 1977, p. 377). This relationship means that taller, and therefore proportionately heavier, runners tend to have a greater gross $\dot{V}O_2$max (ml O_2 • min^{-1}) and a lower mass-corrected $\dot{V}O_2$max (ml O_2 • kg^{-1} •min^{-1}). Another way of looking at this is to say that larger people have to expend less power to move a kilogram of body mass at maximum speed. Further, it means that the overall power output of larger runners should be greater at speeds corresponding to maximal oxygen uptake. Because maximum oxygen uptake usually corresponds to a level of effort that can be sustained for about 10 minutes, it also stands to reason that larger runners should expend less energy to move a kilogram of body mass when racing at distances between 3,000 and 5,000 m. These larger runners would also have more total power available for running at those distances.

Not all physiological systems involved in oxygen uptake need follow this same proportion. For example, because they are volumes, values for various lung capacities that are physiologically related to $\dot{V}O_2max$ should be more closely $\alpha\ l^3$ (or $\alpha\ m^{1/3}$). And, in fact, Cook and Hamann (1961) found that residual volume, vital capacity, and total lung capacity are α $l^{2.97}$, $l^{2.92}$, and $l^{2.91}$, respectively. This nearly geometric scaling of lung capacities must be the case if our arguments for metabolic power being $\alpha\ m^{2/3}$ are to hold.

At first glance, Hill's prediction that power requirements should vary as the cube of speed appears to be off the mark. Numerous studies that determine the rate of oxygen uptake at various speeds report that these variables are linearly related over a range of speeds. Daniels (1985), however, though corroborating this linearity, has observed that total energy expenditure (from both aerobic and anaerobic sources) would actually rise exponentially at speeds corresponding to over about 80% of $\dot{V}O_2max$, the point beyond which anaerobic sources are contributing a significant proportion of total metabolic power (Daniels, 1985). Daniels has suggested that more research is needed to clarify this point. Studies of the mechanical power requirements of running lend some support to his suggestion.

Fukunaga, Matsuo, Yuasa, Fujimatsu, and Asahina (1980) estimated the mechanical power output of running at various speeds. Although one has to use caution when interpreting their data because the calculation of mechanical power is a debatable procedure, the results are intriguing when kept in perspective. Fukunaga and coworkers found that both total power output and power output in the vertical plane appear to be linear functions of speed but that power output in the forward (horizontal) direction increases as the second power of the velocity. The power output in the forward direction also was found to exceed the power output in the vertical plane at speeds over 6 m • s^{-1}. Curiously, this is not far off a speed corresponding to 80% of $\dot{V}O_2max$ for a runner with a $\dot{V}O_2max$ of 80 ml O_2 • kg^{-1} • min^{-1}—a likely value for an elite distance runner. If the body could be shown to be less efficient at performing work to overcome resistance in the horizontal plane it might well be that there is a mechanical rationale for Daniel's observations. Unfortunately, Pugh (1971) found the opposite result when he showed that the apparent efficiency of overcoming wind resistance was greater than that of performing vertical work to move the body's mass up an inclined plane. But Pugh's models of mechanical work were incomplete. And although he observed kinematic changes in his subjects in response to wind resistance, he did not use kinematic data to perform determinations like those of Fukunaga et al. (1980). In the absence of these determinations, we may speculate that Hill may have overestimated the apparent efficiency of performing work against a horizontal force. Future experiments may help us sort out the truth in Hill's proposition.

In addition to the relationship between body mass and $\dot{V}O_2$basal and $\dot{V}O_2$max, Montoye and Ayen (1986) have also shown that submaximal $\dot{V}O_2$ is dependent on body mass. If we use the mean data presented in Table 1 of Montoye and Ayen's paper, in which 1,001 subjects were studied, to generate a power curve for weight on oxygen uptake (ml O_2 • min^{-1}) while walking on the level, we find that $\dot{V}O_2$submax α $m^{.69}$ (r^2 = .98), very close to what we would expect. Taken together these various data support the notion that maximal power and probably submaximal power as well are α $m^{2/3}$. These conclusions can be interpreted differently depending on the distance of the run being considered.

Maximum Speed

Maximum sprint speed, as defined by average speed in 75-m or 100-m dashes, for boys of different sizes was investigated by Asmussen and Christensen (1967). Surprisingly, they found a pronounced improvement in speed with increased body height for the 14-year-olds. This result may be due to differences in sexual maturity evidenced in body height. Support for this explanation comes from their data on 18-year-olds in which the height effect on speed is nearly absent despite the fact that a similar range of statures was included. The observed range in stature of champion sprinters seems to support this notion (Tanner, 1964).

Distance running, the subject of this volume, has quite different requirements than sprint running does. Such factors as sustained power output and efficient movement are more important in events performed at or below $\dot{V}O_2$max. For this reason we must consider other variables that may affect the maximum speed that can be sustained for distances greater than the 400 m or less of sprint events.

Several authors have found that significant proportions of the interindividual variation in performance in distance running can be explained by $\dot{V}O_2$max (Katch & Henry, 1972; Kumagai, Tanaka, Matsuura, Matsuzaka, Hirakoba, & Asano, 1982; Tanaka & Matsuura, 1982). But it is also apparent that the importance of $\dot{V}O_2$max changes with distance. Tanaka and Matsuura (1982) found a lower contribution of mass relative $\dot{V}O_2$max (ml O_2 • kg^{-1} • min^{-1}) to the variation in performance of 10,000-m runners than to that of 5,000-m, 1,500-m, or 800-m runners. And interestingly enough, the correlation between the gross $\dot{V}O_2$max (ml O_2 • min^{-1}) and performance at 10,000 m was not significant, even though it was significant and quite high for the 5,000 m, 1,500 m, and 800 m. These findings seem to support the notion that power output is more important at the relatively shorter distances, which are more power limited when we recall that $P_{max} \propto m^{2/3}$ and $\dot{V}O_2$ max $\propto m^{.67}$. In the longer distances power per kilogram of body mass becomes a contributing factor to success but it is not as important as other factors (Costill & Fox, 1969; Powers, Dodd, Deason, Byrd, & McKnight, 1983). This observation may contribute to fundamental differences in the ''ideal'' body dimensions of athletes who are successful at the longer versus shorter distance events.

In light of Tanaka and Matsuura's (1982) findings, it would seem that runners in the shorter distance races should be bigger, and, in fact, these same authors found significant negative correlations between stature and body weight and performance time in the 1,500 m and 800 m, but not for the longer distances. This indicates that bigger athletes were more successful at the shorter distances (Tanaka & Matsuura, 1982). Success among the 10,000-m runners, on the other hand, was significantly affected only by indicators of a slender build, regardless of stature (Tanaka & Matsuura, 1982). This should not be surprising. And it provides support for the classic notion of the slightly built distance runner. Provided that muscle mass is not significantly diminished, a slender and light build would passively increase $\dot{V}O_2$max and lower the gross power requirement (ml O_2 • min^{-1}) needed to maintain a given speed (Cureton, Sparling, Evans, Johnson, Kong, & Purvis, 1978). Other authors have also found similar trends in the anthropometry of elite runners competing at different distances (Carter, 1970; Tanner, 1964).

To summarize, there is support for the idea that athletes who are bigger (i.e., taller and proportionately heavier) should be more successful at the middle-distance events. In addition, although there is no apparent overall size advantage for the long-distance events, long-distance runners should be as lean as possible for their stature.

Kinetics

In support of the notion that peak muscle force $F \propto l^2 \propto m^{2/3}$, M.H. Lietzke found that records of total weight lifted in strength competitions are proportional to $m^{.67}$ (Lietzke, 1956). This indicates that larger people are capable of developing greater peak muscle forces. This should affect the ground reaction forces at maximum running speed in a similar manner, but no data are available in the literature to confirm this suggestion.

What about the reaction forces during distance running? One can make the case that if metabolic power is a function of the developed muscle force, and if $\dot{V}O_2$submax $\propto m^{2/3}$ then submaximal forces may also be $\propto$ $m^{2/3}$. No data could be found on the relationship between resultant forces and body mass or integrated forces and body mass, but peak vertical forces appear to be size-influenced although not α $m^{2/3}$.

Cavanagh and Lafortune (1980) and a number of other authors (Bates, Osternig, Sawhill, & James, 1983; Hamill, Bates, Knutzen, & Sawhill, 1983; Miller, 1978; Nigg, Denoth, & Neukomm, 1981) have demonstrated that there are commonly two principal peaks in the typical vertical ground reaction force (VGRF) curve: an "impact" peak (F_{z1}) occurring shortly after first contact, and an "active" peak (F_{z2}) (Nigg, Denoth, & Neukomm, 1981) occurring near the middle of stance time. The literature provides ample evidence that vertical ground reaction forces show considerable interindividual and intraindividual variation (Bates et al., 1983; Cavanagh & Lafortune, 1980; Hamill et al., 1983; Kinoshita, Bates, & DeVita, 1985; Nigg, 1985, 1986). One wonders what contribution scale effects and other size-related factors make to this variation.

Frederick and Hagy (1986) examined the influence on VGRF of a number of dimensional parameters including *speed*, *body mass*, *stature* (standing height), and the *reciprocal ponderal index* (which corrects for the isometric scale effect of height and weight).

$$RPI = \text{Stature (inches)} \cdot [\text{body weight (lbs)}^{-0.33}].$$

It should be noted that the RPI is reported in English units because that is how it is defined and generally computed, and because it cannot be directly converted to metric units.

In the Frederick and Hagy study, nine subjects ranging in body mass from 90.9 to 45.5 kg ran repeated trials across a force platform while being filmed to determine kinematic parameters. The subjects ran barefoot trials at each of three speeds: 3.35, 3.83, and 4.47 m $\cdot$ s^{-1}. Force data were analyzed for the magnitude and temporal characteristics of F_{z1} and F_{z2} of the VGRF. Their results support the general conclusion that speed and, indirectly, body mass are significant effectors of the magnitudes of F_{z1}. In addition, other factors that correlate significantly with F_{z1} are connected in some manner to stature: RPI and stature; half-stride length (one step), step length (the distance the body travels during contact), leg length, and vertical hip excursion during a stride cycle.

Body mass was found to correlate highly with F_{z2} ($r = 0.95$). Their results show that 92% of the variability in F_{z2} can be explained by the combination of body mass, stature, RPI, and leg length, with 90% being explained by mass alone.

These data support earlier findings (Nigg & Denoth, 1980) that speed and the effective mass of the leg at contact are important effectors of the magnitude of F_{z1}. In addition, the kinematic and anthropometric parameters that contribute significantly to the variability in F_{z1} and F_{z2} are generally cross-correlated with body size, running speed, or both.

Taken together these data suggest that body mass has a consistent and pronounced influence on the absolute magnitude of the two major peaks in VGRF, but particularly F_{z2}. This is not surprising when one considers that the mass contributing to the development of these forces is either the mass of the body itself in the case of F_{z2} or, in the case of F_{z1}, the effective mass of the leg, which should be a function of body mass. Figure 12.1 shows a plot of F_{z2} against body mass for the nine subjects in the Frederick and Hagy study along with a line representing $F \propto m^1$. Although this is too small a sample for sweeping conclusions, this figure lends some

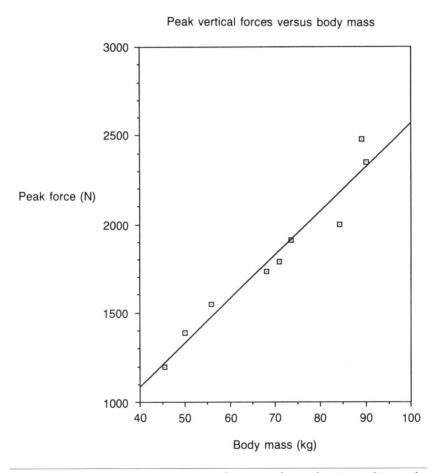

Figure 12.1. Peak active vertical ground reaction forces for nine subjects of various body masses. Each data point represents the average of five trials for each subject at a speed of 3.35 m • s^{-1}. The plotted line is for peak force, $F \propto m^1$. (Partial data from Frederick and Hagy, 1986, p. 45).

support to the idea that the peak active VGRFs required to maintain speed at a particular speed are a function of body mass to the first power. Impact forces, on the other hand, although indirectly related to mass, appear to be more a function of speed and linear dimensions such as stature and stride length.

Hamill et al. (1983) have shown a similar relationship between running speed and F_{z1}. Their subjects ran trials at 4, 5, 6, and 7 m • s^{-1} and showed progressive increases in body mass–specific F_{z1} (N per kg body mass) with speed. In addition, a lesser effect of speed on F_{z2} was observed, which also agrees with the results of this study.

Acceleration

Accelerations (a), according to Hill's model, should be $\alpha\ l^{-1}$. Although it is predicted that accelerations attendant to maximum forces should be $\alpha\ l^{-1}$, that should not necessarily be the case for submaximal accelerations. And, in fact, Clarke, Cooper, Clark, and Hamill (1983) found that longer strides and lower stride frequencies at a given speed caused increases in peak accelerations measured at the tibia.

Decelerations like these, which are associated with peak active VGRFs, should be independent of body size because, if as mentioned above, we can assume that F $\alpha\ m^1 \alpha\ l^3$, and because $a = F/m$, therefore a should be $\alpha l^3/l^3$. So it seems surprising that peak shank deceleration should be proportional to stride length unless we consider the possibility that the mass involved in this acceleration is not the mass of the body but the effective mass (Nigg, 1986).

Plantar Pressures

Frederick and Clarke (1982) predicted that larger runners are subjected to proportionately higher peak VGRFs because of the effects of scale. This is borne out by Frederick and Hagy (1986) who showed that peak active forces are $\alpha\ m^1$. Also influenced by the effects of scale are peak pressures on the plantar surface of the foot. If plantar surface contact area is $\alpha\ l^2$ as would be expected in isometric scaling, and force is proportional to m and therefore $\alpha\ l^3$, then peak pressures should be $\alpha\ l^3/l^2$ or $\alpha\ l$. Because foot length should be $\alpha\ l$ then peak pressures should go up as foot length and shoe size increase. Frederick and Clarke (1982) have argued that this means that shoe manufacturers may have to scale the cushioning in various sizes of a particular model of shoe. At present, the cushioning systems of different sizes of a particular shoe design generally vary only in their area and length dimensions. This means, for example, that if the amount of cushioning that a size 9 shoe provides is adequate, then cushioning may be inadequate in the larger sizes and too great in the smaller sizes. Because the cushioning provided is not just a function of the area of the sole under the foot but is also a function of sole thickness and softness, shoe designers should scale these properties $\alpha\ l$.

Frictional Parameters

Hill (1927) confirmed aerodynamic theory when he observed that the force required to overcome wind resistance is proportional to the projected area that the body presents to the wind. Pugh (1971) found projected area to be a consistent proportion of the total surface area of the body (SA), [0.266 • $SA(m^2)$]. Although total surface area as calculated by the DuBois and DuBois (1916) formula is α (body mass$^{.425}$ • height$^{.725}$), this can be shown to actually work out to $SA\ \alpha\ m^{.667}$. This would mean that wind resistance should be $\alpha\ A$ or $\alpha\ m^{2/3}$, or, in other words, independent of normal body size. Hill argued this point (1950), and the limited research that has been presented to date on this subject seems to support his argument.

Other frictional parameters, however, may not be size-independent. The static translational friction coefficient (μ), for example, is thought to be dependent on the normal force and independent of surface area. Although there is conflicting evidence on the validity of this classical view for certain nonrunning applications (Schlaepfer, Unold, & Nigg, 1983; van Gheluwe, DePorte, & Hebbelinck, 1983), to the extent that it is correct, normal forces and therefore frictional forces should be proportionately greater in larger people and lesser in smaller people. Rotational friction, which is further dependent on the plantar pressure and on the area of contact with the surface, should also tend to be affected by body size. To date no experimental proof of this expected effect of body size on frictional forces has been presented. If it is correct, however, it may have important consequences for the design of sports shoes and playing surfaces for children and women as well as for generally larger people such as American football players.

Summary

The biomechanical consequences of human scale effects are many and varied. Among the important aspects of human running that are thought to be influenced by body dimensions are maximal and submaximal metabolic power, speed in the various distance events, maximum forces and accelerations, plantar pressures, and some frictional forces. These factors may contribute to the relative success of certain sized individuals in running events of various distances. In addition these scale effects should be borne in mind when we are tempted to extrapolate the results of our biomechanical investigations beyond the scope of the dimensional variety of the subject pool.

References

Asmussen, E., & Christensen, E.H. (1967). Kompendium i Legems-övelsernes Specielle Teori, Köbenhavns Universitets Fond til Tilvejebringelse af Läremidler, Köbenhavn

Åstrand, P.O., & Rodahl, K. (1977). *Textbook of Work Physiology* (2nd Ed.). New York: McGraw-Hill.

Bates, B.T., Osternig, L.P., Sawhill, J.A., & James, S.L. (1983). An assessment of subject variability, subject-shoe interaction and the evaluation of running shoes using ground reaction force data. *Journal of Biomechanics*, **16**, 181-192.

Carter, J.E.L. (1970). The somatotypes of athletes: A review. *Human Biology*, **42**, 535-569.

Cavanagh, P.R., & Lafortune, M.A. (1980). Ground reaction forces in distance running. *Journal of Biomechanics*, **13**, 397-406.

Clarke, T.E., Cooper, L.B., Clark, D., & Hamill C. (1983). The effect of varied stride rate and length upon shank deceleration during ground contact in running. *Medicine and Science in Sports and Exercise*, **15**(2), 170.

Cook, C.P., & Hamann, J.F. (1961). Relationship of lung volumes to height in healthy persons between 5 and 38 years. *Journal of Pediatrics*, **59**, 710-714.

Costill, D.L., & Fox, E.L. (1969). Energetics of marathon running. *Medicine and Science in Sports and Exercise*, **1**(1), 81-86.

Cureton, K., Sparling, P., Evans, B.W., Johnson, S.M., Kong, U.D., & Purvis, J.W. (1978). Effects of experimental alterations in excess weight on aerobic capacity and distance running performance. *Medicine and Science in Sports and Exercise*, **10**(3), 194-199.

Daniels, J.T. (1985). A physiologist's view of running economy. *Medicine and Science in Sports and Exercise*, **17**(3), 332-338.

Diagram Group (1980). *Comparisons*. New York: St. Martin's.

DuBois, D., & DuBois, E.F. (1916). Clinical calorimetry: A formula to estimate the approximate surface area if height and weight be known. *Archives of Internal Medicine*, **17**, 863-871.

Frederick, E.C., & Clarke, T.E. (1981). Consequences of scaling on impact loading in running. *Medicine and Science in Sports and Exercise*, **13**(2), 96.

Frederick, E.C., & Clarke, T.E. (1982). Body size and biomechanical consequences. In R.C. Cantu & W.J. Gillespie (Eds.), *Sports medicine, sports science: Bridging the gap* (pp. 47-57). Lexington, MA: Collamore.

Frederick, E.C., & Hagy, J.L. (1986). Factors affecting peak vertical ground reaction forces in running. *International Journal of Sports Biomechanics*, **2**, 41-49.

Fukunaga, T., Matsuo, A., Yuasa, K., Fujimatsu, H., & Asahina, K. (1980). Effect of velocity on external mechanical power output. *Ergonomics*, **23**, 123-136.

Guyton, A.C. (1966). *Textbook of medical physiology* (3rd ed.). Philadelphia: W.B. Saunders.

Hamill, J., Bates, B.T., Kuntzen, K.M., & Sawhill, J.A. (1983). Variations in ground reaction force parameters at different running speeds. *Human Movement Science*, **2**, 47-56.

Hill, A.V. (1927). The air resistance of a runner. *Proceedings of the Royal Society London, Series B*, **102**, 380-385.

Hill, A.V. (1950). The dimensions of animals and their muscular dynamics. *Science Progress*, **38**, 209-230.

Katch, V., & Henry, F.M. (1972). Prediction of running performance from maximal oxygen debt and intake. *Medicine and Science in Sports and Exercise*, **4**(3), 187-191.

Kinoshita, H., Bates, B.T., & DeVita, P. (1985). Intertrial variability for selected running gait parameters. In D.A. Winter, R.W. Norman, R.P. Wells, K.C. Hayes, & A.E. Patla (Eds.), *Biomechanics IX-B* (pp. 499-502). Champaign, IL: Human Kinetics.

Kumagai, S., Tanaka, K., Matsuura, Y., Matsuzaka, A., Hirakoba, K., & Asano, K. (1982). Relationships of anaerobic threshold with the 5km, 10km, and 10 mile races. *European Journal of Applied Physiology*, **49**, 13-23.

Lietzke, M.H. (1956). Relation between weight-lifting totals and body weight. *Science*, **124**, 486-487.

McMahon, T.A. (1973). Size and shape in biology. *Science*, **179**, 1201-1204.

McMahon, T.A. (1975). Using body size to understand the structural design of animals: Quadrupedal locomotion. *Journal of Applied Physiology*, **39**, 619-627.

McMahon, T.A., & Bonner, J.T. (1983). *On size and life*. New York: W.H. Freeman.

Miller, D.I. (1978). Biomechanics of running—what should the future hold? *Canadian Journal of Applied Sports Sciences*, **3**, 229-236.

Montoye, H.J., & Ayen, T. (1986). Body size adjustment for oxygen requirement in treadmill walking. *Research Quarterly*, **57**, 82-84.

Nigg, B.M. (1985). Load in selected sports activities—an overview. In D.A. Winter, R.W. Norman, R.P. Wells, K.C. Hayes, & A.E. Patla (Eds.), *Biomechanics IX-B* (pp. 91-96). Champaign, IL: Human Kinetics.

Nigg, B.M. (Ed.) (1986). *Biomechanics of running shoes*. Champaign, IL: Human Kinetics.

Nigg, B.M., & Denoth, J. (1980). *Sportplatzbelaege* [Playing surfaces]. Zurich: Juris Verlag.

Nigg, B.M., Denoth, J. & Neukomm, P.A. (1981). Quantifying the load on the human body. In A. Morecki & K. Fidelus (Eds.), *Biomechanics VII* (pp. 88-99). Baltimore: University Park.

Powers, S.K., Dodd, S., Deason, R., Byrd, R., & McKnight, T. (1983). Ventilatory threshold, running economy and distance running performance in trained athletes. *Research Quarterly, 54*, 179-182.

Pugh, L.G.C.E. (1971). The influence of wind resistance in running and walking and the mechanical efficiency of work against horizontal or vertical forces. *Journal of Physiology, 213*, 255-276.

Schlaepfer, F., Unold, E., & Nigg, B. (1983). The frictional characteristics of tennis shoes. In B.M. Nigg & B.A. Kerr (Eds.), *Biomechanical aspects of sport shoes and playing surfaces* (pp. 153-160). Calgary, AB: The University of Calgary.

Tanaka, K. & Matsuura, Y. (1982). A multivariate analysis of the role of certain anthropometric and physiological attributes in distance running. *Annals of Human Biology, 9*, 473-482.

Tanner, J.M. (1964). *The physique of the Olympic athlete*. London: George Allen and Unwin.

Taylor, C.R. (1977). Why big animals? *Cornell Veterinarian, 67*, 155-175.

van Gheluwe, B., DePorte, E., & Hebbelinck, M. (1983). Frictional forces and torques of soccer shoes on artificial turf. In B.M. Nigg & B.A. Kerr (Eds.), *Biomechanical aspects of sport shoes and playing surfaces* (pp. 161-168). Calgary, AB: The University of Calgary.

Chapter 13

Gender Differences in Distance Running

Anne E. (Betty) Atwater

As an athletic group, female distance runners have been the subjects of biomechanical research in very few investigations. One of the first studies that involved a biomechanical comparison of elite female and male distance runners was reported in 1977 at a scientific conference focusing on the marathon (Nelson, Brooks, & Pike, 1977). Even with this stimulus, there have been but a few recent investigations directed at providing additional data on the biomechanics of female distance runners. In contrast, extensive studies on various aspects of gait analysis have been performed on male runners over several decades.

The paucity of biomechanical data on female distance runners may partially reflect the fact that distance running has experienced only recent acceptance as a sport for women. Complex combinations of factors including myths, cultural patterns, prejudices, ill-founded arguments, and biological misconceptions served to restrict female participation in distance running, more than in other sports, during the first two thirds of the 20th century. Like Melpomene, a Greek woman who ran unofficially beside male marathon competitors in the first modern Olympics of 1896 because her entry was refused official sanction, a few American women jumped into and unofficially completed marathons to which their entry was barred during the 1960s (Hult, 1986). Coinciding with the increasing popularity of the national jogging movement during the 1970s was the struggle of a growing number of female distance runners for the opportunity to demonstrate their endurance capabilities in the marathon. This decade saw the beginning of several organized marathon runs for women and the opening of previously all-male marathons to women (Kuscsik, 1977). As performance records were lowered by increasingly better trained women runners, the pressure grew for sanctioning of a women's marathon by the International Olympic Committee. Intense lobbying efforts finally overcame early petition defeats and culminated in the addition of the women's marathon to the roster of the 1984 Olympic events (Hult, 1986).

The goal of many female runners today is the development of a well-conditioned body and an efficient running technique to successfully and safely meet the challenge of competitive distance running events. In addition, thousands of women participate in recreational jogging over undefined distances for purposes that may include better health and physical endurance, improved self-image, socialization, and relief from tension or anxiety. As female runners and joggers strive to achieve their personal objectives in this form of physical activity, it is clear that many physical and physiological factors can influence their endurance performance.

The purpose of this chapter is to review, from a biomechanical perspective, physical differences and similarities between the sexes and to examine the extent to which observed differences may affect (a) temporal, kinematic, and kinetic aspects of running technique; and (b) the type and incidence of overuse injuries in female runners. It should be noted that, in all probability, physiological or functional differences between males and females are more important than structural and biomechanical differences where endurance performance is concerned. For a discussion of the interrelationships between physical and physiological factors as they specifically apply to girls and women, the reader is referred to the following sources: Drinkwater (1984, 1986), Fox and Mathews (1981), Wells (1985), and Wilmore (1982).

Performance Records

Comparisons of running performance records for women and men can serve as a background for the discussion of gender-related differences in biomechanics. Table 13.1 lists world records as of mid-1986 for women and men in running events ranging from sprints to the marathon. The ratio expressing the performance of women relative to that of men varies from 89.0% in the 5,000-m event to 92.3% in the 100-m sprint.

During the past 20 years, the improvement of women runners has been much greater in the middle- and long-distance running events than in the sprints. In 1963, the 100-m world record for women was already 89% of the men's record (Wells, 1985). By contrast, two decades ago the world's best time for the women's marathon was only 66.2% of the men's best time. However, this performance ratio improved to 81.2% in 1976 and 90.2% in 1986. Evidence of the rapidly closing gap between male and female world records in the marathon is illustrated in Figure 13.1. Since 1963 the rate of improvement in the women's marathon performance has far exceeded that of the men.

Table 13.1 Comparison of Female and Male World Records: 1986

Running event (meters)	Women (hr.min:s)	Men (hr.min:s)	Performance ratio (percent)
100	10.76	9.93	92.3
200	21.71	19.72	90.8
400	47.60	43.86	92.1
800	1:53.28	1:41.73	89.8
1,500	3:52.47	3:29.46	90.1
3,000	8:22.62	7:32.10	90.0
5,000	14:37.33	13:00.40	89.0
10,000	30:13.74	27:13.81	90.1
Marathon	2.21:06.00	2.07:12.00	90.2

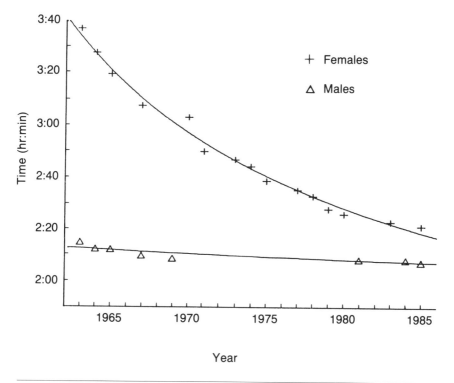

Figure 13.1. Trends in improvement of world best performances for men and women in the marathon.

Physical Similarities and Differences Between the Sexes

Obviously, sports performance is a very complex behavior and many factors interact to produce this performance. In the sport of distance running, which has only recently become popular for women, it is logical to question if the performance differences observed between males and females may be the result of biological differences between the sexes or if they reflect behavioral differences produced by cultural expectations and previous training experiences during the developing years. Although this question may be difficult to answer, some insights may be gained from comparisons of structural data on average males and females versus those who are highly trained distance runners.

It would indeed be simplistic to attempt to attribute athletic success to physical factors alone or to use physique as a key predictor of success in a sport. Instead, body structure might be viewed as either a limiting or a contributing factor to an individual's ultimate athletic accomplishments. An examination of the type and magnitude of physical differences between the sexes should serve to correct some fallacies and clarify some generalizations that have been perpetuated to the disadvantage of female runners. Also, it should serve as a basis for later discussion of possible gender-related differences in running technique and running injuries.

Several writers have expressed the need for caution when making comparisons between average males and females and between highly trained males and females on variables related to body structure and function (Adrian, 1972; Ulrich, 1960; Wells, 1985; Wells & Plowman, 1983; Wilmore, 1982; Wyrick, 1974). Frequency distributions for most physical variables are represented by a bell-shaped "normal" curve, with the average value (mean) at the center of the distribution. The amount of variability about the mean is reflected by the width of this curve (Figure 13.2). If females and males are compared on a given variable, the extent of overlap between the curves for the two groups will depend not only on the magnitude of the difference between their means but also on the amount of variability among the members of each group. For example, a difference as small as two score units between the means of males and females on variable A (Figure 13.2) may be significant because within-group variability is small, whereas this same mean difference between males and females on variable B would not be significant. If within-group variability is large, as reflected in the hypothetical curves for males and females on variable B (Figure 13.2), it is possible for greater differences to exist among members of the same sex than between the average male and the average female of each group. Furthermore, highly trained female athletes within a certain sport may differ, as a group, from average females or from nonathletic females.

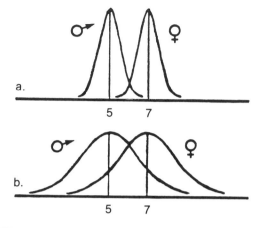

Figure 13.2.   Hypothetical frequency distributions for females and males on two physical variables.

Physical attributes that have been most commonly analyzed for their association with running ability are those pertaining to (a) body size and composition, and (b) selected skeletal dimensions, particularly hip width and shoulder width. In the following sections these physical attributes will be examined to learn the extent to which they differ between average males and females, between trained female and male distance runners, and between female runners of differing ability levels. It should be noted, however, that data on female runners reported in the research literature over the past 20 years reflect a wide range in the quality of the athletes and in the duration and intensity of their training. Not all of the runners studied were of elite caliber; nor were their competitive performances and levels of training equal to those of male distance runners. These factors may have important implications when comparisons are made between the sexes.

Body Size and Composition

At full sexual maturity, the average young adult male is taller and heavier and possesses less adipose tissue than the average young adult female. Reference data (Tables 13.2 and 13.3) for several groups of males and females reveal that the average female is approximately 12 cm shorter than the average male, is 18 kg lighter, and has about 8% to 10% more body fat. Even at identical heights, the average male has a larger total muscle mass and more bone weight than the average female. Higher levels of testosterone in males are believed to be responsible for the greater muscle mass, whereas increases in estrogen contribute to the higher percent body fat in females (Wilmore, 1982).

Table 13.2 Height, Weight, and Body Composition for Female Reference Groups and Distance Runners

Groups	N	Age (years)	Height (cm)	Weight (mass) (kg)	% Fat (%)	Source
Reference females						
College females	64	19-23	165.9 ± 4.3	58.4 ± 6.7	21.5 ± 5.7*	Katch and Michael, 1968
College females	128	21.4 ± 3.8	164.9 ± 6.6	58.6 ± 7.1	25.7 ± 4.5*	Wilmore and Behnke, 1970
Women age 18-55	249	31.4 ± 10.8	165.0 ± 6.0	57.2 ± 7.6	24.1 ± 7.2*	Jackson et al., 1980
Female distance runners[a]						
Elite distance runners	6	15-26	164.0 ± 5.0	50.4 ± 5.1	11.7 ± 4.5*	Brown and Wilmore, 1971
Elite distance runners	11	32.4 ± 4.5	169.4 ± 9.0	57.2 ± 6.7	15.2 ± 8.1*	Wilmore and Brown, 1974
Marathon runners	5	28.6 ± 5.5	162.3 ± 2.8	49.8 ± 4.1	12.5 ± 2.0*	Drinkwater et al., 1977
Distance runners	11	29.2 ± 3.5	165.2 ± 3.4	55.2 ± 5.5	16.6 ± 2.8**	Jordan, 1977
Elite distance runners	42	25.0 ± 9.1	166.8 ± 6.9	54.3 ± 6.4	16.9 ± 6.4*	Wilmore et al., 1977
Trained runners	14	24.1	159.7 ± 6.2	49.3 ± 5.8	15.5 ± 3.7*	Conley et al., 1981
Trained runners	4	28.5 ± 5.3	166.6 ± 5.5	52.0 ± 8.4	11.4 ± 0.7*	Wells et al., 1981
Novice marathon runners	10	29.6 ± 5.6	164.0 ± 6.9	55.7 ± 8.0	18.0 ± 2.9**	Christensen and Ruhling, 1983
Experienced marathoners	13	32.7 ± 4.1	164.7 ± 4.9	57.0 ± 5.4	16.3 ± 1.4**	Christensen and Ruhling, 1983
Distance runners	38	37.8 ± 5.0	165.1 ± 5.4	54.1 ± 4.8	15.5 ± 4.4*	Upton et al., 1983
British marathon runners: Moderate	13	29.7 ± 5.3	161.3 ± 6.6	53.0 ± 5.4	20.8 ± 2.5*	Bale et al., 1985
Good	12	29.8 ± 8.4	163.9 ± 6.0	51.2 ± 2.2	19.8 ± 3.1*	Bale et al., 1985
Elite	11	29.4 ± 7.6	166.4 ± 3.9	54.7 ± 5.6	16.4 ± 2.2*	Bale et al., 1985

[a]Chronological listing of the majority of studies in the past two decades that report data on women distance runners.

*Body density measured by hydrostatic weighing.

**Percent fat estimated from multiple skinfold measurements.

Table 13.3 Height, Weight, and Body Composition for Male Reference Groups and Distance Runners

Groups	N	Age (years)	Height (cm)	Weight (mass) (kg)	% Fat (%)	Source
Reference males						
College males	133	22.0 ± 3.1	177.3 ± 7.2	75.6 ± 11.0	14.4 ± 6.2*	Wilmore and Behnke, 1969
U.S. Marines	297	29.7 ± 8.2	177.1 ± 6.3	77.9 ± 9.8	16.5 ± 6.2*	Wright and Wilmore, 1974
Men age 18-61	308	32.6 ± 10.8	179.2 ± 6.5	74.8 ± 11.8	17.7 ± 8.0*	Jackson and Pollock, 1978
Male distance runners						
Elite marathon runners	114	26.1 ± 4.1	175.7 ± 6.8	64.2 ± 5.1	7.5**	Costill et al., 1970
Elite distance runners	20	26.2 ± 3.0	177.0 ± 6.0	63.1 ± 4.8	4.7 ± 3.1*	Pollock et al., 1977
Trained runners	7	30.9 ± 8.3	180.1 ± 5.1	69.8 ± 5.4	7.3 ± 3.2*	Wells et al., 1981

*Body density measured by hydrostatic weighing.
**Percent fat estimated from multiple skinfold measurements.

Active, athletic individuals of both sexes tend to have a lower percent body fat than their sedentary peers (Fleck, 1983; Sparling, 1980; Wells, 1985; Wilmore, Brown, & Davis, 1977). The average height, weight, and relative body fat reported in the literature for several groups of male and female distance runners are summarized in Tables 13.2 and 13.3. In all of these groups of runners the mean values for percent fat were less than those for the reference groups of the same sex. Body weight also was lower in distance runners than in the reference groups whereas height, although variable across the different studies, was similar to that of the reference groups.

Female distance runners have approximately the same percent fat as reference males (Tables 13.2 and 13.3). At least 10 of the elite female distance runners studied by Wilmore (Wilmore & Brown, 1974; Wilmore et al., 1977) had a percent fat of less than 10% (as measured by hydrostatic weighing), which placed them within the range of values reported for topflight male distance runners. Comparisons among female marathon runners of differing ability levels were conducted by Bale, Rowell, and Colley (1985) and Christensen and Ruhling (1983). In both studies, percent fat tended to decrease as the ability level of the runners increased (Table 13.2); only the British elite runners measured by Bale et al. (1985) had a significantly lower percent body fat than did the groups of good or moderate runners. The novice and experienced women runners studied by Christensen and Ruhling (1983) were actually nonelite marathoners whose average finish times exceeded 4 hours.

Body composition is a fairly important factor influencing performance in distance running. Many nonelite female runners may be at somewhat of a disadvantage when compared to male runners because of the additional fat they must carry during the run. Body fat is essentially dead weight to a runner. Extra body fat can hinder performance and may contribute to certain overuse injuries in the weight-bearing phase of running. Furthermore, a higher percent body fat and a lower total muscle mass contribute to lower measures of maximum aerobic power ($\dot{V}O_2$max). Absolute measures of maximal oxygen consumption summarized from 13 studies by Sparling (1980) showed the male values to be 56% higher, on the average, than the female values. Expressing maximal oxygen consumption relative to body weight reduced the apparent sex differences to 28%, which illustrates that part of the difference in $\dot{V}O_2$max between males and females is related to differences in body size and weight. Even when $\dot{V}O_2$max was expressed relative to fat-free weight instead of total body weight, the values for males were 12% to 15% higher than the values for females (Sparling, 1980). However, as Drinkwater (1973) has pointed out, the latter comparison may be merely an academic exercise because women cannot entirely discard their adipose tissue when participating in sport or physical work.

Regardless of the method used to express $\dot{V}O_2$max, there is considerable overlap of values between males and females. Several studies have clearly demonstrated that highly trained female distance runners often have a larger $\dot{V}O_2$max than most untrained men (Brown & Wilmore, 1971; Christensen & Ruhling, 1983; Wells, Hecht, & Krahenbuhl, 1981; Wilmore & Brown, 1974). Although it is well known that $\dot{V}O_2$max has a high genetic component, an individual's capacity for aerobic energy production may be influenced as much by training as it is by factors such as body size and composition. In fact, it is possible that improved training may play a significant role in reducing the gender differences observed in many physical and physiological variables associated with distance running performance.

Selected Skeletal Dimensions

On the average, girls start and complete their adolescent growth spurt in height about 2 years before boys do. Under the influence of increased amounts of estrogen, the closing of the epiphyses in the long bones of females is completed by approximately 18 years of age. Males usually do not reach full maturation until their early 20s, by which time they are almost 10% taller than the average female. The rate of growth in the hips increases by about the same amount in both males and females, whereas the shoulders and thoracic cage grow more rapidly in males (Sloane, 1980). Therefore, although hip width ends up about the same in both sexes, men tend to have wider shoulders in relation to their hips, whereas women have wider hips compared to their shoulders.

Anthropometric measurements of shoulder width (biacromial breadth [BA]) and pelvic width (biiliocristal breadth [BC])[1] for female and male reference groups and various athletic groups are summarized in Tables 13.4 and 13.5. When data included in these tables is examined, care should be taken to distinguish between absolute values (breadths measured in centimeters) and relative values (breadths expressed as a percentage of height or of other breadths). Reference group data support the previously stated generalizations that absolute pelvic width is very similar in both sexes whereas males have broader shoulders than females. When pelvic width is expressed relative to height (BC/Ht), the ratio is somewhat larger in reference females than in reference males primarily because the females are not as tall. Also, the ratio of pelvic width to shoulder width (BC/BA) is higher in mature females than it is in males, not because of differences in absolute pelvic width but because the male's shoulders are broader (Tables 13.4 and 13.5). The proportional change in the hip-shoulder ratio (BC/BA) for boys and girls after the age of 6 (Figure 13.3) reflects the different rates of growth in these regions of the trunk during adolescence.

---

[1]Biiliocristal breadth is defined as the horizontal distance between the most lateral points on the superior border of the iliac crest of the pelvis (Carter, 1982).

Table 13.4 Hip and Shoulder Widths for Female Reference Groups and Female Athletes

Groups	N	Age (years)	Height (cm)	Biacromial breadth (cm)	BA/Ht %	Biiliocristal breadth (cm)	BC/Ht %	BC/BA %	Source
Reference groups									
Reference woman	—	20-24	163.8	35.0	21.4	28.6	17.5	81.7	Behnke and Wilmore, 1974
College women	128	21.4 ± 3.8	164.9 ± 6.6	36.5 ± 1.7	22.1	28.4 ± 1.7	17.2	77.8	Wilmore and Behnke, 1970
College women	94	20.6 ± 2.6	165.7 ± 6.1	35.5 ± 1.6	21.4	27.5 ± 1.9	16.6	77.5	Carter et al., 1978
Female athletes									
Competitors in 1968 Olympics	143	19.4 ± 4.2	165.2 ± 7.9	37.0 ± 1.9	22.4	26.9 ± 2.2	16.3	72.8	Carter et al., 1978
Teenage cross-country runners	13	16.2 ± 0.4	162.2 ± 1.7	34.0 ± 0.6	21.0	25.2 ± 0.5	15.5	74.1	Burke and Brush, 1979
Track and field athletes in 1976 Olympics	34	21.8 ± 3.7	165.5 ± 7.1	36.3 ± 2.0	21.5	27.2 ± 2.5	16.1	74.9	Carter et al., 1982

Note. BA/Ht = (biacromial breadth/height) × 100; BC/Ht = (biiliocristal breadth/height) × 100; BC/BA = (biiliocristal breadth/biacromial breadth) × 100.

Table 13.5   Hip and Shoulder Widths for Male Reference Groups and Distance Runners

Groups	N	Age (years)	Height (cm)	Biacromial breadth (cm)	BA/Ht %	Biiliocristal breadth (cm)	BC/Ht %	BC/BA %	Source
Reference groups									
Reference man	—	20-24	174.0	40.6	23.3	28.6	16.4	70.4	Behnke and Wilmore, 1974
College men	133	22.0 ± 3.1	177.3 ± 7.2	40.4 ± 2.2	22.8	28.4 ± 1.7	16.0	70.3	Wilmore and Behnke, 1969
College men	153	21.3 ± 2.9	178.6 ± 7.1	40.0 ± 2.1	22.4	27.9 ± 1.7	15.6	69.8	Carter et al., 1982
Male distance runners									
Marathon runners in 1960 Olympics	9	adults	171.1	39.8	23.3	28.3	16.5	71.1	Tanner, 1964
Runners in Olympic training program	13	adults	178.6 ± 4.9	41.2 ± 1.3	23.1	29.9 ± 3.0	16.7	72.5	Adrian and Kreighbaum, 1973
Elite distance runners	20	26.2 ± 3.0	177.0 ± 6.0	39.5 ± 1.8	22.3	28.0 ± 1.4	15.8	70.9	Pollock et al., 1977
Track and field athletes in 1976 Olympics	40	23.8 ± 3.4	179.1 ± 7.7	40.2 ± 2.5	22.4	27.1 ± 2.2	15.1	67.4	Carter et al., 1982

Note. BA/Ht = (biacromial breadth/height) × 100; BC/Ht = (biiliocristal breadth/height) × 100; BC/BA = (biiliocristal breadth/biacromial breadth) × 100.

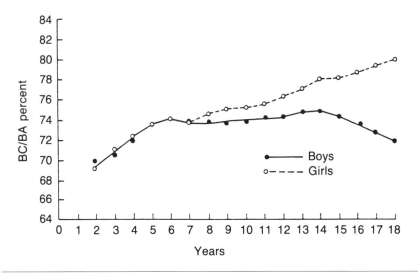

Figure 13.3. Ratio of pelvic width (BC) to shoulder width (BA) in boys and girls during the first 18 years of life. *Note.* From *Growth Diagnosis* by L.M. Bayer and N. Bayley, 1959, Chicago: The University of Chicago. Copyright 1959 by University of Chicago Press. Reprinted by permission.

It is interesting to note that both absolute and relative measurements of pelvic and shoulder width in groups of male distance runners (Table 13.5) are very similar to these same measurements in the reference groups of young adult males. Comparisons of these measurements between female distance runners and reference females are more difficult because very few anthropometric studies have been conducted on adult female distance runners of topflight caliber. The majority of the published anthropometric studies of female athletes have involved Olympic competitors (Carter, 1984; Carter, Hebbelinck, & de Garay, 1978; Carter, Ross, Aubry, Hebbelinck, & Borms, 1982). Few of these subjects could be classified as true distance runners because the longest competitive Olympic running event for women prior to the 1984 Olympics was 1,500 meters. Within the limits of existing data on female runners (Table 13.4), which include 1968 Olympic competitors in track and field as well as other sports (Carter et al., 1978), 1976 Olympic track and field athletes (Carter et al., 1982), and teenage cross-country runners (Burke & Brush, 1979), it appears that these female athletes are more similar to reference males and athletic males than to reference females on both absolute and relative (BC/Ht and BA/BC) measures of pelvic width. In fact, Carter et al. (1982) stated that most male and female groups of Olympic athletes differed by approximately ±1 cm in absolute pelvic width.

The misconception that absolute pelvic width is greater in females than in males is prevalent in the literature dealing with comparisons between athletes of both sexes and with injuries to women athletes. More than 20 references were located in which comments about ''the wider female pelvis'' were made in apparent disregard of existing data. Only a few references contained statements that correctly distinguished between absolute and relative differences in pelvic width of male and female athletic groups and reference groups (James & Brubaker, 1973; Malina et al., 1984; Ross & Ward, 1984; Wilmore, 1982). In the interest of scientific accuracy it is important to make these distinctions.

Another generalization concerning women is that they are more frequently seen with knock-knees when running or standing than are men. Greater pelvic width, increased femoral obliquity, and shorter legs have been mentioned as possible factors contributing to the knock-kneed appearance (Sloane, 1980). Undoubtedly there is considerable individual variation among women—and men—on all of these factors. If an individual had a wider-than-average pelvis and shorter-than-average legs, it is possible that the shaft of the femur may slope inward at a more oblique angle. The abnormal condition known as genu valgum (knock-knees) would occur if the angle between the oblique femur and the essentially vertical tibia were less than 165 degrees, as measured on the lateral side of the extremity (Norkin & Levangie, 1983). If, in fact, there is a greater tendency for knock-knees to exist in females than in males (a fact that has not been substantiated by scientific study in the general population), then this condition is more likely to be associated with relative pelvic width than with absolute pelvic width. The ratio of pelvic width to height (BC/Ht) does tend to be greater in reference females (Table 13.4) primarily because they are shorter in stature than are males. It would be interesting to quantify the degree of relationship among the variables of hip width, leg length, and genu valgum angle. However, athletic females, who typically have a narrower-than-average pelvis, may be no more likely than males to have knock-knees.

The Q (quadriceps) angle is a structural variable that also has been associated with gender differences in hip width. This angle (Figure 13.4), formed by a line (A) representing the resultant vector of the quadriceps muscle pull and a line (B) representing the pull of the patellar tendon, is a measurement used to detect patellar malalignment (Norkin & Levangie, 1983). The Q angle generally measures approximately 10 degrees in men and 15 degrees in women, whereas a measurement in excess of 20 degrees would be considered abnormal (James, 1979). The larger the Q angle, the more likely is the patella to be displaced laterally when quadriceps contraction occurs. This lateral tracking causes abnormal lateral stresses on the patella and can contribute to patellar instability or chondromalacia (topics to be discussed later in this chapter). Among the factors often associated with an increased Q angle are a wider pelvis and genu valgum (Brattstrom, 1964; Cox, 1985). A wide pelvis relative to leg length, and/or genu valgum, could cause a more lateral pull of the

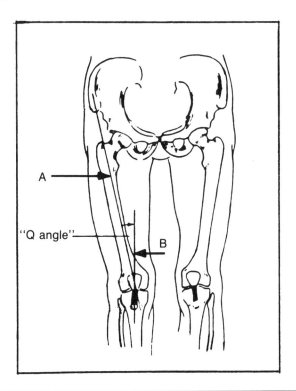

Figure 13.4. The Q angle.

quadriceps on the patella, which may lead to the increased possibility of patellofemoral injuries. Once again, though, an athletic female with relatively narrow hips may be less likely to have such problems if her Q angle is smaller than that of the average female.

Generalizations that females as a group are inferior to males in running events due to their wider pelvis and greater femoral obliquity have been stated repeatedly in the literature without reference to supporting research evidence (Cramer Products, Inc., 1985; Darden, 1979; James & Brubaker, 1973; Rasch & Burke, 1978). Rasch and Burke (1978) claim that the female "suffers certain mechanical disadvantages in running" (p. 456) because her wider pelvis requires a greater lateral shift of the center of gravity to keep her body weight over the supporting foot in each stride. Apparently no biomechanical studies have been conducted to quantify possible associations between hip width and lateral pelvic shift in running, although a few investigations have examined the relationship between hip width and running ability over distances of a mile or less. Correlations between sprinting speed and various measures of hip width were reported to be very low in large groups of nonathletic college women

(Carpenter, 1938; Oyster & Wooten, 1971; Thorsen, 1964). In the most recent of these studies, Oyster and Wooten (1971) found a low but significant correlation ($r = -0.31$) between bitrochanteric diameter and running velocity during the last 35 yards of a 50-yard run performed by 107 untrained college women. The correlation between measures of femoral obliquity and running velocity in this same group was 0.00. Interestingly, Berg and Bell (1980) also found significant correlations, in a group of 33 males, between bitrochanteric width expressed as a percent of height and time to run a 60-yard dash ($r = 0.52$), a 400-yard dash ($r = 0.49$), and a distance of one mile ($r = 0.30$). Taken collectively, these studies suggest that females and males with wide hips may be at a disadvantage in running compared to individuals with narrower hips, particularly in sprinting events. Most likely those female and male runners who are successful at distances greater than one mile tend to have narrower hips or they compensate for this potential structural limitation by superior physiological capabilities.

Gender-Related Biomechanical Aspects of Running

Comprehensive reviews of research literature on the biomechanics of running have been written by Dillman (1975), James and Brubaker (1973), Vaughan (1984), and Williams (1985). In the most recent of these reviews, Williams (1985) stated that ''most measures of biomechanical parameters during running do not show differences based on sex'' (p. 390). However, substantial comparative data that could serve as a basis for gender comparisons of several biomechanical parameters of distance running are sadly lacking because biomechanical studies of female distance runners were essentially nonexistent until approximately 10 years ago. Even since 1975 there has been, on the average, fewer than one published English-language study per year in which trained women distance runners were the primary focus of research or were compared to male runners. In most of these investigations, temporal and kinematic variables were analyzed rather than kinetic factors.

The following review will concentrate on biomechanical aspects of distance running that appear, at first glance, to be gender-related. Other running stride parameters that show little or no difference between females and males, or on which comparable data for distance runners of both sexes is lacking, have been omitted. Although based on only a limited amount of available research data, this review should help to clarify existing knowledge in addition to identifying areas for future research. Where possible, comparisons between male and female distance runners on selected biomechanical variables will be based on the analysis of relative measurements (expressed in proportion to the individual's size or maximum speed) as well as measurements presented in absolute units.

Temporal and Kinematic Factors

Stride length (SL), stride rate (SR), and the temporal components of stride time are the most commonly studied parameters in running. The terms *stride* and *step* often are used interchangeably, which can lead to difficulties in the interpretation of reported data. A stride cycle is defined here as the basic unit of locomotion, consisting of two successive (right and left) steps. The stride is commonly measured from initial ground contact by one foot until subsequent contact by the same (ipsilateral) foot. Two separate support periods (right and left) and two periods of nonsupport constitute the running stride. Support time (ST) is defined as the sum of both support periods in the stride cycle and nonsupport time (NST) represents the sum of the two nonsupport periods.

Average running speeds typical for distance events range from 3.0 m • s⁻¹ (required to complete a marathon in just under 4 hours) to approximately 5.7 m • s⁻¹ (required to complete a 5,000-m race in record time for a female runner). The current world record for women in the marathon (Table 13.1) could be achieved with an average running speed of 5.0 m • s⁻¹. Studies that have reported stride parameters for female and male subjects running at speeds within the 3.5 to 5.7 m • s⁻¹ range are summarized in Tables 13.6 and 13.7. Among the stride parameters reported in these tables are absolute values of stride rate and stride length at running speeds that are similar, in many cases, for groups of female and male runners studied in separate investigations. Also, to facilitate comparisons between female and male runners, relative values of SL, ST, and NST are presented in Tables 13.6 and 13.7: Stride length is expressed in relation to height (SL/Ht), support time (ST%) and nonsupport time (NST%) are expressed as a percent of stride time, and the ratio ST/NST reflects relative amounts of support versus nonsupport time.

Running speed is determined by the product of SL and SR. As average horizontal running velocity increases within the range of speeds from 3 to 10 m • s⁻¹, SL and SR change in a curvilinear manner (Dillman, 1975; Luhtanen & Komi, 1978; Saito, Kobayashi, Miyashita, & Hoshikawa, 1974; Williams, 1985). At the lower velocities (3.0 to 6.5 m • s⁻¹), proportionally greater increases are made in SL than in SR. As running speeds increase beyond 6.5 m • s⁻¹, larger increments are made in SR whereas SL changes relatively little and may even decrease slightly at maximal speeds. In skilled male runners, the dominance of SL as a means of increasing running speed appears to persist up to approximately 6.5 m • s⁻¹ or even 7.0 m • s⁻¹ (Dillman, 1975; Williams, 1985). However, Saito et al. (1974) reported this critical speed to be as low as 5.5 m • s⁻¹ in an untrained male runner. It appears that the critical speed of skilled female runners also may be somewhat lower than 6.5 m • s⁻¹ (Gregor, Rozenek, Brown, & Garhammer, 1979; Lusby, 1983; Miller, Enoka, & McCulloch, 1980), although few investigators have studied skilled female runners to observe the changes in SL and SR that occur as these subjects increase their velocity over a full range of speeds from slow jogging to sprinting.

A given steady-state running speed can be produced by a variety of combinations of SL and SR. For example, a runner who takes short strides may increase SR to keep pace with a taller runner who takes longer strides at a slower rate. Because males typically are taller than females (Tables 13.2 and 13.3), they would be expected to use longer strides and a lower SR when running at the same speed as shorter females. In fact, the absolute average SL of all female groups included in Table 13.6 was shorter than the absolute average SL of male runners (Table 13.7) when both sexes were compared at approximately the same running speeds. Because speed is a product of SL and SR, the absolute SR used by female runners was greater in all cases than that observed in male runners at comparable speeds (Tables 13.6 and 13.7).

Analysis of relative measures of SL has the potential to be more informative as a basis for gender comparisons because it serves to minimize the effect of body size differences between males and females. For instance, it might be logical to hypothesize that runners of both sexes adopt a stride length, for a given speed of running, that is the same in its proportion to their height (Ht) or leg length (LL). However, an underlying relationship between SL and body size measures has not been consistently supported by research evidence. For example, several investigators have found correlations ranging from 0.60 to 0.89 between SL and Ht or between SL and LL for male sprinters (Atwater, 1980; Hoffman, 1971), female sprinters (Hoffman, 1972), male and female physical education students running at speeds from 2.5 to 5.5 m • s⁻¹ (Elliott & Blanksby, 1979), male and female distance runners averaging speeds between 3.4 and 5.4 m • s⁻¹ (Roy, 1981), and good male distance runners at 4.96 m • s⁻¹ (Cavanagh, Pollock, & Landa, 1977). In contrast to these results are correlations that are low or even negative between SL and Ht or between SL and LL reported for male sprinters (Atwater, 1979), collegiate female distance runners at 3.73 and 5.15 m • s⁻¹ (Lusby, 1983), good male distance runners at 3.83 m • s⁻¹ (Cavanagh & Williams, 1982), and elite male distance runners at 4.96 m • s⁻¹ (Cavanagh et al., 1977). Correlations between SL and either Ht or LL tended to be lower when the runners in the group under study were relatively homogeneous in ability and ran at approximately the same speed. In several studies, SL had a slightly higher correlation with LL than with Ht, but the reverse was true in other studies.

Given the conflicting results concerning the relationship of SL to Ht and LL, comparisons between male and female distance runners on these

Table 13.6 Stride Parameters of Female Runners

Subjects/Source	N	Speed (m · s⁻¹)	Stride rate (strides · s⁻¹)	Absolute stride length (m)	Relative stride length (SL/Ht)	Relative support time (ST%)	Relative nonsupport time (NST%)	Ratio (ST/NST)
Elliott and Blanksby, 1979 P.E. college student joggers	10	3.50*	1.44 ± 0.11	2.44 ± 0.20	1.50 ± 0.12			
		4.50*	1.57 ± 0.05	2.88 ± 0.20	1.72 ± 0.10			
		5.50*	1.73 ± 0.08	3.20 ± 0.16	1.92 ± 0.10			
Elliott and Blanksby, 1976 College joggers	8	3.97*	1.40 ± 0.13	2.92 ± 0.38	1.75	58.3	41.7	1.40
	12	5.29*	1.53 ± 0.10	3.47 ± 0.29	2.08	52.4	47.6	1.10
Lusby, 1983 College varsity middle- and long-distance	8	3.73 ± 0.21	1.40 ± 0.06	2.66 ± 0.14	1.58 ± 0.10	72.0 ± 5.7	28.0 ± 5.7	2.63 ± 0.83
		5.15 ± 0.12	1.53 ± 0.04	3.38 ± 0.07	2.01 ± 0.07	61.3 ± 3.2	38.7 ± 3.2	1.59 ± 0.20
Miller et al., 1980 Skilled middle-distance runners	5	3.96 ± 0.15	1.48 ± 0.10	2.66 ± 0.11		67.1 ± 3.4	32.9 ± 3.4	2.07 ± 0.29
		5.76 ± 0.27	1.64 ± 0.10	3.51 ± 0.15		56.2 ± 2.2	43.8 ± 2.2	1.29 ± 0.11

Buckalew et al., 1985	40							
Runners in Olympic marathon trials among top 53 finishers (Time mean=2:37:48)								
Filmed at 9-mile point		4.46 ± 0.15	1.55 ± 0.08	2.89 ± 0.16	1.74 ± 0.10	64.6	35.4	1.89 ± 0.39
Filmed at 16-mile point		4.43 ± 0.17	1.55 ± 0.07	2.86 ± 0.14	1.72 ± 0.10	71.9	28.1	2.62 ± 0.73
Filmed at 20-mile point		4.33 ± 0.23	1.55 ± 0.07	2.79 ± 0.15	1.68 ± 0.10	69.2	30.8	2.34 ± 0.48
Filmed at 24-mile point		3.73 ± 0.22	1.52 ± 0.06	2.45 ± 0.14	1.48 ± 0.10	70.8	29.2	2.42 ± 0.55
Nelson et al., 1977	21	4.83	1.53 ± 0.13	3.19 ± 0.23	1.96 ± 0.11	55.4 ± 3.6	44.6	1.24
Elite distance runners		5.15	1.56 ± 0.14	3.33 ± 0.25	2.04 ± 0.12	54.0 ± 3.7	46.0	1.17
studied at selected		5.33	1.59 ± 0.14	3.41 ± 0.26	2.08 ± 0.12	53.3 ± 3.7	46.7	1.14
paced velocities		5.68	1.63 ± 0.14	3.53 ± 0.26	2.16 ± 0.12	52.2 ± 3.7	47.8	1.09

Note. A blank space indicates that no data were available for this variable. A stride is defined as two successive steps, measured from initial ground contact by one foot until that same (ipsilateral) foot contacts the ground again.
*Treadmill running.

Table 13.7   Stride Parameters of Male Runners

Subjects/Source	N	Speed (m · s⁻¹)	Stride rate (strides · s⁻¹)	Absolute stride length (m)	Relative stride length (SL/Ht)	Relative support time (ST%)	Relative nonsupport time (NST%)	Ratio (ST/NST)
Elliott and Blanksby, 1979 P.E. college student joggers	10	3.50*	1.36 ± 0.08	2.60 ± 0.16	1.44 ± 0.12			
		4.50*	1.45 ± 0.06	3.12 ± 0.12	1.74 ± 0.06			
		5.50*	1.54 ± 0.08	3.58 ± 0.20	2.00 ± 0.10			
Elliott and Blanksby, 1976 College joggers	12	3.70*	1.41 ± 0.05	2.62 ± 0.29	1.49	63.4	36.6	1.73
	11	5.41*	1.52 ± 0.09	3.57 ± 0.15	2.03	54.4	45.6	1.19
Richards, 1980	11							
Marathon runners (average finish time just under 3 hours)								
Filmed at 6-mile point		4.36	1.45	2.99		72.1	27.9	2.50
Filmed at 12-mile point		4.63	1.45	3.20		65.0	35.0	1.85
Filmed at 18-mile point		4.02	1.42	2.83		68.5	31.5	2.17
Filmed at 24-mile point		3.54	1.43	2.47		67.9	32.1	2.12

	n							
Cavanagh et al., 1977								
Elite middle- and long-distance runners	14	4.96*	1.59	3.12 ± 0.34	1.75	63.1	36.9	1.71
Good middle and long-distance runners	8	4.96*	1.52	3.28 ± 0.32		60.6	39.4	1.54
Nelson et al., 1977	10	4.83	1.46 ± 0.04	3.32 ± 0.10	1.88 ± 0.07	59.6 ± 4.2	40.4	1.46
College distance runners		5.15	1.49 ± 0.04	3.45 ± 0.09	1.96 ± 0.07	58.4 ± 4.0	41.6	1.40
studied at selected		5.33	1.52 ± 0.05	3.53 ± 0.10	2.00 ± 0.08	57.7 ± 3.9	42.3	1.36
paced velocites		5.68	1.56 ± 0.05	3.66 ± 0.12	2.06 ± 0.09	56.5 ± 3.7	43.5	1.30

Note. A blank space indicates that no data were available for this variable. A stride is defined as two successive steps, measured from initial ground contact by one foot until that same (ipsilateral) foot contacts the ground again.
*Treadmill running.

parameters should be interpreted cautiously. The question of gender differences in relative SL was investigated in two of the studies for which data are reported in Tables 13.6 and 13.7. Both studies set identical pre-determined running speeds for female and male subjects. Elliott and Blanksby (1979) used four different treadmill speeds (the three fastest of which are reported in Tables 13.6 and 13.7) in the analysis of college female and male joggers. Nelson et al. (1977) studied elite female distance runners at seven overground speeds (the four slowest of which are included in Tables 13.6 and 13.7) that were comparable to the running speeds of college male distance runners. However, the results regarding relative SL differed in the two studies. Relative stride lengths (SL/Ht and SL/LL) were not significantly different at any speed for the men and women runners analyzed by Elliott and Blanksby (1979), whereas the female runners studied by Nelson et al. (1977) had relative strides (SL/Ht) significantly longer than the relative strides of males running at the same speeds. Possible reasons for these conflicting results might be the different speeds of running employed in the two investigations or the different ability levels of the groups involved. Clearly, additional research is needed before conclusions are drawn regarding gender differences in relative stride lengths of distance runners.

Although differences exist between male and female runners on measures of absolute SR (Tables 13.6 and 13.7), these differences have rarely been analyzed in relation to the Ht or LL of the runners. One such study that examined these relationships was performed on sprinters rather than distance runners. Hoffman (1971, 1972) reported that male sprinters had a higher inverse relationship between SR and Ht and between SR and LL (-0.81 and -0.76, respectively) than did female sprinters (-0.50 and -0.51, respectively). More typically, the parameter of SR has been analyzed by taking its reciprocal, which is the time per stride, and subdividing this stride time into the two components of support time (ST) and nonsupport time (NST). Of particular interest here are comparisons between female and male runners on measures of absolute and relative ST and NST.

As increases in running speed occurred within the range of 3.5 to 5.7 m • s⁻¹ for groups of female and male runners included in Tables 13.6 and 13.7, absolute ST tended to decrease whereas NST changed very little. These data fit the general observation by Dillman (1975) that decreases in stride time accompanying increases in running speed are due primarily to a reduction in ST. Relative support time (ST%) ranged between 60% and 70% of stride time for runners of both sexes at speeds from 3.5 to approximately 4.7 m • s⁻¹ and generally decreased to 50% to 60% of stride time at speeds from 4.7 to 5.7 m • s⁻¹ (Tables 13.6 and 13.7). Changes in relative nonsupport time (NST%) complemented those occurring in ST%. The ratio ST/NST decreased for both males and females as running speed increased but remained greater than 1.00, indicating that ST still exceeded NST even at running speeds approaching 6.0 m • s^{-1}.

The study by Nelson et al. (1977) is deserving of comment because it is the only investigation presented in Tables 13.6 and 13.7 that included an analysis of stride time components for females and males running at the same average speeds. The 21 elite female distance runners in this study were found to have significantly longer absolute and relative values of NST and shorter absolute and relative values of ST than did the male college distance runners to whom they were compared. The authors concluded that these female runners obtained their longer NST by projecting the center of gravity at a higher angle and with greater vertical velocity in each stride takeoff. In contrast to these results are the temporal component data reported by Lusby (1983) for female distance runners whose average velocity of 5.15 m • s^{-1} was identical to one of the velocities employed in the study by Nelson et al. (1977). These female runners had somewhat shorter absolute and relative NST values and longer absolute and relative ST values than did the males in the Nelson et al. study (Tables 13.6 and 13.7). This same pattern of a shorter NST% and a longer ST% in females than in males was also observed in the latter third of marathon running (Buckalew, Barlow, Fischer, & Richards, 1985; Richards, 1980). In light of the rather incomplete and conflicting data regarding gender differences in the temporal components of running, it is clear that further research on this topic is needed.

One promising approach that should be considered for future investigations of gender-related biomechanical aspects of distance running involves the concept of relative velocity. The values of most running stride parameters have been shown to change as speed increases from a slow jog up to maximum velocity. When comparisons are to be made between groups of runners who differ significantly in maximum running speed, it may be informative to calculate the values of stride parameters at velocities that are relative to the maximum speed for each individual. This approach was used by Nelson et al. (1977), whose results were presented for several absolute running velocities (Tables 13.6 and 13.7) as well as for relative velocities of 60%, 70%, 80%, and 90% of maximum velocity. At comparable relative running speeds, the differences between male and female distance runners decreased for SL and SR parameters but increased for the temporal components of SR. Nelson et al. (1977) concluded that basing biomechanical comparisons on running speeds that are proportional to the individual's maximum speed did not eliminate differences between the male and female runners studied in that investigation.

Another potentially fruitful approach to the comparison of males and females on running stride parameters involves the calculation of dimensionless numbers (ratios) representing SL, SR, and running speed. The equations for these dimensionless stride parameters are described by

Alexander (1977) and incorporate values for h (height of the hip joint from the ground) and g (the acceleration during free fall). Using dimensionless numbers, Alexander (1977) has elucidated relationships between SL, SR, and speed that hold for mammals of all sizes. The advantage of calculating dimensionless stride parameters for male and female runners would be to facilitate the analysis of gender differences apart from the influence of body size differences.

Kinetic Factors

To date, the study of running biomechanics has been approached far more frequently from a kinematic point of view (involving analysis of time and space factors of the motion) than from a kinetic perspective (involving analysis of the forces causing the motion). Within the area of kinetics, some of the topics that have been examined by investigators studying the running gait have included muscle activity, resultant joint forces and moments, external forces and pressures acting on the foot, mechanical energy and power, and air resistance (see reviews by Miller, 1978; Vaughan, 1984; Williams, 1985). Only a small percentage of the published kinetic analyses of running has used females as subjects, and still fewer of these investigations have directly compared male and female runners on various kinetic parameters. In several of the studies that included runners of both sexes as subjects, results were presented for the combined sample and no separate analysis of the data, by gender, was attempted. Therefore, the entire area of running kinetics appears to be a promising one for future research on gender differences.

The following study is one of the few kinetic analyses of running in which female and male subjects have been compared. Matsuo and Fukunaga (1983) investigated the effect of age and sex on external mechanical energy in running. Subjects were 10 males and 10 females at each age from 8 to 20 years and 10 male adult sprinters. The mechanical energy of the center of gravity of the body was measured by means of force platforms as the subjects ran at maximum velocity. No significant differences in the regression equations for mechanical work versus running velocity were found between subjects of different age, sex, and athletic background.

Gender Differences in Overuse Injuries of Runners

Overuse athletic injuries typically occur in sports that require repetitive movements. Injuries of this type are caused by repeated submaximal stresses on a certain anatomical structure that outpace the structure's capacity to respond to the stress and repair itself. Due to the high-mileage endurance training programs adopted by most competitive distance runners, overuse injuries in this group of athletes are most frequently seen in the knee, lower leg, and foot (Clement, Taunton, Smart, & McNicol, 1981). Of foremost interest in this section are questions concerning possible gender differences in overuse running injuries. Specifically, do female runners have a higher incidence of overuse injuries than male runners? If so, what factors are associated with observed sex differences in the type and rate of injury? Although Nilson (1986) recently discussed the full scope of injuries in female distance runners, the following sections will highlight only the relationships between certain running injuries and factors such as gender, anatomical structure, and training level.

Incidence and Type of Injury

The incidence of chronic overuse injuries in distance runners compared to other athletic groups is poorly documented. Also, reported differences in the injury rates for female and male runners are sometimes difficult to evaluate when studies are based on clinical records of only those patients who seek medical treatment. For example, some studies indicate that fewer injuries occurred in female runners than in male runners (James, Bates, & Osternig, 1978; Pagliano & Jackson, 1980). However, the percentage of injured females and males relative to the percentage of each sex in the running population was not stated. Informative data on gender differences in running-related injuries would be best provided by epidemiological studies in which the numbers of injured female and male runners are reported in proportion to the populations at risk.

Only a few published epidemiological investigations were located in which males and females were compared on the basis of incidence of injury. In one of these studies (Koplan, Powell, Sikes, Shirley, & Campbell, 1982), registrants in the 10-km Peachtree Road Race were surveyed, by mail, one year after the 1980 race; they were asked to describe their running habits and injuries during the preceding year. Of the 1,250 females and 1,250 males randomly selected to receive the questionnaire, 730 (58.4%) of the females and 693 (55.4%) of the males responded. Telephone interviews with a randomly selected group of nonrespondents revealed no significant difference between respondents and nonrespondents on rates of injuries. Among the respondents, 37% of the men and 38% of the women incurred a musculoskeletal injury (attributed to running) during the previous year that was severe enough to require a decrease in weekly mileage. The risk of injury increased for runners of both sexes as weekly running mileage increased. No association was found between injuries and age, running speed, body mass index (weight/height^2), or years of running.

Information from another epidemiological study of running injuries, presented by Blair at a national meeting, was included in an article by Powell, Kohl, Caspersen, and Blair (1986). A mail survey was used to question 720 regular runners, who were members of a fitness and exercise club, about their running-related injuries during the previous year. The results of this study were similar to those found by Koplan et al. (1982) in that 24% of the 438 respondents had experienced injuries severe enough to curtail their running, and the incidence of injury was directly associated with running mileage per week. Age, gender, years of running, and speed were not related to the incidence of injury once weekly mileage was taken into account.

Female and male runners tend to have the same types of injuries, and no single overuse injury is unique to the runners of one sex. Among the runners who reported injuries in the study by Koplan et al. (1982), the new injuries per person-year of running were essentially the same for men and women at 10 different body sites. Knee injuries were most frequent whereas the incidence of foot injuries ranked second. Clement et al. (1981) also reported similar types of overuse running injuries for 987 men and 663 women whose clinical records were the subject of a retrospective survey. A high percentage of all injuries in these runners occurred at the knee (42.4% in men, 41.1% in women), and the next most frequently injured locations were the lower leg (27% in men, 28.6% in women) and the foot (19.6% in men, 16.1% in women).

Although these limited data indicate that the type and rate of injuries do not appear to differ between groups of average female and male runners, it is premature to make a similar statement concerning elite female and male distance runners. Epidemiological studies, such as those reported by Koplan et al. (1982) and Powell et al. (1986), have not been conducted to clarify gender differences in the type and incidence of injuries to competitive runners at the elite level. It is possible that the percentage of elite distance runners who sustain injuries is just as high as, if not higher than, the percentage of average runners who are injured because the elite runners must undertake extensive training programs to achieve world-class times. For instance, Nilson (1986) reported data on the injury history in those women who qualified for the first U.S. women's Olympic marathon trials. Of the 210 women who completed medical history questionnaires, 44% described a history of musculoskeletal problems and 29 of the 267 qualifiers were unable to compete in the trials because of injury. However, there is yet no evidence to indicate that elite male runners differ from the females in the type or rate of injury.

Anatomical and Mechanical Factors
Related to Certain Running Injuries

The probable cause of overuse injuries in runners can be attributed broadly to extrinsic factors and intrinsic (anatomical) factors. Among the extrinsic factors identified by James et al. (1978), training errors were associated with the largest percentage (60%) of running injuries. The most common training error was excessive mileage; other training mistakes included rapid changes in the training routine and increases in the time spent running on hard surfaces or hills (James et al., 1978). Inappropriate selection and maintenance of running shoes was another external factor thought to contribute to running injuries (James et al., 1978).

Several intrinsic or anatomical factors have been implicated in overuse problems of runners, although there is a lack of data relating specific structural or functional variations to the mechanisms of overuse injuries (Williams, 1985). For example, abnormalities such as excessive pronation of the foot or malalignment of the femur, tibia, and foot have been observed in a large percentage of injured runners, but these conditions could not always be correlated with a single, specific diagnosis (James et al., 1978). Minor abnormalities, of no consequence in other sports, may become significant factors in the development of injuries in distance runners, who strike the ground with each foot approximately 70 to 100 times per minute at a force between two and three times body weight. A portion of the impact force is attenuated by running shoes whereas the remainder is transmitted upward to other anatomical structures. Thus, the combination of accumulated impact loadings and deviations in anatomical structure may contribute to many of the overuse injuries experienced by distance runners.

Two types of overuse injuries, (a) chondromalacia and (b) stress fractures, have been selected for attention here because they are among the few running injuries singled out by some writers as differing in incidence between males and females due to anatomical or mechanical causes.

Chondromalacia. In a review of knee injuries in runners, Rubin and Collins (1980) stated that the most common condition causing knee pain is chondromalacia patellae. This condition, which typically is diagnosed during the teenage and young adult years, affects the cartilage on the articulating surface of the patella, causing it to progressively degenerate. The prevalent view in medical literature prior to the late 1960s was that chondromalacia occurred predominantly in knock-kneed, overweight, teenage females (Hughston, 1968; Outerbridge & Dunlop, 1975). Wide hips and an increased Q angle were generally believed to predispose females to this injury. Hughston (1968) was one of the first to dispel this erroneous belief by showing that male athletes were three times more likely than female athletes to experience symptomatic recurrent subluxation or dislocation of the patella, a condition most frequently associated with quadriceps malalignment and chondromalacia (Outerbridge & Dunlop, 1975).

The current opinion in medical literature, which is supported primarily by clinical observations rather than by epidemiological data, is that chondromalacia occurs almost twice as often in females as in males within the general population, but more frequently in men than in women when only athletes are studied (Levine, 1979). For example, DeHaven, Dolan, and Mayer (1979) conducted a prospective study of 100 athletes in whom the clinical diagnosis of chondromalacia was made and found that males outnumbered females three to two. Also, Rubin and Collins (1980) reported that 172 of the 2,137 new patients seen during a 20-month period at the Cleveland Clinic Foundation Sports Injury Clinic were diagnosed as having chondromalacia. Of these, 139 were males and 33 were females.

Despite the growing evidence that chondromalacia is no more likely to occur in female athletes than in male athletes, some authors have perpetuated the generalization that women athletes are more susceptible than males to patellofemoral pain and chondromalacia primarily because of their wider hips and greater Q angle (Haycock & Gillette, 1976; Hunter, 1984; Jones, 1980; Potera, 1986). Data presented in Tables 13.4 and 13.5 and discussed earlier in this chapter demonstrate that neither absolute nor relative hip width is greater in female athletes than in males. The Q angle, though somewhat larger on the average in females than in males (James, 1979), has not been shown to differentiate between male and female athletes. Furthermore, an increased Q angle is only one of several factors that have been proposed to explain the etiology of chondromalacia. Among the structural variations that reportedly predispose the patellofemoral joint to chondromalacia are an underdeveloped lateral lip of the lateral femoral condyle, a small or high-riding patella, weakness of the vastus medialis muscle, and lower extremity malalignment (Insall, 1979; Levine, 1979; Norkin & Levangie, 1983; Rubin & Collins, 1980). The latter factor, lower extremity malalignment, may include one or more variations in anatomical structure in joints or bones from the hip to the foot (James et al., 1978; Leach, 1982). An increased Q angle often, but not always, reflects this malalignment (Insall, Falvo, & Wise, 1976; Levine, 1979).

Although knee pain is the most frequent complaint of runners (James et al., 1978), it should be noted that chondromalacia is but one of several conditions that collectively constitute a patellofemoral stress syndrome (Cox, 1985; Hunter, 1984). The predisposing structural factors most frequently associated with peripatellar pain can occur in both sexes. It has been suggested that the lack of proper conditioning may be a more important factor related to knee injuries, especially in women, than are anatomical or structural deviations (Micheli, 1982). Until recent years, the average female athlete has had a history of less strength training and less rigorous physical conditioning than the male athlete (Beck & Wildermuth, 1985), which may have contributed to a higher incidence of patellar injuries (Ryan, 1975). With greater emphasis now being placed on the importance of preparticipation fitness and strength training programs, it is apparent that knee injuries specifically, and sports injuries in general, are more likely to be sport-related rather than gender-related (Clarke & Buckley, 1980; Whiteside, 1980).

Stress Fractures. Stress or fatigue fractures in runners typically occur under conditions of repetitive cyclic loading of weight-bearing bones. According to Wolff's law, bone remodels in response to the mechanical demands placed upon it (Frankel & Nordin, 1980). During repeated sub-maximal loading of bone, the normal remodeling process involves a balance of bone resorption and bone formation. If sufficient time is allowed for the remodeling to occur, modest increases in the amount or frequency of loading can produce bone hypertrophy and greater bone strength (McBryde, 1982). If, however, the frequency of loading precludes the remodeling necessary to prevent bone failure, a stress fracture may result (Frankel & Nordin, 1980).

The frequency of stress fractures among running injuries diagnosed in a clinical setting varies in different studies. Stress fractures have been ranked as the second most common type of running injury, or 14.4% of 376 injuries (McBryde, 1982), the fifth most common, or 6% of 232 injuries (James et al., 1978), and the eighth most common, or 4.2% of 224 injuries (Gudas, 1980). Although McBryde (1985) has stated that the relative incidence and distribution of stress fractures in women runners parallels that in men, others have reported that stress fractures occur with greater frequency in female runners than in male runners (Lombardo & Benson, 1982; Micheli, 1982; Noakes, Smith, Lindenberg, & Wills, 1985; Paty & Swafford, 1984; Prescott, 1983; Taunton, Clement, & Webber, 1981). How-ever, none of these authors presented epidemiological data comparing female and male runners on the frequency of stress fractures at various skeletal sites.

One study, in which the incidence and location of stress fractures were compared in females and males, did include distance runners among the subjects examined. Orava, Puranen, and Ala-Ketola (1978) described a series of 142 stress fractures in athletes and noncompetitive sports par-ticipants over an 8-year period. The number of patients with stress frac-tures made up about 1% of the total number of athletes attending the clinic. Middle- and long-distance runners in the study incurred 68 stress fractures, noncompetitive joggers sustained 21 fractures, and the remaining

53 fractures occurred in other athletes such as skiers, orienteering runners, gymnasts, and cyclists. Women had 25 (17.6%) of the 142 stress fractures and men had 117 (82.4%) of the total, but the distribution of female and male patients by sport was not identified. The authors noted that only 13 of the 142 stress fractures were a consequence of training other than running. By far the majority of stress fractures (88.7%) occurred in bones below the knee joint, 7.8% were located in the femur and pelvis, and 3.5% were found in the vertebrae and upper extremities. The distribution of these injuries was very similar for subjects of both sexes. The only locations at which a slightly larger percentage of stress fractures occurred in females than in males were the femoral neck and the tarsal navicular (Orava et al., 1978).

Other studies also have identified the bones of the lower leg and foot as the most frequent sites of stress fractures in runners (Gudas, 1980; McBryde, 1982; Taunton et al., 1981) and in military recruits whose training included daily marching and running (Meurman, 1980; Protzman & Griffis, 1977; Reinker & Ozburne, 1979; Wilson & Katz, 1969). Unfortunately, gender differences in this most frequent category of stress fractures have not been analyzed in detail. Taunton et al. (1981), who reviewed records of 62 runners (25 men and 37 women) treated for stress fractures during a 9-month period, reported that 93.5% of the fractures occurred in the lower leg and foot. Although the authors of this study commented on the predominance of stress fractures in women, they presented data only for the combined group of females and males when describing the sites of stress fractures, alignment variances in the lower extremity, and the activity level of the runners. Reports on stress fractures in the lower extremities of female and male military recruits have been presented relative to the larger group of subjects at risk of injury, but they often lack specificity as to the exact amount and type of exercise that precipitated the injuries. The incidence of stress fractures occurring primarily in the lower leg and foot was more than 2 times greater in female than in male military recruits during a 1976-1977 basic training cycle at Fort Jackson (Reinker & Ozburne, 1979), and almost 10 times greater in female than in male cadets participating in a summer training program at West Point in 1976 (Protzman & Griffis, 1977).

The increased frequency of stress fractures reported in women recruits during the last half of the 1970s, after they were first admitted to military training programs, may have been largely due to their lower fitness level and general lack of prior physical conditioning (Cox & Lenz, 1979, 1984; Hunter, 1984; Kowal, 1980; Tomasi, Peterson, Pettit, Vogel, & Kowal, 1977). Similar observations have been made regarding the higher likelihood of stress fractures in novice female runners who undertook vigorous training programs without adequate physical preparation (Jackson & Strizak, 1982; McBryde, 1985; Micheli, 1979). Likewise, rapid increases in the intensity and amount of running within the first few months after commencement of training were reported to be the major causes of stress fractures in two groups of male and female runners (Orava et al., 1978; Taunton et al., 1981). These proposed causes of stress fractures are all illustrations of a repetitive loading mechanism in which the remodeling process of bone may have been outpaced by the fatigue process.

Gender differences in the incidence or site of stress fractures have been attributed to anatomical factors in only a few studies. Taunton et al. (1981) stated that the predominance of stress fractures in women was the result of a wider pelvis and genu valgum, but no data were presented to support this claim. A wide pelvis was also viewed by Kowal (1980) as a factor contributing to the increased risk of injury in female military recruits, although hip width was not among the measures recorded in this study. McBryde (1982) observed that the width of the woman's pelvis "tends to produce a slightly different foot plant in the running gait" (p. 25), yet he added that this difference was relatively unimportant as a contributing factor to bony stress syndromes.

Stress fractures of the pubic ramus and the femur comprise one category of injuries for which gender differences appear to exist without a satisfactory explanation of the causes. Although stress fractures in these bones are relatively infrequent in runners (Orava et al., 1978) and in military recruits (Meurman, 1980), their incidence among females is reportedly growing (Pavlov, Nelson, Warren, Torg, & Burstein, 1982). The inferior pubic ramus near the symphysis pubis was the site of stress fractures observed by Pavlov et al. (1982) in six females and two males who were long-distance or marathon runners and in three females who were joggers. Noakes et al. (1985) described stress fractures at the same location in four female competitive distance runners, three of whom were elite marathoners, and in eight male distance runners. Additional cases of pubic ramus stress fractures were reported in one male and five female distance runners (Prescott, 1983), in a female jogger (Latshaw, Kantner, Kalenak, Baum, & Corcoran, 1981), and in a male cross-country runner (Tehranzadeh, Kurth, Elyaderani, & Bowers, 1982). In the majority of these runners, the stress fracture occurred during a period of intensified training.

Pavlov et al. (1982) pointed out that the pubic ramus stress fractures differed from other stress or fatigue fractures in joggers and runners because they occurred in a bone that was exposed to tensile stress as opposed to compressive stress. Investigators who have reported the same type of stress fracture in female and male military recruits (Hughes, 1983; Meurman, 1980; Ozburn & Nichols, 1981) called attention to the fact that the hip adductor muscles have their proximal attachment on the inferior pubic ramus and are actively involved in the motions of running and marching. Ozburn and Nichols (1981), who described a series of stress fractures of the inferior pubic ramus in 67 female and three male military recruits, observed that 14 of these patients (including one male) also had a periosteal stress reaction on the femur at the distal attachment site of the adductor brevis and longus muscles. These facts suggest several possible reasons why females might be more susceptible than males to stress fractures in the pubic ramus and femur: Either the tensile stresses caused by repeated adductor muscle contractions are higher in women than in men, or the female pelvis is more vulnerable to these tensile stresses, or the gaits of men and women differ in a way that affects the fatigue mechanism.

No etiologic factor has yet been identified as the primary contributor to gender differences in the frequency of stress fractures of the pubic ramus and the femur. However, Ozburn and Nichols (1981) proposed that stress fractures at these two sites were related more frequently to short stature and length of marching stride than to a lack of physical conditioning. They reviewed the records of 70 female and male military recruits who developed inferior pubic ramus stress fractures, or adductor insertion stress reactions, or both during 12 weeks of basic training. The average height of the 67 female recruits in this group was 63.7 in., whereas the three male recruits had an average height of 65 in. All of these recruits said they experienced difficulty in keeping up with other recruits (males and females of different heights were in the same platoon) during marching drills, and many described "taking giant steps all day" (p. 332). It is interesting to note that the definition of a marching step in the 1976 Army Field Manual was "30 inches heel to heel for the marching *man* [italics added]" (Ozburn & Nichols, 1981, p. 332). None of the recruits were considered to be in poor physical condition, and 57% were previous high school athletes.

The possibility exists that female distance runners, in whom stress fractures of the inferior pubic ramus were diagnosed, may have developed habits of overstriding when running alone or with a taller companion. Of course, this same link between pubic stress fractures, stature, and stride length also should be investigated in male runners. Overstriding may repeatedly cause increased tension stress on both the pubic and femoral attachments of the adductor muscles. In all likelihood, though, the increased incidence of pubic ramus stress fractures in female runners and military recruits may be related to a combination of factors, including those previously mentioned as well as training errors, anatomical alignment deviations, decreased bone girth or density, and individual idiosyncrasies in running technique. This topic is a fertile one for future investigations.

Summary and Concluding Statement

Contrary to popular belief, research indicates that there are few substantiated biomechanical differences between female and male distance runners. Current performance records of women in middle- and long-distance running events have improved to within 10% of the men's records. Although differences exist between the average female and the average male on several structural parameters, female distance runners are more similar to average and athletic males than to the average female on measures of percent body fat, absolute pelvic width, and pelvic width relative to height and to shoulder width. However, they usually are not as heavy or as lean as elite male distance runners. There is no evidence to support generalizations that female distance runners differ from male distance runners in lower extremity alignment even though the Q angle (a measure of lower extremity alignment) is slightly greater, on the average, in females than in males.

The temporal and kinematic parameters of stride length and stride rate tend to differ between male and female distance runners only when expressed as absolute values and not when expressed in relative form. Specifically, male distance runners usually have longer absolute stride lengths and lower stride rates when running at the same speed as female distance runners, primarily because they are taller. When stride length is expressed relative to height or leg length and the components of stride rate (support time and nonsupport time) are expressed in relation to stride time, differences between the sexes decrease. No evidence yet exists to show that any kinematic aspect of running is altered by sex-related structural differences. However, the number of available studies on which gender comparisons of distance running kinematics can be based is small, and results have been conflicting. Analyses comparing female and male distance runners on kinetic factors are even more rare. Until additional research is available, it would be reasonable to conclude that few biomechanical parameters of distance running can be differentiated on the basis of the sex of the performer.

Overuse injuries can occur in any athlete (male or female) who exceeds a level of activity that will allow the body time to respond to this stress and repair itself. Epidemiological studies, in which the numbers of injured female and male distance runners are reported in proportion to the populations at risk, have shown no differences between females and males in either the incidence or the site of injury. When two specific types of overuse injuries, chondromalacia and stress fractures, are singled out for closer analysis, there is inconclusive evidence that these injuries are found more frequently in female runners than in male runners. Although clinical data indicates that chondromalacia may occur almost twice as often in females as in males in the general population, the reverse is true when only athletes are studied. Stress fractures at most skeletal sites do not seem to differ for males and females, yet those in the pelvis and femur have been noted with increased frequency in women distance runners. There is no evidence that the hip width of female runners is associated with the incidence of either chondromalacia or stress fractures. Variations in the alignment of the lower extremities, including an increased Q angle, have been linked with chondromalacia, stress fractures, and several other injuries in both female and male distance runners. One factor that appears to be related to the incidence of overuse injuries in females is their level of physical conditioning. As women distance runners take greater advantage of fitness and strength training programs, their potential for stress-related injuries will in all likelihood decrease.

Distance running, for either competitive or recreational goals, is a relatively new sport for women. Thus, only a limited amount of research is currently available to serve as the basis for a review of gender differences centered around biomechanical factors. Undoubtedly this situation will be remedied in the future as unresolved topics are studied by biomechanists working in conjunction with exercise physiologists and medical personnel. Until additional data are provided to clarify gender differences in distance running, it would be wise to adopt the attitude illustrated by the null hypothesis; that is, no difference is assumed to exist unless objective evidence shows the difference to be a significant and meaningful one. Such an approach may diminish the prevalence of misconceptions regarding the distance running capabilities of women.

Acknowledgments

Appreciation is extended to Fred B. Roby for his critical reading of the manuscript and to Deborah Jane Power for her assistance with the literature search.

References

Adrian, M. (1972). Sex differences in biomechanics. In D.V. Harris (Ed.), *Women in sport: A national research conference* (Vol.2, pp. 389-397). University Park, PA: The Pennsylvania State University.

Adrian, M., & Kreighbaum, E. (1973). Mechanics of distance-running during competition. In S. Cerquiglini, A. Venerando, & J. Wartenweiler (Eds.), *Biomechanics III* (pp. 354-358). Baltimore: University Park.

Alexander, R.M. (1977). Terrestrial locomotion. In R.M. Alexander & G. Goldspink (Eds.), *Mechanics and energetics of animal locomotion* (pp. 168-203). New York: John Wiley & Sons.

Atwater, A.E. (1979). Kinematic analysis of striding during the sprint start and mid-race sprint. *Medicine and Science in Sports*, **11**(1), 85.

Atwater, A.E. (1980). Kinematic analysis of sprinting. In J.M. Cooper & B. Haven (Eds.), *Proceedings of the Biomechanics Symposium, Indiana*

University, October 26-28, 1980 (pp. 303-314). Bloomington, IN: The Indiana State Board of Health.

Bale, P., Rowell, S., & Colley, E. (1985). Anthropometric and training characteristics of female marathon runners as determinants of distance running performance. *Journal of Sports Sciences, 3*, 115-126.

Bayer, L.M., & Bayley, N. (1959). *Growth diagnosis.* Chicago: University of Chicago.

Beck, J.L., & Wildermuth, B.P. (1985). The female athlete's knee. *Clinics in Sports Medicine: Symposium on the Knee, 4*(2), 345-366.

Behnke, A.R., & Wilmore, J.H. (1974). *Evaluation and regulation of body build and composition.* Englewood Cliffs, NJ: Prentice-Hall.

Berg, K., & Bell, C.W. (1980). Physiological and anthropometric determinants of mile run time. *Journal of Sports Medicine and Physical Fitness, 20*, 390-396.

Brattstrom, H. (1964). Shape of the intercondylar groove normally and in recurrent dislocation of patella: A clinical and x-ray anatomical investigation. *Acta Orthopaedica Scandinavica, 68*(Suppl.), pp. 7-33.

Brown, C.H., & Wilmore, J.H. (1971). Physical and physiological profiles of champion women long distance runners. *Medicine and Science in Sports, 3*(1), page h.

Buckalew, D.P., Barlow, D.A., Fischer, J.W., & Richards, J.G. (1985). Biomechanical profile of elite women marathoners. *International Journal of Sport Biomechanics, 1*(4), 330-347.

Burke, E.J., & Brush, F.C. (1979). Physiological and anthropometric assessment of successful teenage female distance runners. *Research Quarterly, 50*(2), 180-187.

Carpenter, A. (1938). Strength, power and "femininity" as factors influencing athletic performance of college women. *Research Quarterly, 9*, 120-127.

Carter, J.E.L. (Ed.) (1982). *Physical structure of Olympic athletes. Part I: The Montreal Olympic Games Anthropological Project* (Vol. 16). Basel, Switzerland: S. Karger.

Carter, J.E.L. (1984). Age and body size of Olympic athletes. In J.E.L. Carter (Ed.), *Medicine and sport science* (Vol. 18, pp. 53-79). Basel, Switzerland: S. Karger.

Carter, J.E.L., Hebbelinck, M., & de Garay, A. (1978). Anthropometric profiles of Olympic athletes at Mexico City. In F. Landry & W.A.R. Orban (Eds.), *Biomechanics of sports and kinanthropometry* (Vol. 6, pp. 305-312). Miami, FL: Symposia Specialists.

Carter, J.E.L., Ross, W.D., Aubry, S.P., Hebbelinck, M., & Borms, J. (1982). Anthropometry of Montreal Olympic athletes. In J.E.L. Carter (Ed.), *Medicine and sport* (Vol. 16, pp. 25-52), Basel, Switzerland: S. Karger.

Cavanagh, P.R., & Williams, K.R. (1982). The effect of stride length variation on oxygen uptake during distance running. *Medicine and Science in Sports and Exercise*, **14**(1), 30-35.

Cavanagh, P.R., Pollock, M.L., & Landa, J. (1977). A biomechanical comparison of elite and good distance runners. In P. Milvy (Ed.), *The marathon: Physiological, medical, epidemiological, and psychological studies* (pp. 328-345). Annals of the New York Academy of Sciences (Vol. 301). New York: The New York Academy of Sciences.

Christensen, C.L., & Ruhling, R.O. (1983). Physical characteristics of novice and experienced women marathon runners. *British Journal of Sports Medicine*, **17**, 166-171.

Clarke, K.S., & Buckley, W.E. (1980). Women's injuries in collegiate sports. *American Journal of Sports Medicine*, **8**(3), 187-191.

Clement, D.B., Taunton, J.E., Smart, G.W., & McNicol, K.L. (1981). A survey of overuse running injuries. *Physician and Sportsmedicine*, **9**(5), 47-58.

Conley, D.L., Krahenbuhl, G.S., Burkett, L.W., & Millar, A.L. (1981). Physiological correlates of female road racing performances. *Research Quarterly for Exercise and Sport*, **52**(4), 441-448.

Costill, D.L., Bowers, R., & Kammer, W.F. (1970). Skinfold estimates of body fat among marathon runners. *Medicine and Science in Sports*, **2**(2), 93-95.

Cox, J.S. (1985). Patellofemoral problems in runners. *Clinics in Sports Medicine: Symposium on Running*, **4**(4), 699-715.

Cox, J.S., & Lenz, H.W. (1979). Women in sports: The naval academy experience. *American Journal of Sports Medicine*, **7**, 355-357.

Cox, J.S., & Lenz, H.W. (1984). Women midshipmen in sports. *American Journal of Sports Medicine*, **12**, 241-243.

Cramer Products, Inc. (1985, October). The female athlete: A look at the facts. *The First Aider*, pp. 8-10.

Darden, E. (1979). Are women really the weaker sex? *Young Athlete*, **2**(10), 60-61.

DeHaven, K.E., Dolan, W.A., & Mayer, P.J. (1979). Chondromalacia patellae in athletes. *American Journal of Sports Medicine*, **7**(1), 5-11.

Dillman, C.J. (1975). Kinematic analyses of running. In J.H. Wilmore & J.F. Keogh (Eds.), *Exercise and sport sciences reviews* (Vol. 3, pp. 193-218). New York: Academic Press.

Drinkwater, B.L. (1973). Physiological responses of women to exercise. In J.H. Wilmore (Ed.), *Exercise and sport sciences reviews* (Vol. 1, pp. 125-153). New York: Academic Press.

Drinkwater, B.L. (1984). Women and exercise: Physiological aspects. In R.L. Terjung (Ed.), *Exercise and sport sciences reviews* (Vol. 12, pp. 21-51). Lexington, MA: Collamore.

Drinkwater, B.L. (Ed.) (1986). *Female endurance athletes*. Champaign, IL: Human Kinetics.

Drinkwater, B.L., Kupprat, I.C., Denton, J.E., & Horvath, S.M. (1977). Heat tolerance of female distance runners. In P. Milvy (Ed.), *The marathon: Physiological, medical, epidemiological, and psychological studies* (pp. 777-792). Annals of the New York Academy of Sciences (Vol. 301). New York: The New York Academy of Sciences.

Elliott, B.C., & Blanksby, B.A. (1976). A cinematographic analysis of overground and treadmill running by males and females. *Medicine and Science in Sports*, **8**(2), 84-87.

Elliott, B.C., & Blanksby, B.A. (1979). Optimal stride length considerations for male and female recreational runners. *British Journal of Sports Medicine*, **13**, 15-18.

Fleck, S.J. (1983). Body composition of elite American athletes. *American Journal of Sports Medicine*, **11**(6), 398-403.

Fox, E.L., & Mathews, D.K. (1981). *The physiological basis of physical education and athletics* (3rd Ed.). New York: Saunders College.

Frankel, V.H., & Nordin, M. (1980). *Basic biomechanics of the skeletal system*. Philadelphia: Lea & Febiger.

Gregor, R.J., Rozenek, R., Brown, C.H., & Garhammer, J. (1979). Variations in running stride mechanics as a function of velocity in elite distance runners. *Medicine and Science in Sports*, **11**(1), 85.

Gudas, C.J. (1980). Patterns of lower-extremity injury in 224 runners. *Comprehensive Therapy*, **6**, 50-59.

Haycock, C.E., & Gillette, J.V. (1976). Susceptibility of women athletes to injury: Myths vs. reality. *Journal of the American Medical Association*, **236**, 163-165.

Hoffman, K. (1971). Stature, leg length, and stride frequency. *Track Technique*, **46**, 1463-1469.

Hoffman, K. (1972). Stride length and frequency of female sprinters. *Track Technique*, **48**, 1522-1524.

Hughes, L.Y. (1983). Pelvic fractures in soldiers (Letters to the Editor). *Physician and Sportsmedicine*, **11**(9), 16.

Hughston, J.C. (1968). Subluxation of the patella. *Journal of Bone and Joint Surgery*, **50-A**(5), 1003-1026.

Hult, J.S. (1986). The female American runner: A modern quest for visibility. In B.L. Drinkwater (Ed.), *Female endurance athletes* (pp. 1-39), Champaign, IL: Human Kinetics.

Hunter, L.Y. (1984). Women's athletics: The orthopedic surgeon's viewpoint. *Clinics in Sports Medicine: Symposium on the Athletic Woman*, **3**(4), 809-827.

Insall, J. (1979). ''Chondromalacia patellae'': Patellar malalignment syndrome. *Orthopedic Clinics of North America: Symposium on Disorders of the Knee Joint*, **10**(1), 117-127.

Insall, J., Falvo, K.A., & Wise, D.W. (1976). Chondromalacia patellae: A prospective study. *Journal of Bone and Joint Surgery*, **58-A**(1), 1-8.

Jackson, A.S., & Pollock, M.L. (1978). Generalized equations for predicting body density of men. *British Journal of Nutrition, 40,* 497-504.

Jackson, A.S., Pollock, M.L., & Ward, A. (1980). Generalized equations for predicting body density of women. *Medicine and Science in Sports and Exercise, 12*(3), 175-182.

Jackson, D.W., & Strizak, A.M. (1982). Stress fractures in runners, excluding the foot. In R.P. Mack (Ed.), *Symposium on the foot and leg in running sports* (pp. 109-122). St. Louis: C.V. Mosby.

James, S.L. (1979). Chondromalacia of the patella in the adolescent. In J.C. Kennedy (Ed.), *The injured adolescent knee* (pp. 205-251). Baltimore: Williams & Wilkins.

James, S.L., Bates, B.T., & Osternig, L.R. (1978). Injuries to runners. *American Journal of Sports Medicine, 6*(2), 40-50.

James, S.L., & Brubaker, C.E. (1973). Biomechanical and neuromuscular aspects of running. In J.H. Wilmore (Ed.), *Exercise and sport sciences reviews* (Vol. 1, pp. 189-216). New York: Academic Press.

Jones, R.E. (1980). Common athletic injuries in women. *Comprehensive Therapy, 6*(9), 47-49.

Jordan, D.B. (1977). Analysis of exercise stress test responses of adult women marathon runners. *Journal of Sports Medicine and Physical Fitness, 17,* 59-64.

Katch, F.I., & Michael, E.D. (1968). Prediction of body density from skinfold and girth measurements of college females. *Journal of Applied Physiology, 25*(1), 92-94.

Koplan, J.P., Powell, K.E., Sikes, R.K., Shirley, R.W., & Campbell, C.C. (1982). An epidemiologic study of the benefits and risks of running. *Journal of the American Medical Association, 248,* 3118-3121.

Kowal, D.M. (1980). Nature and causes of injuries in women resulting from an endurance training program. *American Journal of Sports Medicine, 8*(4), 265-269.

Kuscsik, N. (1977). The history of women's participation in the marathon. In P. Milvy (Ed.), *The marathon: Physiological, medical, epidemiological, and psychological studies* (pp. 862-876). Annals of the New York Academy of Sciences (Vol. 301). New York: The New York Academy of Sciences.

Latshaw, R.F., Kantner, T.R., Kalenak, A., Baum, S., & Corcoran, J.J. (1981). A pelvic stress fracture in a female jogger: A case report. *American Journal of Sports Medicine, 9*(1), 54-56.

Leach, R. (1982). Running injuries of the knee. In R.D'Ambrosia & D. Drez (Eds.), *Prevention and treatment of running injuries* (pp. 55-75). Thorofare, NJ: Charles B. Slack.

Levine, J. (1979). Chondromalacia patellae. *Physician and Sportsmedicine, 7*(8), 41-49.

Lombardo, S.J., & Benson, D.W. (1982). Stress fractures of the femur in runners. *American Journal of Sports Medicine*, **10**(4), 219-227.

Luhtanen, P., & Komi, P.V. (1978). Mechanical factors influencing running speed. In E. Asmussen & E. Jorgensen (Eds.), *Biomechanics VI-B* (pp. 23-29). Baltimore: University Park.

Lusby, L.A. (1983). *Speed-related position-time profiles of arm motion in trained women distance runners*. Unpublished master's thesis, University of Arizona, Tucson.

Malina, R.M., Little, B.B., Bouchard, C., Carter, J.E.L., Hughes, P.C.R., Kunze, D., & Ahmed, L. (1984). Growth status of Olympic athletes less than 18 years of age. In J.E.L. Carter (Ed.), *Medicine and sport science* (Vol. 18, pp. 183-201). Basel, Switzerland: S. Karger.

Matsuo, A., & Fukunaga, T. (1983). The effect of age and sex on external mechanical energy in running. In H. Matsui & K. Kobayashi (Eds.), *Biomechanics VIII-B* (pp. 676-680). Champaign, IL: Human Kinetics.

McBryde, A.M. (1982). Stress fractures in runners. In R. D'Ambrosia & D. Drez (Eds.), *Prevention and treatment of running injuries* (pp. 21-42). Thorofare, NJ: Charles B. Slack.

McBryde, A.M. (1985). Stress fractures in runners. *Clinics in Sports Medicine: Symposium on Running*, **4**(4), 737-752.

Meurman, K.O.A. (1980). Stress fractures of the pubic arch in military recruits. *British Journal of Radiology*, **53**, 521-524.

Micheli, L.J. (1979). Injuries to female athletes. *Surgical Rounds*, **2**(5), 44-55.

Micheli, L. (1982). Female runners. In R. D'Ambrosia & D. Drez (Eds.), *Prevention and treatment of running injuries* (pp. 125-134). Thorofare, NJ: Charles B. Slack.

Miller, D.I. (1978). Biomechanics of running—what should the future hold? *Canadian Journal of Applied Sport Sciences*, **3**, 229-236.

Miller, D.I., Enoka, R.M., & McCulloch, R.G. (1980). Influence of speed on thigh and knee kinematics of female distance runners. Unpublished manuscript.

Nelson, R.C., Brooks, C.M., & Pike, N.L. (1977). Biomechanical comparison of male and female distance runners. In P. Milvy (Ed.), *The marathon: Physiological, medical, epidemiological, and psychological studies* (pp. 793-807). Annals of the New York Academy of Sciences (Vol. 301). New York: The New York Academy of Sciences.

Nilson, K.L. (1986). Injuries in female distance runners. In B.L. Drinkwater (Ed.), *Female endurance athletes* (pp. 149-161). Champaign, IL: Human Kinetics.

Noakes, T.D., Smith, J.A., Lindenberg, G., & Wills, C.E. (1985). Pelvic stress fractures in long distance runners. *American Journal of Sports Medicine*, **13**(2), 120-123.

Norkin, C.C., & Levangie, P.K. (1983). *Joint structure and function: A comprehensive analysis*. Philadelphia: F.A. Davis.

Orava, S., Puranen, J., & Ala-Ketola, L. (1978). Stress fractures caused by physical exercise. *Acta Orthopaedica Scandinavica*, **49**, 19-27.

Outerbridge, R.E., & Dunlop, J.A.Y. (1975). The problem of chondromalacia patellae. *Clinical Orthopaedics and Related Research*, **110**, 177-196.

Oyster, N., & Wooten, E.P. (1971). The influence of selected anthropometric measurements on the ability of college women to perform the 35-yard dash. *Medicine and Science in Sports*, **3**(3), 130-134.

Ozburn, M.S., & Nichols, J.W. (1981). Pubic ramus and adductor insertion stress fractures in female basic trainees. *Military Medicine*, **146**(5), 332-334.

Pagliano, J., & Jackson, D. (1980). The ultimate study of running injuries. *Runner's World*, **15**, 42-50.

Paty, J.G., & Swafford, D. (1984). Adolescent running injuries. *Journal of Adolescent Health Care*, **5**(2), 87-90.

Pavlov, H., Nelson, T.L., Warren, R.F., Torg, J.S., & Burstein, A.H. (1982). Stress fractures of the pubic ramus. *Journal of Bone and Joint Surgery*, **64-A**(7), 1020-1025.

Pollock, M.L., Gettman, L.R., Jackson, A., Ayres, J., Ward, A., & Linnerud, A.C. (1977). Body composition of elite class distance runners. In P. Milvy (Ed.), *The marathon: Physiological, medical, epidemiological, and psychological studies* (pp. 361-370). Annals of the New York Academy of Sciences (Vol. 301). New York: The New York Academy of Sciences.

Potera, C. (1986). Women in sports: The price of participation. *Physician and Sportsmedicine*, **14**(6), 149-153.

Powell, K.E., Kohl, H.W., Caspersen, C.J., & Blair, S.N. (1986). An epidemiological perspective on the causes of running injuries. *Physician and Sportsmedicine*, **14**(6), 100-114.

Prescott, L. (1983). Pelvic stress fractures more common in women. *Physician and Sportsmedicine*, **11**(5), 25-26.

Protzman, R.R., & Griffis, C.G. (1977). Stress fractures in men and women undergoing military training. *Journal of Bone and Joint Surgery*, **59-A**(6), 825.

Rasch, P.J., & Burke, R.K. (1978). *Kinesiology and applied anatomy*. Philadelphia: Lea & Febiger.

Reinker, K.A., & Ozburne, S. (1979). A comparison of male and female orthopaedic pathology in basic training. *Military Medicine*, **144**(8), 532-536.

Richards, J.G. (1980). Mechanical analysis of gait during a marathon. In J.M. Cooper & B. Haven (Eds.), *Proceedings of the Biomechanics Symposium, Indiana University, October 26-28, 1980* (pp. 275-285). Bloomington, IN: The Indiana State Board of Health.

Ross, W.D., & Ward, R. (1984). Proportionality of Olympic athletes. In J.E.L. Carter (Ed.), *Medicine and sport science* (Vol. 18, pp. 110-143). Basel, Switzerland: S. Karger.

Roy, B. (1981). Temporal and dynamic factors of long distance running. In A. Morecki, K. Fidelus, K. Kedzior, & A. Wit (Eds.), *Biomechanics VII-B* (pp. 219-225). Baltimore: University Park.

Rubin, B.D., & Collins, H.R. (1980). Runner's knee. *Physician and Sportsmedicine*, **8**(6), 49-58.

Ryan, A.J. (1975). Women in sports—are the 'problems' real? *Physician and Sportsmedicine*, **3**(5), 49-56.

Saito, M., Kobayashi, K., Miyashita, M., & Hoshikawa, T. (1974). Temporal patterns in running. In R.C. Nelson & C.A. Morehouse (Eds.), *Biomechanics IV* (pp. 106-111). Baltimore: University Park.

Sloane, E. (1980). *Biology of women*. New York: John Wiley and Sons.

Sparling, P.B. (1980). A meta-analysis of studies comparing maximal oxygen uptake in men and women. *Research Quarterly for Exercise and Sport*, **51**(3), 542-552.

Tanner, J.M. (1964). *The physique of the Olympic athlete*. London: George Allen and Unwin.

Taunton, J.E., Clement, D.B., & Webber, D. (1981). Lower extremity stress fractures in athletes. *Physician and Sportsmedicine*, **9**(1), 77-86.

Tehranzadeh, J., Kurth, L.A., Elyaderani, M.K., & Bowers, K.D. (1982). Combined pelvic stress fracture and avulsion of the adductor longus in a middle-distance runner. *American Journal of Sports Medicine*. **10**(2), 108-111.

Thorsen, M. (1964). Body structure and design: Factors in the motor performance of college women. *Research Quarterly*, **35**(3, Part 2), 418-432.

Tomasi, L.F., Peterson, J.A., Pettit, G.P., Vogel, J.V., & Kowal, D.M. (1977). Women's response to army training. *Physician and Sportsmedicine*, **5**(6), 32-37.

Ulrich, C. (1960). Women and sport. In W.R. Johnson (Ed.), *Science and medicine of exercise and sports* (pp. 508-516). New York: Harper & Row.

Upton, S.J., Hagan, R.D., Rosentswieg, J., & Gettman, L.R. (1983). Comparison of the physiological profiles of middle-aged women distance runners and sedentary women. *Research Quarterly for Exercise and Sport*, **54**(1), 83-87.

Vaughan, C.L. (1984). Biomechanics of running gait. *CRC Critical Reviews in Biomedical Engineering*, **12**(1), 1-48.

Wells, C.L. (1985). *Women, sport & performance: A physiological perspective*. Champaign, IL: Human Kinetics.

Wells, C.L., Hecht, L.H., & Krahenbuhl, G.S. (1981). Physical characteristics and oxygen utilization of male and female marathon runners. *Research Quarterly for Exercise and Sport*, **52**(2), 281-285.

Wells, C.L., & Plowman, S.A. (1983). Sexual differences in athletic performance: Biological or behavioral? *Physician and Sportsmedicine*, **11**(8), 52-63.

Wells, C.L., Hecht, L.H., & Krahenbuhl, G.S. (1981). Physical characteristics and oxygen utilization of male and female marathon runners. *Research Quarterly for Exercise and Sport*, **52**(2), 281-285.

Whiteside, P.A. (1980). Men's and women's injuries in comparable sports. *Physician and Sportsmedicine*, **8**(3), 130-140.

Williams, K.R. (1985). Biomechanics of running. In R.L. Terjung (Ed.), *Exercise and sport sciences reviews* (Vol. 13, pp. 389-441). New York: Macmillan.

Wilmore, J.H. (1982). *Training for sport and activity*. Boston: Allyn and Bacon.

Wilmore, J.H., & Behnke, A.R. (1969). An anthropometric estimation of body density and lean body weight in young men. *Journal of Applied Physiology*, **27**(1), 25-31.

Wilmore, J.H. & Behnke, A.R. (1970). An anthropometric estimation of body density and lean body weight in young women. *American Journal of Clinical Nutrition*, **23**(3), 267-274.

Wilmore, J.H., & Brown, C.H. (1974). Physiological profiles of women distance runners. *Medicine and Science in Sports*, **6**(3), 178-181.

Wilmore, J.H., Brown, C.H., & Davis, J.A. (1977). Body physique and composition of the female distance runner. In P. Milvy (Ed.), *The marathon: Physiological, medical, epidemiological, and psychological studies* (pp. 764-776). Annals of the New York Academy of Sciences (Vol. 301). New York: The New York Academy of Sciences.

Wilson, F.S., & Katz, F.N. (1969). Stress fractures. An analysis of 250 consecutive cases. *Radiology*, **92**, 481-486.

Wright, H.F., & Wilmore, J.H. (1974). Estimation of relative body fat and lean body weight in a United States Marine Corps population. *Aerospace Medicine*, **45**(3), 301-306.

Wyrick, W. (1974). Biophysical perspectives. In E.W. Gerber, J. Felshin, P. Berlin, & W. Wyrick (Eds.), *The American woman in sport* (pp. 403-529). Reading, MA: Addison-Wesley.